# Effective Techniques for Bioinformatic Exploration

Paulo Fazendeiro
*University of Beira Interior, Portugal*

Carmelina Leite
*Portuguese Catholic University, Portugal*

IGI Global
Publishing Tomorrow's Research Today

Published in the United States of America by
IGI Global
701 E. Chocolate Avenue
Hershey PA, USA 17033
Tel: 717-533-8845
Fax: 717-533-8661
E-mail: cust@igi-global.com
Web site: https://www.igi-global.com



Library of Congress Cataloging-in-Publication Data

Names: Fazendeiro, Paulo, 1972- editor. | Leite, Carmelina Figueiredo Vieira, 1978- editor.
Title: Effective techniques for bioinformatic exploration / edited by Paulo Fazendeiro, Carmelina Leite.
Description: Hershey, PA : Medical Information Science Reference, [2025]| Includes bibliographical references and index. | Summary: "This project is dedicated to a comprehensive examination of advanced concepts within bioinformatics. We will explore cutting-edge concepts, address challenges, and apply insights to develop practical solutions. The primary focus includes investigating emerging technologies, overcoming complex data analysis hurdles, and translating gained insights into tangible outcomes"-- Provided by publisher.
Identifiers: LCCN 2024038321 (print) | LCCN 2024038322 (ebook) | ISBN 9798369331927 (h/c) | ISBN 9798369366493 (s/c) | ISBN 9798369331934 (eISBN)
Subjects: LCSH: Bioinformatics.
Classification: LCC QH324.2 .E34 2024 (print) | LCC QH324.2 (ebook) | DDC 570.285--dc23/eng/20240926
LC record available at https://lccn.loc.gov/2024038321
LC ebook record available at https://lccn.loc.gov/2024038322

Vice President of Editorial: Melissa Wagner
Managing Editor of Acquisitions: Mikaela Felty
Managing Editor of Book Development: Jocelynn Hessler
Production Manager: Mike Brehm
Cover Design: Phillip Shickler

British Cataloguing in Publication Data
A Cataloguing in Publication record for this book is available from the British Library.

# Table of Contents

# Detailed Table of Contents



Bioinformatics is a modern field of research that encompasses the use of computational techniques for analyzing biological data. Bioinformatics is directly connected to the central dogma of molecular biology, such as the analysis of DNA, RNA, and proteins. The first studies date back to the 1960s when standards were defined for the representation of proteins in files. However, it gained popularity in the early 2000s with the publication of the draft human genome. Nowadays, bioinformatics has stood out as an interdisciplinary science that uses growing computational power to help obtain new scientific discoveries. In this chapter, you will learn a little more about the fundamentals of bioinformatics.

Genomics is an exciting and rapidly evolving field at the intersection of biology, computer science, and statistics, which has made remarkable strides in recent years thanks to advancements in high-throughput sequencing technologies. This field allows us to analyze and understand the genetic code of organisms, from humans to microorganisms, shedding light on everything from the diagnosis and treatment of genetic disorders to the identification of disease-causing genes in pathogens, to the mysteries of biodiversity. With its wide range of applications, genomics is transforming our understanding of life itself.

Within this chapter, the authors offer a concise introduction to the concept of structural bioinformatics. In addition to proteins, DNA, and RNA, peptides are another type of biomolecule covered in this chapter. Peptides are short chains of amino acids (generally, 2-50) connected by peptide bonds. Due to their versatility, they play essential roles in several biological processes in living beings, acting as hormones, enzymes, antibiotics, structural support, etc. Additionally, the chapter will address in detail the interaction between RNA, peptides, and protein, the application of structural bioinformatics tools for studying these complexes, and their implications in research related to interactions, the development of new drugs, future prospects, and challenges related to the topic.

Advancements in computational methods have significantly transformed the traditional drug discovery process, which has been historically lengthy and costly, often taking over a decade and billions of dollars. Virtual screening (VS) is a computational technique used in drug discovery to identify potential drug candidates by screening large chemical libraries. This technique has transformed drug discovery by enabling rapid and cost-effective screening of chemical libraries, reducing time and resources. This leverages algorithms, such as similarity-based, quantitative, and machine learning, to prioritize compounds binding to target proteins. Additionally, databases provide vast compound collections for screening. However, experimental validation is essential to ensure accuracy. In this chapter, the authors delve into the concepts and methodologies of this technique, focusing on its crucial role in drug discovery.

The multidisciplinary field of bioinformatics integrates information technology, biology, and computer science to analyze and comprehend biological data. It entails the creation and use of computational methods and tools to comprehend biological relationships and processes. The incorporation of bioinformatics into the field of food and nutritional science is examined in this chapter, with a focus on how it has revolutionized both research and practical applications. A thorough review of food science and nutritional research is given in the introduction, laying the groundwork for understanding how bioinformatics advances both disciplines. The foundations of bioinformatics are examined, together with significant databases and computing instruments required for data processing and interpretation. This chapter explores the various uses of bioinformatics in food and nutritional research, highlighting developments that have advanced these fields.

# Preface

In today's rapidly advancing scientific landscape, bioinformatics has emerged as a cornerstone of modern biology, offering a vital intersection between biology, computer science, mathematics, and statistics. The growing importance of this discipline can be seen in a wide array of fields, from healthcare and medicine to agriculture, environmental science, and biotechnology. This book, *Effective Techniques for Bioinformatic Exploration*, seeks to provide readers with a comprehensive guide to the foundational and cutting-edge techniques used in bioinformatics, while also addressing its increasing relevance across diverse scientific domains.

The need for bioinformatics expertise has never been more pressing. With the huge amount of biological data generated by high-throughput technologies, the challenge is no longer merely to collect data but to analyze and interpret it in ways that are meaningful and actionable. This book is designed to guide students, researchers, educators, and professionals alike through the core principles and latest innovations in bioinformatics, with a special focus on practical applications that can be translated into real-world solutions. Whether you are a newcomer to the field or an experienced practitioner looking to expand your knowledge, this book offers insights that will help you navigate the ever-evolving bioinformatics landscape.

The book is organized into distinct chapters, each focusing on a critical area of bioinformatics.

Chapter 1 provides a concise introduction to bioinformatics as a transdisciplinary science. The chapter emphasizes the field's integration of biology, computer science, mathematics, and statistics to organize biological data, analyze genetic sequences, and model biological systems. It lays the foundation for the book by highlighting bioinformatics' key objectives, including gene expression analysis, protein interaction studies, and artificial intelligence applications.

In Chapter 2, we delve into genomics, one of the key components of the multi-omics landscape. The rise of high-throughput technologies has revolutionized biological research by generating vast datasets that enable the exploration of genetic information at unprecedented scales. This chapter explores the role of bioinformatic tools in handling genomic data, discussing their applications in understanding diseases, agricultural advancements, and environmental science. The chapter emphasizes the need for interdisciplinary collaboration and technological innovation to harness the full potential of genomics.

Chapter 3 introduces immunoinformatics, a paradigm shift in vaccine development. This chapter highlights how computational approaches can accelerate immunogen design and optimize vaccine formulations. The chapter also discusses the emergence of Immunoinformatics 2.0, driven by artificial intelligence (AI) and machine learning, which promises to enhance vaccine precision and efficiency. By examining pathogen-host interactions and modeling immune responses, this chapter underscores the field's critical role in modern vaccine development.

In Chapter 4, we shift focus to computational modeling and machine learning, crucial for analyzing biological data. This chapter presents machine learning techniques as powerful tools for tackling problems related to DNA, RNA, and protein sequences. With growing data complexity, computational modeling becomes increasingly vital for solving intricate biological questions, guiding researchers in the development of predictive models and understanding macromolecular interactions.

Following the discussion on machine learning, Chapter 5 explores deep learning and its profound impact on bioinformatics. By leveraging AI techniques, deep learning models have significantly advanced fields like protein structure prediction and drug discovery. This chapter discusses the pivotal role of state-of-the-art models, in redefining our understanding of biological systems. Deep learning's contributions to personalized medicine and novel molecular design are also addressed, demonstrating its transformative influence.

Chapter 6 examines structural bioinformatics, focusing on the applications of molecular modeling, docking, and molecular dynamics. This chapter presents how computational techniques provide insights into macromolecular functions, protein-ligand interactions, and drug design. Tools such as AlphaFold and SWISS-MODEL are explored, along with the challenges and opportunities that arise from the integration of multi-omics data in structural analysis. The chapter highlights structural bioinformatics' vital role in precision medicine and biomolecular research.

Chapter 7 takes readers through the importance of virtual screening in drug discovery. This computational technique has revolutionized the pharmaceutical industry by rapidly screening large chemical libraries to identify potential drug candidates. The chapter explores various methodologies, including ligand-based, structure-based, and fragment-based screening, emphasizing the role of bioinformatics in transforming the traditionally time-consuming process of drug discovery into a more efficient and cost-effective approach.

Chapter 8 expands the discussion into bioinformatics' contributions to food and nutritional science. While bioinformatics holds tremendous potential in this field, challenges remain in terms of data quality, standardization, and interdisciplinary collaboration. This chapter discusses the real-world limitations and ethical considerations associated with using bioinformatics in food science, focusing on the need for more robust computational techniques and ethical frameworks to guide future research.

Finally, Chapter 9 explores the intersection of bioinformatics and ecology through ecoinformatics. As the field continues to evolve, ecoinformatics plays a crucial role in biodiversity conservation and ecological research. By integrating big data and advanced computational tools, this chapter shows how bioinformatics contributes to monitoring ecosystems and managing biodiversity more effectively. The chapter underscores the potential of ecoinformatics to address environmental challenges through innovative data-management strategies and collaboration.

In conclusion, this book covers a broad spectrum of bioinformatics applications, emphasizing the importance of computational approaches in solving some of the most pressing biological challenges of our time. Each chapter illustrates a unique facet of bioinformatics, providing readers with a holistic understanding of how data-driven science is reshaping our understanding of biology, health, and the environment. The chapters are designed to engage both newcomers to the field and experienced researchers, offering insights that span from theoretical foundations to cutting-edge applications.

As a final remark let us emphasize that the creation of this book has been a collaborative journey, which would not have been possible without the contributions of many dedicated individuals. First and foremost, we wish to express our deepest gratitude to the esteemed authors and researchers who have contributed their expertise to this volume. Their dedication to advancing the field of bioinformatics and their commitment to sharing their knowledge has greatly enriched the content of this book.

We are also deeply appreciative of the peer-reviewers whose insightful feedback helped to improve the quality of each chapter. Their attention to detail and thoughtful suggestions ensured the accuracy and relevance of the information presented.

Special thanks go to the editorial team at IGI Global, whose professionalism and support made the development and publication of this book possible. We are especially grateful to our Development Editor, Nina Eddinger, whose tireless efforts and guidance throughout the process have been instrumental in bringing this project to fruition. This work was also supported by FCT - Fundação para a Ciência e Tecnologia, I.P. by project reference UIDB/50008/2020, and DOI identifier https://doi.org/10.54499/UIDB/50008/2020.

We are pleased to present this book to you and hope that it proves to be a useful resource for your journey in bioinformatics.

**Paulo Fazendeiro**
*Instituto de Telecomunicações, Portugal & University of Beira Interior, Portugal*

**Carmelina Leite**
*Portuguese Catholic University, Portugal*

# Chapter 1
# A Concise Overview of Bioinformatics

**Alessandra Lima da Silva**
https://orcid.org/0000-0002-2698-6889
*Universidade Federal de Minas Gerais, Brazil*

**Diego Mariano**
https://orcid.org/0000-0002-5899-2052
*Universidade Federal de Minas Gerais, Brazil*

**Mariana Parise**
*Instituto de Ciências Exatas, Brazil*

**Angie L. A. Puelles**
*Instituto de Ciências Exatas, Brazil*

**Tatiane Senna Bialves**
https://orcid.org/0000-0002-3827-2859
*Instituto de Ciências Exatas, Brazil*

**Luana Luiza Bastos**
https://orcid.org/0000-0002-6932-0438
*Instituto de Ciências Exatas, Brazil*

**Lucas Santos**
*Instituto de Ciências Exatas, Brazil*

**Rafael Pereira Lemos**
*Instituto de Ciências Exatas, Brazil*

## ABSTRACT

*Bioinformatics is a modern field of research that encompasses the use of computational techniques for analyzing biological data. Bioinformatics is directly connected to the central dogma of molecular biology, such as the analysis of DNA, RNA, and proteins. The first studies date back to the 1960s when standards were defined for the representation of proteins in files. However, it gained popularity in the early 2000s with the publication of the draft human genome. Nowadays, bioinformatics has stood out as an interdisciplinary science that uses growing computational power to help obtain new scientific discoveries. In this chapter, you will learn a little more about the fundamentals of bioinformatics.*

DOI: 10.4018/979-8-3693-3192-7.ch001

## 1. WHAT IS BIOINFORMATICS

Bioinformatics is a discipline that transcends the boundaries of traditional academic disciplines, integrating and mobilizing knowledge from various areas associated with life science and computational technologies, thus demonstrating its transdisciplinarity. Moreover, bioinformatics is an interdisciplinary field mediated by biology, computer science, mathematics, and statistics. The main objectives of bioinformatics include storage and organization of biological metadata, analysis of genetic and protein sequences, genomic comparison and evolution (complete or incomplete in the form of drafts), gene expression and proteomics (microarrays, RNA sequencing, or molecule interactions), and also biological systems modeling (development of models and algorithms, virtual screening, or artificial intelligence).

## 2. A BRIEF HISTORY OF BIOINFORMATICS

As the processing power of computers increased after the Second World War, their use began to be considered for solving biological problems. In the 1960s, several experimental techniques were used to obtain collections of amino acid sequences. At that time, scientists were using computational methods without the need for supercomputers or computer networks to understand protein sequences. This led to a set of interesting problems that were impossible to solve without the processing power of computers (Hagen, 2000). This is partly responsible for giving rise to a fundamental area of research in biology and how information is processed in living systems, such as evolutionary biology and how genes are transmitted (Gauthier et al., 2019).

Margaret Dayhoff, Richard Eck, and Robert Ledley were pivotal figures in bioinformatics' early development, with Dayhoff's contributions being particularly notable. She is often credited as a pioneer in developing the first amino-acid substitution matrix and other foundational bioinformatics tools. Margaret Dayhoff became known as the mother of Bioinformatics. In the mid-1960s, she published the "Atlas of Protein Sequence and Structure", a collection of protein sequences known at the time, which received several updates in the following years (Dayhoff, 1972). When representing protein sequences and other polypeptide molecules, each amino acid residue was represented by a letter. For example, alanine is represented by the letter "A," and tryptophan is represented by the letter "W". The representation of a single letter facilitates comparison between different sequences (alignment). Furthermore, based on analyzing several amino acid sequences, Dayhoff and colleagues proposed the PAM (Point Accepted Mutation) substitution matrix, which estimated the possibility of mutations accepted by natural selection (Sasidharan;

Chothia, 2007). These early contributions laid the groundwork for contemporary bioinformatics, illustrating the discipline's significance in understanding protein evolution and biological data analysis.

Computing revolutionized in the 1970s with the emergence of integrated circuit microprocessors. Furthermore, the internet took its first steps in these years (Mariano et al., 2020). Thus, with the increasing volume of accumulated DNA and protein sequence data, mainly obtained from Sanger DNA sequencing techniques, algorithms have become necessary to analyze this data. Needleman and Wunsch proposed the first sequence alignment algorithms based on a dynamic programming method.

Also, in the 1990s, with the popularization of the internet, several research groups had access to available biological databases, such as Genbank. Furthermore, tools such as BLAST (Basic Local Alignment Search Tool) allowed the comparison of several sequences in parallel (Altschul et al., 1990). Below, we can see an example of an alignment between two myoglobin sequences extracted from a human from a mouse (Figure 1):

*Figure 1. The human myoglobin (P02144) and mouse myoglobin (P04247) are aligned*

```
Range 1: 1 to 154 Graphics

Score          Expect  Method                         Identities      Positives       Gaps
264 bits(674)  4e-97   Compositional matrix adjust.   129/154(84%)    141/154(91%)    0/154(0%)

Query  1    MGLSDGEWQLVLNVWGKVEADIPGHGQEVLIRLFKGHPETLEKFDKFKHLKSEDEMKASE  60
            MGLSDGEWQLVLNVWGKVEAD+ GHGQEVLI LFK HPETL+KFDKFK+LKSE++MK SE
Sbjct  1    MGLSDGEWQLVLNVWGKVEADLAGHGQEVLIGLFKTHPETLDKFDKFKNLKSEEDMKGSE  60

Query  61   DLKKHGATVLTALGGILKKKGHHEAEIKPLAQSHATKHKIPVKYLEFISECIIQVLQSKH  120
            DLKKHG TVLTALG ILKKKG H AEI+PLAQSHATKHKIPVKYLEFISE II+VL+ +H
Sbjct  61   DLKKHGCTVLTALGTILKKKGQHAAEIQPLAQSHATKHKIPVKYLEFISEIIIEVLKKRH  120

Query  121  PGDFGADAQGAMNKALELFRKDMASNYKELGFQG  154
             GDFGADAQGAM+KALELFR D+A+ YKELGFQG
Sbjct  121  SGDFGADAQGAMSKALELFRNDIAAKYKELGFQG  154
```

However, bioinformatics experienced exponential growth only in the 2000s with the publication of the draft human genome (Venter et al., 2001). In this context, several other genome projects began to be developed, and computer specialists were increasingly requested to help analyze this data. Nowadays, we see an evolution in video cards, which has increasingly allowed more robust experiments that involve more calculations, such as molecular dynamics simulations and analyses involving artificial intelligence (Mariano et al., 2020).

These initial efforts in bioinformatics prepared the field for what we know. Over time, bioinformatics grew from the observation of molecules in biology to the use of computer tools such as data analysis and machine learning. Consequently, the change has shown how the field has moved from a general view of biological information to a detailed study of genes and their evolution, highlighting the importance of genetic research for bioinformatics and its fundamental role in modern biology.

Nowadays, bioinformatics has emerged as an interdisciplinary field combining biology, computer science, mathematics, and statistics to address complex biological questions and process large amounts of data from genomic research. Furthermore, artificial intelligence and machine learning algorithms have made a huge impact on several subfields of bioinformatics (Dos Santos; Mariano; De Melo-Minardi, 2024). We could cite, for example, de novo protein modeling techniques using deep learning algorithms such as those proposed by the Alphafold algorithm (Jumper et al., 2021). However, before discussing this, we need to know more about genome projects and the rise of so-called omic sciences.

## 3. THE HUMAN GENOME PROJECT

Sequencing is the process of reading DNA fragments. Understanding the sequence of DNA is vital to understanding not only how organisms function but the entire principle of life. The first successful sequencing of the insulin protein, achieved using Sanger sequencing, occurred between 1945 and 1955, marking a significant milestone in protein biochemistry (Hagen, 2000). At the end of the 1980s, bioinformatics began to focus mainly on genes and their sequences (Wang; Li; Zhang, 2022). In the following years, several milestones in sequencing led to the Human Genome Project, that is, understanding the sequence of the more than three million base pairs of the human genome (Figure 2).

*Figure 2. Sequencing technologies timeline (Yang et al., 2020) (CC-BY 4.0)*

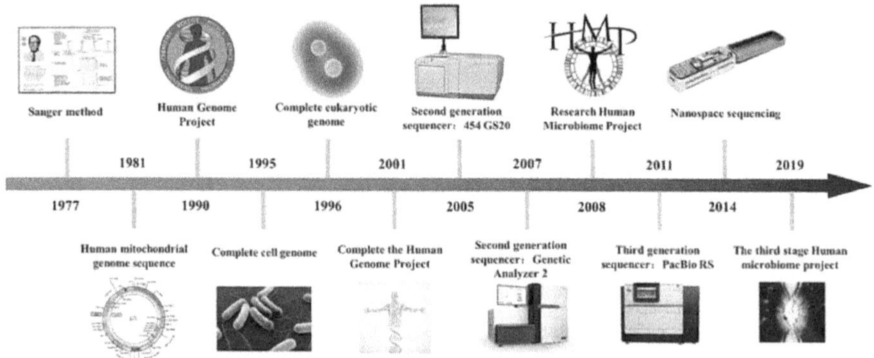

The Human Genome Project (HGP) was a large-scale international collaboration that aimed to map, read, and understand the entire set of genes in the human genome. The human genome consists of the complete set of hereditary genetic material in our cells, containing all the information necessary for the functioning, development, and maintenance of the human body. The HGP was officially launched in 1990 and completed in 2003. In 2001, a draft of the human genome was published by both government and private initiatives (Lander et al., 2001; Venter et al., 2001).

The Human Genome Project is an example of large-scale genetic research that has produced rich biological knowledge. Additionally, bioinformatics has played a vital role in managing, analyzing, and understanding this data, significantly improving our understanding of the human genome's structure, functionality, and evolutionary path. Through the HGP and similar sequencing efforts, bioinformatics has enabled the discovery of genes, the analysis of genetic variations, and the investigation of their implications for health and disease. It has transformed public health, improving disease detection, accelerating vaccine development, and combating antimicrobial resistance more precisely (Gwinn; Maccannell; Khabbaz, 2017).

Bioinformatics proved indispensable to the HGP, offering crucial tools and methods for analyzing the vast genetic information discovered. Completing the HGP in 2003 was a milestone in genomics and medical research, enriching our knowledge of human genetic diversity and disease mechanisms (Depristo et al., 2011).

*Figure 3. Bioinformatics played a pivotal role in the success of the Human Genome Project (HGP), a landmark scientific endeavor completed in 2003 that mapped and sequenced the entire human genome (Collins; Fink, 1995; Hood; Rowen, 2013; Lander et al., 2001; Venter et al., 2001)*

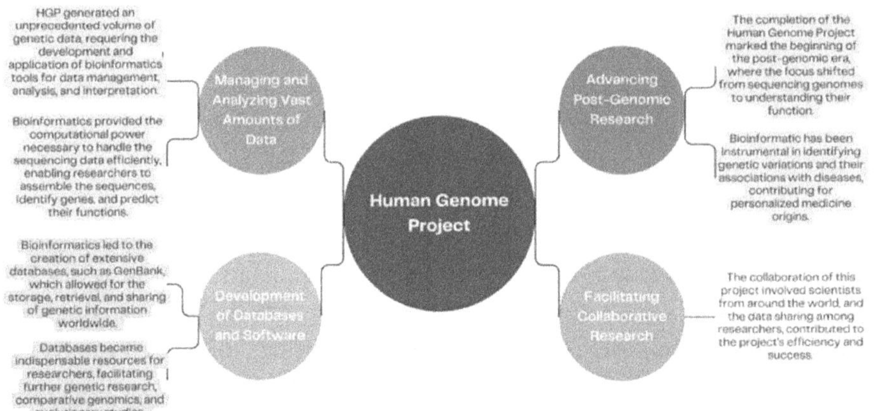

This wealth of data produced by HGP has facilitated predictions about gene regulation, expression, translation, protein structure, mechanisms of action, and evolutionary changes. In recent decades, advances in sequencing technologies and the creation of various tools have enabled the collection and analysis of large amounts of genomic, transcriptomic, proteomic, and metabolomic data (Lu; Lu; Cao, 2012).

The conclusion of the Human Genome Project emphasized the necessity for enhanced technologies and larger datasets to address complex biological inquiries. Despite this, significant challenges, such as limited throughput and high sequencing costs, persisted. In the mid-2000s, the introduction of the initial high-throughput sequencing platform marked a substantial decrease of 50,000-fold in sequencing costs compared to the Human Genome Project. This development earned the moniker "Next-Generation Sequencing" (NGS) (Wetterstrand, 2021).

## 4. NEXT-GENERATION SEQUENCING (NGS)

Despite providing advances in genomic research, Sanger sequencing (Sanger; Nicklen; Coulson, 1977) presents limitations in terms of low scalability and incurs high costs, compared to new techniques, has less capacity for high-throughput sequencing (Reuter; Spacek; Snyder, 2015). In this sense, the emergence of Next-Generation Sequencing (NGS) occurred with the advent of next-generation sequencing technologies that revolutionized the area of genomics (Metzker, 2010).

One of the primary breakthroughs offered by NGS was its capacity to produce vast quantities of data at a reduced cost, often surpassing one billion short reads per instrument run. A broader implication of NGS lies in its potential for resequencing human genomes, which can enhance our comprehension of the impact of genetic variations on health and disease (Metzker, 2010).

The progress in these areas is indeed thrilling; however, they come with their own set of limitations. With the advent of new technologies, both issues are magnified, and novel challenges emerge. While NGS platforms offer immense amounts of data, they also come with higher error rates (approximately 0.1–15%) and shorter read lengths (ranging from 35 to 700 bp for short-read methods) than traditional Sanger sequencing platforms. This necessitates meticulous scrutiny of the findings, especially regarding variant discovery and clinical applications (Liu et al., 2012; Metzker, 2005). Despite these limitations, NGS represents significant advances in genetic sequencing, providing a wide range of applications in biomedical research, clinical diagnosis, and biotechnology.

An example of sequencing using an NGS platform is the Shotgun technique. In this methodology, several DNA fragments are broken at random points. Next, the DNA is read using modern sequencers. Finally, the final genome sequence is assembled through computational analysis (Figure 4).

*Figure 4. The shotgun sequencing (Wikimedia Commons | National Institutes of Health)*

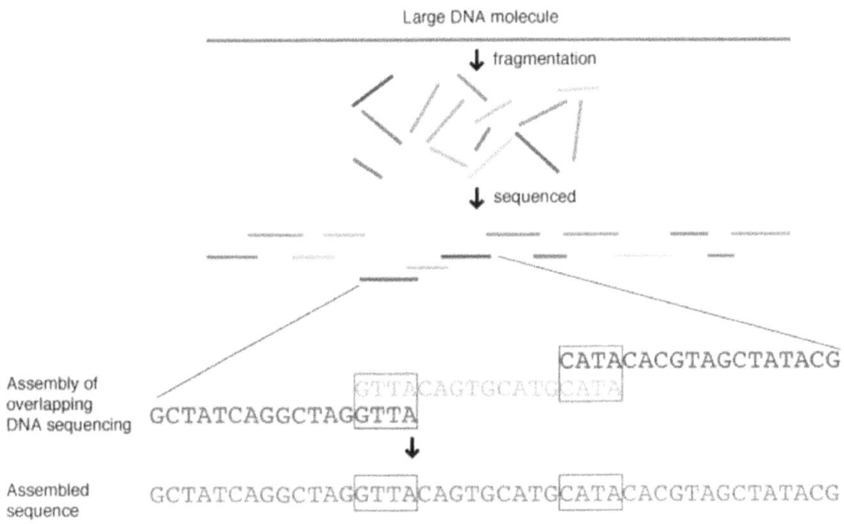


Three essential methodologies in the field of NGS are fundamental for advancing molecular biology:

- **Sequencing by Synthesis (SBS):** this NGS technique involves cycles of DNA synthesis, where each cycle adds a base to the sequencing. The detection of the incorporated base is done using methods like fluorescence, enabling the reading of the nucleotide sequence. Platforms like Illumina (Figure 5) and SOLiD use SBS for generating high-quality data on a large scale (Liu et al., 2012; Metzker, 2005, 2010);
- **Ligation-Based Sequencing (LBS):** in LBS, DNA or RNA fragments are ligated to a solid surface, followed by amplification and sequencing detection. This technique is known for its ability to sequence long regions of DNA or RNA without the need for cloning, and it is used in platforms like PacBio for long-range and high-fidelity analyses (Eid et al., 2009);
- **Nanopore Sequencing:** Nanopore Sequencing allows DNA or RNA molecules to pass through nanopores, where base identification occurs by changes

in electrical current during translocation. This approach offers the advantage of direct and real-time reading of sequences without amplification, and it is implemented in devices like MinION and GridION from Oxford Nanopore Technologies for applications in whole genome sequencing and detection of epigenetic modifications (Ip et al., 2015; Song et al., 1996).

*Figure 5. Example of next-generation sequencing (Illumina NextSeq 500) (Wikimedia Commons (CC-BY 2.0))*

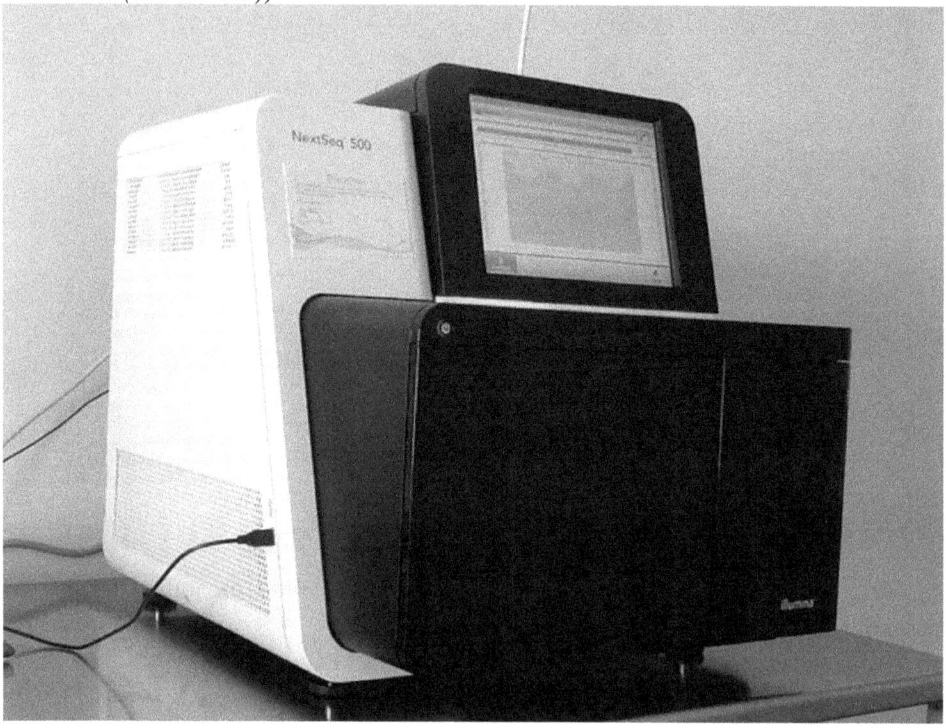


## 5. THE OMICS SCIENCES AND BIOINFORMATICS

Evolution in sequencing technologies has led to the emergence of specific research fields for each biological molecule, such as DNA, RNA, and proteins. The scientific field that studies the comprehensive assessment of these sets of molecules is considered "Omics" science (Hasin; Seldin; Lusis, 2017). The choice of this particular term (Omics) is due to the suffix of the main sub-areas of study, such as genomics, transcriptomics, and proteomics.

Genomics comprises the study of whole genomes, namely, the complete collection of an organism's genetic information. It aims to understand the function, evolution, structure, and mapping (Casas; Saborido-Rey, 2023). As discussed previously, the genome is the set of genes and other hereditary information passed from generation to generation in organisms. The genome comprises the DNA molecule in many organisms, although certain viruses have an RNA genome. On the other hand, Transcriptomics focuses on the complete set of RNA molecules that have been expressed in a cell, tissue, or organism (Lowe et al., 2017), including transcripts' expression levels, location, functions, degradation, splicing, modifications, amongst others (Skerrett-Byrne Anthony et al., 2023). The study of RNA has gained prominence in recent times, in particular, due to the role of non-coding RNAs in various cellular functions. Lastly, Proteomics focuses on the set of protein contents, analyzing its biochemical properties, expression, structure, interactions, and connections (Zubair et al., 2022). Proteins are the macromolecules essential for life. Its role is closely related to its three-dimensional structure, which will be further discussed in the next section. Other examples of omics include metagenomics, which encompasses the study of metagenome samples, and metabolomics, which encompasses the study of metabolites.

Although advances in these fields have largely impacted medicine, biology, and drug discovery (Al-Amrani et al., 2021; Jäkel; Williams, 2020; Pattan et al., 2021; Van De Sande et al., 2023), even greater breakthroughs are expected when combining data from different omics sciences (multi-omics). Integrated omics approaches aim to understand how molecules interact to orchestrate phenotypes (Hawkins; Hon; Ren, 2010; Subramanian et al., 2020). When considering complex diseases involving several relevant tissues, development over time, and both genetic and environmental factors, the results from a multi-omics approach aid in elucidating how the disease develops or how molecules interact to generate a phonetic state (Hasin; Seldin; Lusis, 2017). (Graw et al., 2021; Hasin; Seldin; Lusis, 2017; Kwoji et al., 2023; Subramanian et al., 2020) present approaches and applications of multi-omics.

The advances in both platform technologies and bioinformatics tools have contributed to the generation of enormous biological datasets and the understanding of cellular and molecular systems (Boja et al., 2014; Casas; Saborido-Rey, 2023). Several challenges, such as genome assembly, biomedical data classification, protein structure prediction, and protein-protein interactions prediction, have inspired the development of bioinformatics tools in recent years. Although tools have been developed and tested, several solutions can be improved and optimized (Iqbal; Kumar, 2023; O'Donoghue, 2021; Yang et al., 2021). Not only data analysis and interpretation, but also publicly available biological data repositories such as Genbank (Benson et al., 2012), Gene Expression Omnibus (GEO) (Clough; Barrett, 2016), Uniprot (Consortium, 2015), and PDB (Berman et al., 2000) have also boosted biological research and product development. For an overview of bioinformatics databases, techniques, and tools, see (Iqbal et al, 2023).

## 6. STRUCTURAL BIOINFORMATICS

As discussed, Bioinformatics is a broad field encompassing the development and application of computational algorithms to analyze biological data, such as genetic information, protein amino acid sequences, and protein structures. It is divided into two main sub-areas: (a) traditional bioinformatics, which deals mainly with sequence analysis, and (b) structural bioinformatics, which deals with structures of biomolecules. A prominent area within this domain is Structural Bioinformatics, which focuses on understanding the three-dimensional architecture of biological macromolecules, such as proteins, nucleic acids, and macromolecular complexes (Iacoangeli et al., 2022).

Biological macromolecules are fundamental to various biological and chemical processes that sustain life and the functioning of organisms. Their structure is essential for their stabilization, development of their biological functions, and interaction with other macromolecules. Proteins, for example, are essential for virtually all cellular functions, from catalyzing biochemical reactions to providing structural support and transporting substances (Nelson; Lehninger; Cox, 2008). The three-dimensional structure of proteins is a determinant of their function. Proteins are formed by amino acids linked together by peptide bonds. The three-dimensional structure of proteins is vital for their biological functions. It presents four levels: primary (linear sequence of amino acids), secondary (repetitive arrangements such as alpha helices and beta sheets), tertiary (complete three-dimensional arrangement), and quaternary (association of multiple subunits). The specific three-dimensional conformation of a protein is crucial for its interaction with other molecules and biological activity. Changes in protein structure can lead to dysfunction and disease (Nelson; Lehninger; Cox, 2008).

In summary, the structure of macromolecules plays a central role in understanding and manipulating biological processes from the cellular to the molecular level. This structure determines the properties and functions of molecules, being vital for the maintenance of life and for the variety of shapes and functions in living organisms. In this context, structural bioinformatics emerges as an essential tool to deepen understanding of biological processes, facilitating the analysis and modeling of molecular structures (Lin; Li; Lin, 2020).

Through advanced computational techniques like molecular modeling, molecular dynamics simulation, and protein structure analysis, Structural Bioinformatics enables scientists to visualize and thoroughly investigate the molecular composition of these biomolecules. This approach is crucial for unraveling fundamental biological mechanisms, understanding diseases, and developing new pharmaceutical therapies based on the molecular structure of therapeutic targets. The emergence of the internet and the increase in computational capacity in the 1990s further propelled the field of Structural Bioinformatics. At the same time, sophisticated software programs emerged for visualization, analysis, and modeling of structures (Anfinsen, 1973).

Protein structure databases, such as the Protein Data Bank (PDB), were established to store and share information on biomolecular structures. Major advances in bioinformatics have led to the development of databases such as the Research Collaboratory for Structural Bioinformatics Protein Data Bank (RCSB PDB), which offers extensive resources for exploring the 3D structures of biological macromolecules (Figure 6). These resources are invaluable for fundamental and applied research in various scientific fields, including biology, biomedicine, biotechnology, and bioengineering.

*Figure 6. Protein data bank statistics: growth of released structures per year. Increasing numbers of entries available per year, especially after 2000. Despite the constant increase in information in databases, bioinformatics tools are continually being improved to provide more accurate predictions about the functionality of proteins (https://www.rcsb.org/stats/growth/growth-released-structures)*

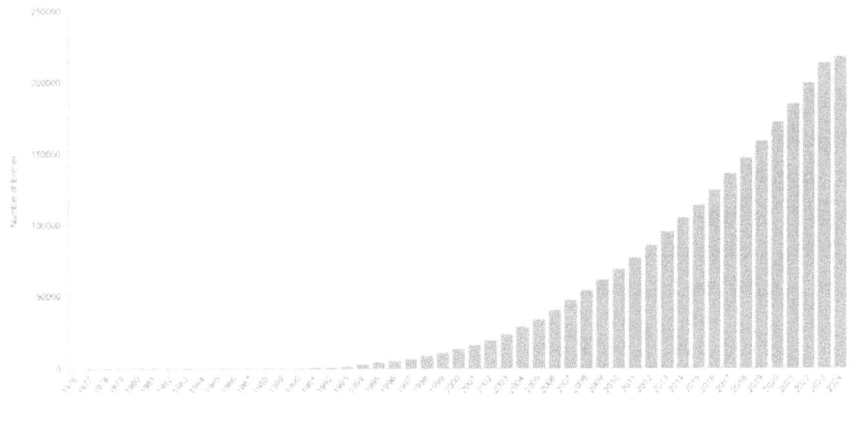

In the 21st century, Structural Bioinformatics has become an indispensable tool in various biological and medical applications, from drug design to protein engineering and understanding genetic diseases. The field of research continues to evolve rapidly, driven by advances in biotechnology, artificial intelligence, and quantum computing. This evolution is crucial to facing complex challenges in biology and the various areas in which it can be applied. Some of the methodologies used in this field of research include molecular docking, virtual drug screening, molecular dynamics simulations, and molecular modeling (PAIVA et al., 2022). In this context, a modern tool that uses a deep learning approach has stood out for having solved one of the most complex problems in structural bioinformatics: *de novo* protein modeling. In the next section, you will learn more about the Alphafold algorithm.

## 7. THE REVOLUTION OF ALPHAFOLD

When Christian Anfinsen won the Nobel Prize in Chemistry in 1972, he postulated, in his acceptance lecture, that the sequence of proteins should fully determine their structure (Anfinsen, 1973). This hypothesis, dubbed the "holy grail" of molecular biology (The Times, 2024), went on for nearly 50 years, contrasting the classical methods of structural biology that focus on the experimental resolution of 3D protein structures. However, according to the 1969 Levinthal Paradox, enumerating all possible configurations of a protein structure by brute force calculation would take longer than the age of the universe (Levinthal, 1968; Zwanzig; Szabo; Bagchi, 1992).

In 2019, the Google subsidiary DeepMind released the AlphaFold algorithm, which won the CASP (Critical Assessment of Protein Structure Prediction) competition $13^{th}$ edition (Moult et al., 1995; Senior et al., 2020). With improved methods and combining machine learning, neural networks, and deep learning, the new AlphaFold2 won the following CASP14 by a landslide, achieving a never seen before-free modeling median score of 87 (on a 0-100 scale) (Google Deepmind, 2020; Jumper et al., 2021; Pereira et al., 2021). At the same time, the human proteome structural coverage was expanded to 98.5% (from 17%) using AlphaFold2, in which 58% of the residues achieved a confident prediction (Jumper et al., 2021).

*Figure 7. Alphafold obtains the 3D structure from sequences using a deep learning approach (Mariano, 2022) (CC-BY 4.0)*

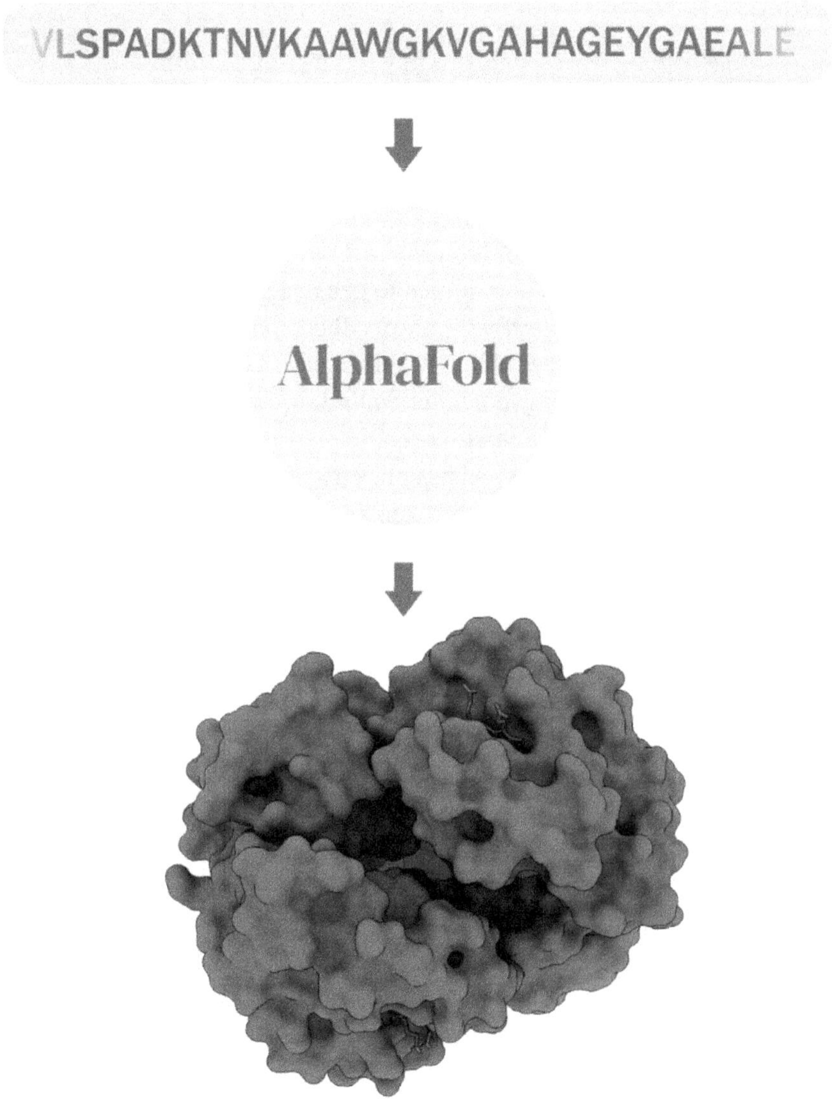


The so-called "AlphaFold Revolution" sparked a new gold rush in structural biology (Callaway, 2020; Thornton; Laskowski; Borkakoti, 2021), and soon after the CASP14, almost the whole Uniprot database (comprised of over 200 million protein sequences) was structurally solved using AlphaFold2, and released in an open-access tool for the scientific community (available at https://alphafold.ebi.ac.uk).

Several tools using the methodology were also released, like the ColabFold, which uses Google Colaboratory free cloud computing service to run a more accessible (computationally and user-friendly) version of AlphaFold2 (Mirdita et al., 2022).

## 8. ARTIFICIAL INTELLIGENCE IN BIOINFORMATICS

Finally, we must highlight the role that artificial intelligence is playing in modern bioinformatics. Although the term *Artificial Intelligence* (AI) was first introduced in 1955 by American mathematician John McCarthy, historical perspectives on AI reveal that its various definitions have aimed to gauge a system's success based on how closely it can mimic *human performance* or adhere to *rationality* (Russell; Norvig, 2016). (Dennett, 2004) further elucidates the concept of artificial intelligence, describing it as "*the art of crafting machines capable of executing tasks that, when performed by humans, necessitate intelligence*".

In the field of Molecular Biology, AI has played a pivotal role for over a decade, facilitated by the exponential growth of heterogeneous biological data originating from various sources, including omics, imaging, and biomedical signals (Min; Lee; Yoon, 2017). Recent approaches based on Deep Learning (DL) have presented innovative solutions to long-standing challenges in structural biology, such as the problem of protein folding based on a given sequence (Alquraishi, 2019; Anishchenko et al., 2021; Senior et al., 2020). In this context, big techs such as DeepMind and Meta AI have made significant contributions to the development of AlphaFold (Jumper et al., 2021; Senior et al., 2020) and ESMFold (Lin et al., 2023), respectively. These tools would become the state of the art when it comes to comparative modeling and achieving unprecedented accuracy in predicting protein structures. This shows that AI's utility extends beyond mere comprehension to creating intelligent systems, as highlighted by (Russell; Norvig, 2016).

One aspect of the applications that contributed to their performance is the employment of Deep Learning (DL) algorithms such as Convolutional Neural Networks (Lecun et al., 1989), which identify complex patterns across large datasets, being widely used in predicting the structure or function of proteins (Gao et al., 2020; Torrisi; Pollastri; LE, 2020). The progress of generative AI has introduced a specialized class of generative models, which have overlapped with established DL architectures. Architectures, including Variational Autoencoders (VAEs), Generative Adversarial Networks (GANs), and, more recently, diffusion models, have shown potential in modeling data distributions from training sets to generate comparable results. Developing novel large-scale protein language models (PLMs) pre-trained from self-supervised learning on billions of protein sequences has been the core of computational strategies for the high-precision prediction of 3D protein structures and de novo protein design. These models demonstrate remarkable potential in creating customized molecules and exploring the space of sequences and their structural and functional properties (Callaway, 2023).

In conclusion, there is great potential in using artificial intelligence, machine learning, and deep learning techniques in bioinformatics. Several areas can benefit from this, from the production of medicines and biotechnological products to the discovery of how diseases work. What else can we get from AI techniques? Only time can tell us.

## REFERENCES


Al-Amrani, S., Al-Jabri, Z., Al-Zaabi, A., Alshekaili, J., & Al-Khabori, M. (2021). Proteomics: Concepts and applications in human medicine. *World Journal of Biological Chemistry*, 12(5), 57–69. DOI: 10.4331/wjbc.v12.i5.57 PMID: 34630910

Alquraishi, M. (2019, November 15). AlphaFold at CASP13. *Bioinformatics (Oxford, England)*, 35(22), 4862–4865. DOI: 10.1093/bioinformatics/btz422 PMID: 31116374

Altschul, S. F., Gish, W., Miller, W., Myers, E. W., & Lipman, D. J. (1990). Basic local alignment search tool. *Journal of Molecular Biology*, 215(3), 403–410. DOI: 10.1016/S0022-2836(05)80360-2 PMID: 2231712

Anfinsen, C. B. (1973, July 20). Principles that Govern the Folding of Protein Chains. *Science*, 181(4096), 223–230. DOI: 10.1126/science.181.4096.223 PMID: 4124164

Anishchenko, I., Pellock, S. J., Chidyausiku, T. M., Ramelot, T. A., Ovchinnikov, S., Hao, J., Bafna, K., Norn, C., Kang, A., Bera, A. K., DiMaio, F., Carter, L., Chow, C. M., Montelione, G. T., & Baker, D. (2021, December). De novo protein design by deep network hallucination. *Nature*, 600(7889), 547–552. DOI: 10.1038/s41586-021-04184-w PMID: 34853475

Benson, D. A., Cavanaugh, M., Clark, K., Karsch-Mizrachi, I., Lipman, D. J., Ostell, J., & Sayers, E. W. (2012). GenBank. *Nucleic Acids Research*, 41(D1, n. D1), D36–D42. DOI: 10.1093/nar/gks1195 PMID: 23193287

Berman, H. M. (2000, January 1). The Protein Data Bank. *Nucleic Acids Research*, 28(1), 235–242. DOI: 10.1093/nar/28.1.235 PMID: 10592235

Boja, E. S. (2014). Integration of omics sciences to advance biology and medicine.

Callaway, E. (2020, November 30). 'It will change everything': DeepMind's AI makes gigantic leap in solving protein structures. *Nature*, 588(7837), 203–204. DOI: 10.1038/d41586-020-03348-4 PMID: 33257889

Callaway, E. (2023). AI tools are designing entirely new proteins that could transform medicine. *Nature*, 619(7969), 236–238. DOI: 10.1038/d41586-023-02227-y PMID: 37433935

Casas, L., & Saborido-Rey, F. (2023). A review of genomics methods and bioinformatics tools for the analysis of close-kin mark-recapture.

Clough, E., & Barrett, T. (2016). The *gene expression omnibus* database. *Statistical Genomics: Methods and Protocols*, 93–110.

Collins, F. S., & Fink, L. (1995). The Human Genome Project. *Alcohol Health and Research World*, 19(3), 190–195. PMID: 31798046

Consortium. (2015, January 28). UniProt: A hub for protein information. *Nucleic Acids Research*, 43(D1, n. D1), D204–D212. DOI: 10.1093/nar/gku989 PMID: 25348405

Dayhoff, M. O. (1972). *Atlas of protein sequence and structure*. National Biomedical Research Foundation.

Dennett, D. (2004). The age of intelligent machines: Can machines think.

Depristo, M. A., Banks, E., Poplin, R., Garimella, K. V., Maguire, J. R., Hartl, C., Philippakis, A. A., del Angel, G., Rivas, M. A., Hanna, M., McKenna, A., Fennell, T. J., Kernytsky, A. M., Sivachenko, A. Y., Cibulskis, K., Gabriel, S. B., Altshuler, D., & Daly, M. J. (2011, May). A framework for variation discovery and genotyping using next-generation DNA sequencing data. *Nature Genetics*, 43(5), 491–498. DOI: 10.1038/ng.806 PMID: 21478889

Dos Santos, L. M., Mariano, D., & De Melo-Minardi, R. C. (2024). O impacto da Inteligência artificial nas ciências da vida através da Bioinformática. *Revista da Universidade Federal de Minas Gerais, 30*, e47996.

Eid, J., Fehr, A., Gray, J., Luong, K., Lyle, J., Otto, G., Peluso, P., Rank, D., Baybayan, P., Bettman, B., Bibillo, A., Bjornson, K., Chaudhuri, B., Christians, F., Cicero, R., Clark, S., Dalal, R., deWinter, A., Dixon, J., & Turner, S. (2009, January 2). Real-time DNA sequencing from single polymerase molecules. *Science*, 323(5910), 133–138. DOI: 10.1126/science.1162986 PMID: 19023044

Gao, W., Mahajan, S. P., Sulam, J., & Gray, J. J. (2020). Deep learning in protein structural modeling and design. *Patterns (New York, N.Y.)*, 1(9), 100142. DOI: 10.1016/j.patter.2020.100142 PMID: 33336200

Gauthier, J. (2019). A brief history of bioinformatics. *Briefings in Bioinformatics, 20*(6).

Google Deepmind. (n.d.). *AlphaFold: a solution to a 50-year-old grand challenge in biology*. https://deepmind.google/discover/blog/alphafold-a-solution-to-a-50-year-old-grand-challenge-in-biology/

Graw, S., Chappell, K., Washam, C. L., Gies, A., Bird, J., Robeson, M. S., & Byrum, S. D. (2021). Multi-omics data integration considerations and study design for biological systems and disease. *Molecular Omics*, 17(2), 170–185. DOI: 10.1039/D0MO00041H PMID: 33347526

Gwinn, M., Maccannell, D. R., & Khabbaz, R. F. (2017, February 22). Integrating Advanced Molecular Technologies into Public Health. *Journal of Clinical Microbiology*, 55(3), 703–714. DOI: 10.1128/JCM.01967-16 PMID: 28031438

Hagen, J. B. (2000). The origins of bioinformatics. *Nature Reviews. Genetics*, 1(3), 231–236. DOI: 10.1038/35042090 PMID: 11252753

Hasin, Y., Seldin, M., & Lusis, A. (2017, May 5). Multi-omics approaches to disease. *Genome Biology*, 18(1), 83. DOI: 10.1186/s13059-017-1215-1 PMID: 28476144

Hawkins, R. D., Hon, G. C., & Ren, B. (2010). Next-generation genomics: An integrative approach. *Nature Reviews. Genetics*, 11(7), 476–486. DOI: 10.1038/nrg2795 PMID: 20531367

Hood, L., & Rowen, L. (2013, September 13). The Human Genome Project: Big science transforms biology and medicine. *Genome Medicine*, 5(9), 79. DOI: 10.1186/gm483 PMID: 24040834

Iacoangeli, A., Marcatili, P., Deane, C., Lesk, A. M., Pastore, A., & Teichmann, S. A. (2022, April). Editorial: A Journey Through 50 Years of Structural Bioinformatics in Memoriam of Cyrus Chothia. *Frontiers in Molecular Biosciences*, 9, 28. DOI: 10.3389/fmolb.2022.885318

Ip, C. L. C., Loose, M., Tyson, J. R., de Cesare, M., Brown, B. L., Jain, M., Leggett, R. M., Eccles, D. A., Zalunin, V., Urban, J. M., Piazza, P., Bowden, R. J., Paten, B., Mwaigwisya, S., Batty, E. M., Simpson, J. T., Snutch, T. P., Birney, E., Buck, D., & Olsen, H. E. (2015, October 15). MinION Analysis and Reference Consortium: Phase 1 data release and analysis. *F1000 Research*, 4, 1075. DOI: 10.12688/f1000research.7201.1 PMID: 26834992

Iqbal, N., & Kumar, P. (2023). From data science to bioscience: Emerging era of bioinformatics applications, tools and challenges. *Procedia Computer Science*, 218, 1516–1528. DOI: 10.1016/j.procs.2023.01.130

Jäkel, S., & Williams, A. (2020). What have advances in transcriptomic technologies taught us about human white matter pathologies? *Frontiers in Cellular Neuroscience*, 14, 238. DOI: 10.3389/fncel.2020.00238 PMID: 32848627

Jumper, J. (2021). Highly accurate protein structure prediction with AlphaFold. *Nature, 596*(7873), 583–589.

Kwoji, I. D. (2023). 'Multi-omics' data integration: applications in probiotics studies. *NPJ Science of Food, 7*(1), 25.

Lander, E. S., Linton, L. M., Birren, B., Nusbaum, C., Zody, M. C., Baldwin, J., Devon, K., Dewar, K., Doyle, M., FitzHugh, W., Funke, R., Gage, D., Harris, K., Heaford, A., Howland, J., Kann, L., Lehoczky, J., LeVine, R., McEwan, P., & Morgan, M. J. (2001, February). Initial sequencing and analysis of the human genome. *Nature*, 409(6822), 860–921. DOI: 10.1038/35057062 PMID: 11237011

Lecun, Y., Boser, B., Denker, J. S., Henderson, D., Howard, R. E., Hubbard, W., & Jackel, L. D. (1989, December 1). Backpropagation Applied to Handwritten Zip Code Recognition. *Neural Computation*, 1(4), 541–551. DOI: 10.1162/neco.1989.1.4.541

Levinthal, C. (1968). Are there pathways for protein folding? *Journal de Chimie Physique*, 65, 44–45. DOI: 10.1051/jcp/1968650044

Lin, X., Li, X., & Lin, X. (2020, January). A Review on Applications of Computational Methods in Drug Screening and Design. *Molecules (Basel, Switzerland)*, 25(6), 1375. DOI: 10.3390/molecules25061375 PMID: 32197324

Lin, Z., Akin, H., Rao, R., Hie, B., Zhu, Z., Lu, W., Smetanin, N., Verkuil, R., Kabeli, O., Shmueli, Y., dos Santos Costa, A., Fazel-Zarandi, M., Sercu, T., Candido, S., & Rives, A. (2023, March 17). Evolutionary-scale prediction of atomic-level protein structure with a language model. *Science*, 379(6637), 1123–1130. DOI: 10.1126/science.ade2574 PMID: 36927031

Liu, L. (2012, July 5). Comparison of Next-Generation Sequencing Systems. *BioMed Research International*, 2012, e251364. PMID: 22829749

Lowe, R., Shirley, N., Bleackley, M., Dolan, S., & Shafee, T. (2017). Transcriptomics technologies. *PLoS Computational Biology*, 13(5), e1005457. DOI: 10.1371/journal.pcbi.1005457 PMID: 28545146

Lu, D., Lu, T., & Cao, S. (2012). Individualized cancer chemotherapy by detecting cancer biomarkers. *Metabolomics*, 2(5), e121.

Mariano, D. (2022). Revista BIOINFO. In *AlphaFold e a busca pelo Santo Graal da Biologia Molecular. 2*. Alfahelix.

Mariano, D. (2020). A brief history of bioinformatics told by data visualization. In *Brazilian Symposium On Bioinformatics*. Springer. DOI: 10.1007/978-3-030-65775-8_22

Metzker, M. L. (2005, January 12). Emerging technologies in DNA sequencing. *Genome Research*, 15(12), 1767–1776. DOI: 10.1101/gr.3770505 PMID: 16339375

Metzker, M. L. (2010, January). Sequencing technologies — The next generation. *Nature Reviews. Genetics*, 11(1), 31–46. DOI: 10.1038/nrg2626 PMID: 19997069

Min, S., Lee, B., & Yoon, S. (2017, September 1). Deep learning in bioinformatics. *Briefings in Bioinformatics*, 18(5), 851–869. PMID: 27473064

Mirdita, M., Schütze, K., Moriwaki, Y., Heo, L., Ovchinnikov, S., & Steinegger, M. (2022). ColabFold: Making Protein folding accessible to all. *Nature Methods*, 19(6), 679–682. DOI: 10.1038/s41592-022-01488-1 PMID: 35637307

Moult, J., Pedersen, J. T., Judson, R., & Fidelis, K. (1995). A large-scale experiment to assess protein structure prediction methods. *Proteins*, 23(3), ii–iv. DOI: 10.1002/prot.340230303 PMID: 8710822

Nelson, D. L., Lehninger, A. L., & Cox, M. M. (2008). *Lehninger Principles of Biochemistry*. W. H. Freeman.

O'Donoghue, S. I. (2021). Grand challenges in bioinformatics data visualization. *Frontiers in Bioinformatics*, 1, 669186. DOI: 10.3389/fbinf.2021.669186 PMID: 36303723

Paiva, V. (2022). Protein structural bioinformatics: An overview. *Computers in Biology and Medicine, 147*.

Pattan, V., Kashyap, R., Bansal, V., Candula, N., Koritala, T., & Surani, S. (2021). Genomics in medicine: A new era in medicine. *World Journal of Methodology*, 11(5), 231–242. DOI: 10.5662/wjm.v11.i5.231 PMID: 34631481

Pereira, J., Simpkin, A. J., Hartmann, M. D., Rigden, D. J., Keegan, R. M., & Lupas, A. N. (2021). High-accuracy protein structure prediction in CASP14. *Proteins*, 89(12), 1687–1699. DOI: 10.1002/prot.26171 PMID: 34218458

Reuter, J. A., Spacek, D. V., & Snyder, M. P. (2015, May 21). High-Throughput Sequencing Technologies. *Molecular Cell*, 58(4), 586–597. DOI: 10.1016/j.molcel.2015.05.004 PMID: 26000844

Russell, S. J., & Norvig, P. (2016). *Artificial intelligence: a modern approach*. Pearson.

Sanger, F., Nicklen, S., & Coulson, A. R. (1977, December). DNA sequencing with chain-terminating inhibitors. *Proceedings of the National Academy of Sciences of the United States of America*, 74(12), 5463–5467. DOI: 10.1073/pnas.74.12.5463 PMID: 271968

Sasidharan, R., & Chothia, C. (2007). The selection of acceptable protein mutations. *Proceedings of the National Academy of Sciences of the United States of America*, 104(24), 10080–10085. DOI: 10.1073/pnas.0703737104 PMID: 17540730

Senior, A. W., Evans, R., Jumper, J., Kirkpatrick, J., Sifre, L., Green, T., Qin, C., Žídek, A., Nelson, A. W. R., Bridgland, A., Penedones, H., Petersen, S., Simonyan, K., Crossan, S., Kohli, P., Jones, D. T., Silver, D., Kavukcuoglu, K., & Hassabis, D. (2020, January). Improved protein structure prediction using potentials from deep learning. *Nature*, 577(7792), 706–710. DOI: 10.1038/s41586-019-1923-7 PMID: 31942072

Skerrett-Byrne Anthony, D. (2023). Transcriptomics. In Bradshaw, R. A., Hart, G. W., & Stahl, P. D. (Eds.), *Encyclopedia of Cell Biology* (2nd ed., pp. 363–371). Academic Press. DOI: 10.1016/B978-0-12-821618-7.00157-7

Song, L., Hobaugh, M. R., Shustak, C., Cheley, S., Bayley, H., & Gouaux, J. E. (1996). Structure of staphylococcal α-hemolysin, a heptameric transmembrane pore. *Science*, 274(5294), 1859–1865. DOI: 10.1126/science.274.5294.1859 PMID: 8943190

Subramanian, I., Verma, S., Kumar, S., Jere, A., & Anamika, K. (2020). Multi-omics data integration, interpretation, and its application. *Bioinformatics and Biology Insights*, 14, 1177932219899051. DOI: 10.1177/1177932219899051 PMID: 32076369

The Times. (2024). Deepmind finds biology's 'holy grail' with answer to protein problem. Available at https://www.thetimes.co.uk/article/deepmind-finds-biology-s-holy-grail-with-answer-to-protein-problem-htg6s7qlq

Thornton, J. M., Laskowski, R. A., & Borkakoti, N. (2021, October). AlphaFold heralds a data-driven revolution in biology and medicine. *Nature Medicine*, 27(10), 1666–1669. DOI: 10.1038/s41591-021-01533-0 PMID: 34642488

Torrisi, M., Pollastri, G., & Le, Q. (2020). Deep learning methods in protein structure prediction. *Computational and Structural Biotechnology Journal*, 18, 1301–1310. DOI: 10.1016/j.csbj.2019.12.011 PMID: 32612753

Van De Sande, B., Lee, J. S., Mutasa-Gottgens, E., Naughton, B., Bacon, W., Manning, J., Wang, Y., Pollard, J., Mendez, M., Hill, J., Kumar, N., Cao, X., Chen, X., Khaladkar, M., Wen, J., Leach, A., & Ferran, E. (2023, June). Applications of single-cell RNA sequencing in drug discovery and development. *Nature Reviews. Drug Discovery*, 22(6), 496–520. DOI: 10.1038/s41573-023-00688-4 PMID: 37117846

Venter, J. C. (2001). The sequence of the human genome. *Science, 291*(5507), 1304–1351.

Wang, J., Li, Z., & Zhang, J. (2022). Visualizing the knowledge structure and evolution of bioinformatics. *BMC Bioinformatics, 23*(8), 404.

Yang, A., Zhang, W., Wang, J., Yang, K., Han, Y., & Zhang, L. (2020, September). Review on the Application of Machine Learning Algorithms in the Sequence Data Mining of DNA. *Frontiers in Bioengineering and Biotechnology*, 8, 4. DOI: 10.3389/fbioe.2020.01032 PMID: 33015010

Yang, C., Chowdhury, D., Zhang, Z., Cheung, W. K., Lu, A., Bian, Z., & Zhang, L. (2021). A review of computational tools for generating metagenome-assembled genomes from metagenomic sequencing data. *Computational and Structural Biotechnology Journal*, 19, 6301–6314. DOI: 10.1016/j.csbj.2021.11.028 PMID: 34900140

Zubair, M., Wang, J., Yu, Y., Faisal, M., Qi, M., Shah, A. U., Feng, Z., Shao, G., Wang, Y., & Xiong, Q. (2022). Proteomics approaches: A review regarding an importance of proteome analyses in understanding the pathogens and diseases. *Frontiers in Veterinary Science*, 9, 1079359. DOI: 10.3389/fvets.2022.1079359 PMID: 36601329

Zwanzig, R., Szabo, A., & Bagchi, B. (1992, January). Levinthal's paradox. *Proceedings of the National Academy of Sciences of the United States of America*, 89(1), 20–22. DOI: 10.1073/pnas.89.1.20 PMID: 1729690

# Chapter 2
# Genomics:
## A Piece of the Multi-Omics Puzzle

**Seyyed Mohammad Amin Mousavi Sagharchi**
*Department of Molecular Biology, Islamic Azad University, Tehran, Iran*

**Mohsen Sheykhhasan**
https://orcid.org/0000-0002-4610-2278
*Cellular and Molecular Research Center, Qom University of Medical Sciences, Iran*

**Atousa Ghorbani**
*Department of Biology, Islamic Azad University, Tehran, Iran*

**Elina Afrazeh**
*Department of Microbiology and Marine Biology, Khorramshahr University of Marine Sciences and Technology, Iran*

**Naresh Poondla**
*Icahn School of Medicine at Mount Sinai, USA & Center for Global Health Research, Saveetha Medical College & Hospital, India*

**Naser Kalhor**
*Academic Center for Education, Culture, and Research, Qom, Iran*

**Hamid Tanzadehpanah**
*Antimicrobial Resistance Research Center, Mashhad University of Medical Sciences, Iran*

**Hanie Mahaki**
*Vascular & Endovascular Surgery Research Center, Mashhad University of Medical Sciences, Iran*

**Hamed Manoochehri**
*The Persian Gulf Marine Biotechnology Research Center, Bushehr University of Medical Sciences, Iran*

## ABSTRACT

*Genomics is an exciting and rapidly evolving field at the intersection of biology, computer science, and statistics, which has made remarkable strides in recent years thanks to advancements in high-throughput sequencing technologies. This field*

DOI: 10.4018/979-8-3693-3192-7.ch002

*allows us to analyze and understand the genetic code of organisms, from humans to microorganisms, shedding light on everything from the diagnosis and treatment of genetic disorders to the identification of disease-causing genes in pathogens, to the mysteries of biodiversity. With its wide range of applications, genomics is transforming our understanding of life itself.*

## 1. INTRODUCTION

Genomics and bioinformatics have revolutionized the field of molecular biology and medicine by enabling scientists to analyze and interpret vast amounts of genetic data. The integration of these two disciplines has led to significant advancements in understanding the genetic basis of diseases, drug development, and personalized medicine (Yehudit Hasin, Marcus Seldin, & Aldons Lusis, 2017). By studying different organisms' genomes, we can access any raw data about under-study organisms. Further, we can process the data with bioinformatic tools to make information and finally predict the appearance and features of organisms (Huang et al., 2023). Whenever "omics" was added to a biology word, this concept contains all molecular analysis of it. Other omics sciences such as proteomics as an analysis of proteins, reactomics, transcriptomics as an examiner of RNA situation qualitatively and quantitatively, and translatomics as an investigation of translation process can link with genomic and bioinformatics (Y. Hasin, M. Seldin, & A. Lusis, 2017; Zhao, Qin, Nikolay, Spahn, & Zhang, 2019). By handling these links we can figure out molecular pathways and discover new methods and ways to improve the metabolites and other cellular functions (Figure 1) (Subramanian, Verma, Kumar, Jere, & Anamika, 2020). In microorganisms, we can make some changes in the genome and improve the secondary metabolites; in multicellular organisms, we can treat genetic disorders with this great knowledge. In the first step, investigation of the gene to product or genotype to phenotype is crucial to understanding genomics. Central dogma describes clearly this process.

*Figure 1. Overview of multi-omics and following the DNA to metabolite in central dogma*

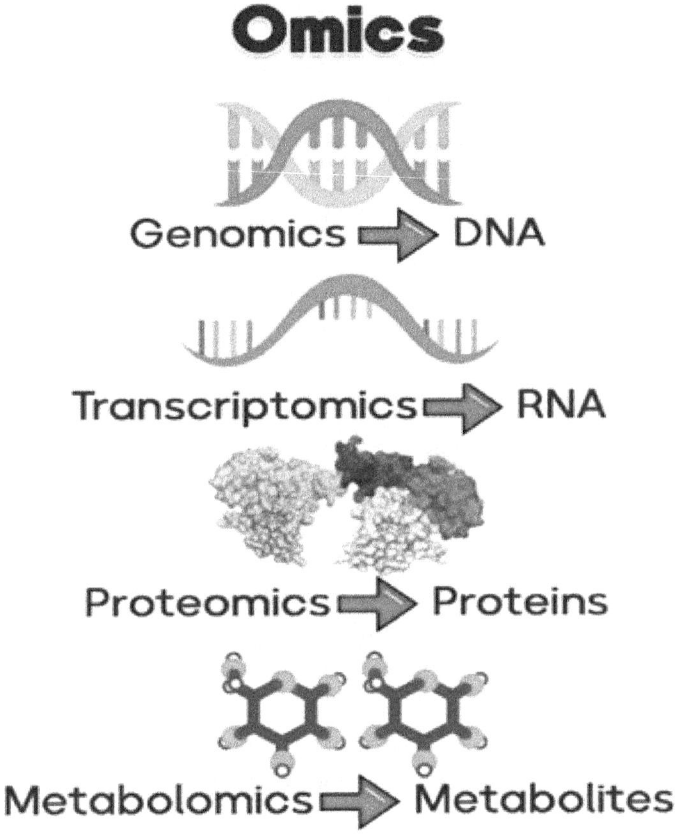


Based on central dogma in molecular biology, a metabolite originates from a related gene or gene cluster. A gene must be transcripted, and translated to make a protein or other metabolites. So, we should follow the main path of central dogma to find out the mechanism of cells to produce products (Thieffry & Sarkar, 1998). DNA after transcription makes types of RNAs (e.g., messenger RNA (mRNA), transfer RNA (tRNA), and ribosomal RNA (rRNA)) each one of them has a voice in different applications. For example, mRNA is converted to peptides by ribosomes and tRNA, finally making a string of polypeptides (Morange, 2006). In the transcription process, RNA polymerases play the main role. These factors in addition to other factors such as boxes and upstream transcription factors, initial the transcription from DNA to make RNA. Genes and factors are the main regulators for this purpose. A large RNA-protein complex known as "spliceosome" catalyzes the process of intron's removal from nuclear pre-mRNA (Lamond, 1993). RNA is under modification by spliceosomes, and new RNAs appear in non-coding or coding RNAs. Coding RNAs and non-coding RNAs are different in function. Coding RNAs such as mRNA make proteins after translation and have a role in the main functions of cells. However noncoding RNAs act role as cellular regulators without converting to proteins (Li & Liu, 2019). Coding RNAs are placed in ribosomes for translation and converting to peptides by helping a non-coding RNA "tRNA". Peptides are linked to each other by an amide type of covalent chemical bond known as peptide bind between the carboxyl group and amino acid group of the next peptide (Lee & Young, 2000). Polypeptides link to each other and make the primary structure of protein. All of this process is controlled and initiated by DNA and in other ways from genes. So, the investigation of genes and their nucleotides is necessary for prediction of proteins and other functions of a gene (Jowett, 1999). Genomics is the area for frontiers in molecular biology and is a broad field for the analysis of genes by different tools.

A gene is the sequence of nucleotides that are linked to each other by phosphor-di-ester binds that contains a nucleoside, and phosphate (Figure 2). This structure is repeated similarly but there is a difference in nitrogenic bases in each nucleotide. Nitrogenic bases are classified into two types based on chemical structure: 1. Purine 2. Pyrimidine. Purine bases are adenine and guanine with two aromatic rings. Pyrimidines are cytosine, thymine, and uracil with one aromatic ring. We know these bases as A, G, C, T, and U. U is presented only in the RNAs (Durland & Ahmadian-Moghadam, 2024). Order of these bases in genome determines the features and details of genes and their phenotypes.

Different methods are designed for genome/genes sequencing. The simplest method is the Sanger method which applies electrophoresis to sequencing (Crossley et al., 2020). The advent of high-throughput sequencing technologies has revolutionized genomics research, enabling scientists to rapidly sequence entire genomes and analyze the vast amounts of data generated. This has led to significant advancements in fields such as personalized medicine, evolutionary biology, and agriculture. Different sequencing methods make us possible to identify genes, investigate the genome of each organism, and find out how phenotype appears.

*Figure 2. A nucleotide contains a purine base*

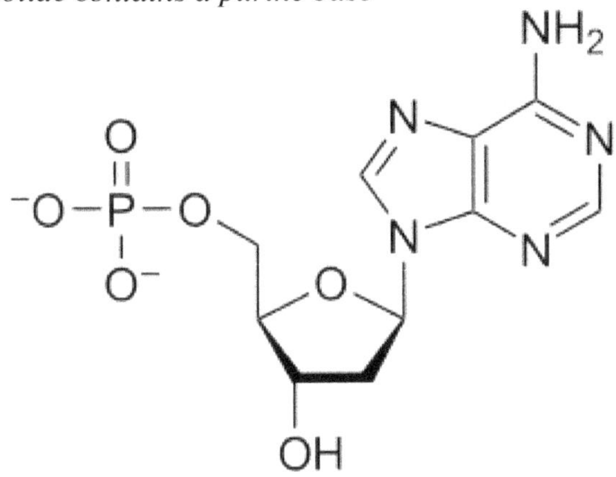


Whole genome sequencing, the process of determining the complete DNA sequence of an organism, has revolutionized the field of biology and medicine. Sequencing occurs by different techniques such as whole genome sequencing (WGS), next-generation sequencing (NGS), third-generation sequencing (TGS), etc. Genome sequencing provides a comprehensive map of an organism's genetic code, revealing the sequence of nucleotide bases that make up its DNA. By sequencing DNA, replication, and translation factors (e.g., ribozymes, and polymerases) and blasting them we can find mutations and unwanted molecular changes. Variant calling is a powerful tool for detecting genetic variations, which include single nucleotide polymorphisms (SNPs) and insertions/deletions (indels). This method helps researchers identify and analyze these genetic differences for various applications, such as understanding disease susceptibility and tracing evolutionary relationships (Chan, 2005; Y. Chen et al., 2021). As mentioned, there are many methods for the sequencing of DNA, also every sequenced DNA/cDNA is recorded on online databases such as the GenBank of National Center for Biotechnology Information (NCBI), Gene cards, KEGG gene database, etc. These online databases give us a huge amount of data about the genomes of organisms and microorganisms. Online databases in genomics offer comprehensive annotations of genomic elements, including genes, regulatory regions, non-coding RNAs, and repetitive sequences, providing valuable insights into the functional and structural aspects of the genome. Studying and sequencing of genes in DNA allows us to map the pathway of genes to products (Hutchins, 2014). So, Databases integrate data from epigenomic studies, such as DNA methylation patterns and histone modifications, enabling researchers to investigate the epigenetic regulation of gene expression and its impact on cellular function. One of the key goals of genomics research is to identify genetic variations that are associated with specific disorders or diseases. By comparing the genomes of individuals with and without a particular trait or disease, researchers can pinpoint genetic variants that may play a role in its development. This information can then be used to develop targeted therapies or interventions to improve health outcomes (Lesk, 2017).

In this chapter, the authors do their best to describe all aspects of genomics and related approaches from sequencing to online databases, and applications in different fields such as medicine, agriculture, and industrial purposes. Genomics has great potential to meet our needs in different fields. To benefit from this valuable science, we should learn all aspects of this science. Sequencing, reaching to data, analysis, and applications are four majors we will investigate in our chapter.

## 2. GENOME SEQUENCING

Genome sequencing is the revolutionary process of uncovering the entire DNA sequence of an organism's genome (Koboldt, Steinberg, Larson, Wilson, & Mardis, 2013). Its significance spans across numerous disciplines, owing to its profound implications and wide-ranging applications (Koboldt et al., 2013). In the realm of medicine, genome sequencing is pivotal in the realm of personalized medicine, allowing for tailored treatment strategies and targeted therapies by identifying genetic variations associated with diseases (Brittain, Scott, & Thomas, 2017). It aids in diagnosing genetic disorders, predicting disease susceptibility, and guiding preventative measures. Additionally, in pharmacogenomics, genome sequencing optimizes drug effectiveness and minimizes adverse reactions based on an individual's genetic profile (P. D. Gupta, 2015). For example, NGS was reported as a reliable and efficient tool for sequencing and identifying genetic variation in genes that refer to drug pharmacokinetics and pharmacodynamics (Schwarz, Gulilat, & Kim, 2019). A prominent instance is the use of warfarin, a widely prescribed anticoagulant. Variations in the CYP2C9 and VKORC1 genes cause considerable differences in warfarin's efficacy among individuals. Genome sequencing enables physicians to tailor warfarin dosages according to a patient's genetic profile, thereby minimizing adverse side effects and enhancing treatment results (Johnson & Cavallari, 2015). Another exemplary case is Ivacaftor, a drug for treating cystic fibrosis. Genome sequencing has revealed that around 5% of cystic fibrosis patients carry a unique G551D mutation in the CFTR gene (Kotha & Clancy, 2013). Ivacaftor has shown remarkable effectiveness in treating this specific patient subset, underscoring the potential of genome sequencing in stratifying patient populations for precision medicine. This precision approach allows for tailored treatment plans based on an individual's genetic makeup, improving overall treatment outcomes and quality of life for patients (Feng, Grosse, Green, Fink, & Sawicki, 2018). Within agriculture, genome sequencing plays a crucial role in enhancing crop productivity and livestock breeding (Gao, 2021). By pinpointing genes linked to desirable traits such as yield, disease resistance, and nutritional content, it facilitates the development of genetically improved crops and livestock breeds, thereby contributing to agricultural sustainability (Abdul Aziz, Brini, Rouached, & Masmoudi, 2022). In evolutionary biology, genome sequencing offers insights into the genetic underpinnings of evolutionary processes, species divergence, and adaptation to diverse environments (Zheng et al., 2023). Comparative genomics enables the exploration of genetic similarities and differences among species, illuminating evolutionary relationships and mechanisms driving genetic diversity (Alföldi & Lindblad-Toh, 2013; Bornstein, Gryan, Chang, Marchler-Bauer, & Schneider, 2023).

## 2.1. Fundamentals of Genome Sequencing

Genome sequencing, the process of precisely determining the sequence of nucleotides within an organism's entire genome, is a monumental scientific endeavor with profound implications across various domains (Jovic et al., 2022). Its historical trajectory traces back to the early days of molecular biology, where pioneering efforts in the 1970s laid the groundwork for subsequent breakthroughs. Notably, the development of Sanger sequencing in the late 1970s represented a transformative leap forward in DNA sequencing, offering unprecedented accuracy and efficiency in determining DNA sequence fragments (Heather & Chain, 2016). The initiation of the Human Genome Project (HGP) in 1990 was a crucial moment in genome sequencing history. The HGP aimed to map and sequence the entire human genome, sparking global collaborative efforts (Hood & Rowen, 2013). By 2003, the HGP successfully completed this monumental task, presenting the initial comprehensive sequence of the human genome. This achievement not only transformed our comprehension of human biology but also set the stage for a new era of genomic research and its practical applications (Collins & Fink, 1995). Key concepts in genome sequencing include DNA sequencing, genome assembly, genome mapping, and genetic variation analysis. DNA sequencing is fundamental to genome sequencing, involving the determination of nucleotide sequences in DNA molecules (Logsdon, Vollger, & Eichler, 2020). Genome assembly is the process of reconstructing a complete genome sequence from the fragmented DNA data produced by sequencing technologies. The goal is to generate an accurate reference genome that captures an organism's full genetic blueprint. This is challenging work, as researchers must address gaps in the input data, mitigate biases in the assembly process, and navigate complex genomic regions. But advances in sequencing and computational methods are making genome assembly better, faster, and cheaper - driving progress in our understanding of genes, genomes, and evolution. Overall, genome assembly is a critical enabling step that allows us to leverage the wealth of genomic data for diverse downstream applications and analyses. It's an active area of research and innovation in the field of genomics (Hyungtaek Jung et al., 2020; Yang et al., 2021).

Genome assembly is the computational process of reconstructing the complete genome sequence from short DNA sequence reads obtained through sequencing technologies. Genome mapping involves identifying the physical locations of genes and other genomic features on chromosomes (Ghurye & Pop, 2019). Genetic variation analysis involves studying the differences in DNA sequences among individuals of the same species. This includes identifying and characterizing variations like insertions and deletions (indels), SNPs, and larger structural variations. Understanding genetic variation is essential for investigating the genetic underpinnings of health and disease, as well as for tracing evolutionary histories. However, accurately detecting and interpreting genetic variation requires high-quality genome assemblies and advanced sequencing technologies. The challenge lies in capturing the full range of genetic variation comprehensively, including rare variants and complex structural changes, which are often overlooked or challenging to analyze with current sequencing and assembly methods (Flynn & Lappalainen, 2022; Yu Lei et al., 2022).

Various sequencing technologies have emerged to facilitate genome sequencing, each utilizing distinct principles and methodologies. Sanger sequencing, known for its accuracy and reliability, uses the chain termination method. NGS technology like Illumina sequencing and Ion Torrent sequencing allow high-throughput sequencing of DNA fragments simultaneously (H. Satam et al., 2023). TGS technology such as PacBio sequencing and Oxford Nanopore sequencing provide long-read sequencing capabilities, enabling direct sequencing of single DNA molecules. These sequencing technologies have transformed genomics research, offering unparalleled insights into genetic information (Athanasopoulou, Boti, Adamopoulos, Skourou, & Scorilas, 2021). From uncovering the genetic roots of diseases to elucidating evolutionary relationships among species, genome sequencing continues to propel scientific discovery and innovation, shaping our understanding of biology and its applications in medicine, agriculture, and beyond (E. D. Green et al., 2020).

## 2.2. Tools for Genome Sequencing

Sequencing platforms have seen significant advancements, each with unique advantages and limitations. Sanger sequencing, based on chain termination, provides longer reads suitable for targeted sequencing and small projects (Cheng, Fei, & Xiao, 2023). Despite its usefulness, Sanger sequencing lacks the throughput of modern methods. NGS platforms like Illumina and Ion Torrent allow simultaneous sequencing of many DNA fragments, offering high throughput at a reasonable cost (N. Gupta & Verma, 2019). While NGS platforms have shorter read lengths, they are widely used in genomics, transcriptomics, and epigenomics due to their scalability and cost-effectiveness (H. Satam et al., 2023). TGS, seen in PacBio and Oxford Nanopore sequencing, enables long-read sequencing and real-time sequencing without the need for amplification or fragmentation. Although TGS platforms have lower accuracy than NGS, their longer reads surpass Sanger and NGS, crucial for tasks needing continuous sequences (Athanasopoulou et al., 2021). Comparing sequencing platforms reveals features like accuracy, throughput, read length, and cost. Sanger sequencing, known for precision, has limited throughput, suitable for focused projects with small samples (Cheng et al., 2023). NGS platforms offer moderate to high accuracy and throughput, with shorter reads but superior scalability. TGS platforms provide long-read sequencing and real-time capabilities, though with reduced accuracy that is improving with technology (Cheng et al., 2023). While Sanger sequencing is costlier per base, NGS and TGS platforms offer more cost-effective options for high-throughput sequencing, depending on initial instrument costs. Decoding sequencing data relies on specialized software for alignment, variant calling, genome annotation, and visualization (Cheng et al., 2023). Alignment tools like Bowtie and BWA align reads accurately to a reference genome. Variant calling software such as GATK and Samtools identify genetic variants like SNPs and structural variations. Genome annotation tools like ANNOVAR and Ensembl annotate genomic features, providing context to genetic variants (Schilbert, Rempel, & Pucker, 2020). Visualization tools like IGV and UCSC Genome Browser help researchers visually analyze reads, annotations, and variants, aiding data interpretation and hypothesis development (Nusrat, Harbig, & Gehlenborg, 2019).

## 2.3. Types of Sequencing and Strategies

DNA sequencing is a powerful analytical technique that allows scientists to determine the precise order of nucleotides - the building blocks of DNA - within a given genetic sample. By obtaining this detailed DNA sequence information, researchers can gain valuable insights into the genetic information encoded within biological systems (Alex, Brundha, & Prathap, 2020). DNA sequencing has evolved significantly over the years, with different generations of technologies emerging. Second-generation sequencing was built around the use of PCR (polymerase chain reaction) techniques (Lu Zhang et al., 2021). This represented an advancement from the first-generation approach, which was centered on shotgun sequencing relying on traditional Sanger sequencing. Interestingly, the Sanger sequencing method using dideoxy chain termination is still considered the gold standard in terms of sequencing accuracy. In fact, shotgun sequencing - the foundational first-generation approach - actually boasts higher accuracy than both second-generation and TGS techniques. Shotgun sequencing also has the advantage of being able to produce long read lengths, from 700 to 1,000 base pairs, and it can handle repeated DNA sequences very well. However, the tradeoff is that it can only process one DNA template at a time, making it a slower and more time-consuming process compared to later advancements (Alex et al., 2020; Rebets, Kormanec, Luzhetskyy, Bernaerts, & Anné, 2017; Lu Zhang et al., 2021). Sanger sequencing, a key genomics technique, relies on chain termination. DNA synthesis stops at specific nucleotide positions, creating fragments of different lengths. These fragments are separated and analyzed to determine the DNA sequence (Heather & Chain, 2016). While Sanger sequencing is valuable for gene exploration, its limited throughput and high cost per base hinder its use in large-scale genomic projects (Giani, Gallo, Gianfranceschi, & Formenti, 2020). NGS referred to as massively parallel sequencing, enables the sequencing of a human's entire genome in less than a day. NGS has revolutionized genomics by enabling the discovery of disease-related genes and pathways. It allows for faster diagnosis and targeted treatments for genetic diseases and cancers. RNA-Seq, an NGS-based method, provides detailed gene expression data, surpassing microarrays. NGS also detects a wide range of genetic differences. While cost-effective and efficient, integrating NGS into routine clinical practice still faces challenges (Alex et al., 2020; Heena Satam et al., 2023). Metagenomic sequencing involves analyzing DNA from microbial communities in environmental samples, supporting microbiome and environmental studies (Akaçin, Ersoy, Doluca, & Güngörmüşler, 2022). Metagenomics extracts genetic material directly from natural environments, bypassing the need to culture microorganisms. This allows for comprehensive genetic analysis of microbial communities. It is divided into functional metagenomics, which discovers new bioactive substances and microbial genes, and sequencing metagenomics, which studies microbial diversity and abundance (L Zhang et al., 2021). Applications span environmental studies (oceans, soils), medical diagnostics (identifying pathogens), and industrial uses (novel enzymes for manufacturing). The advantages of metagenomics include accelerating drug development and identifying unculturable microorganisms. However, there are challenges such as incomplete DNA extraction, low screening efficiency, and limited microbial databases. Despite these hurdles, metagenomics holds immense potential for advancements in various fields, driven by ongoing improvements in high-throughput and sensitive screening technologies (Chiu & Miller, 2019; Pereira, 2019; L Zhang et al., 2021).

Single-cell RNA sequencing (scRNA-Seq) involves isolating single cells, reverse transcription, cDNA synthesis, library construction, high-throughput sequencing, and data analysis (S. Wang et al., 2023). This method helps identify diverse microbial species and their roles in ecosystems. Single-cell sequencing methods are powerful for studying cellular diversity and rare cell populations in complex biological systems (Jovic et al., 2022). Single-cell sequencing finds applications in cancer research, developmental biology, and neuroscience, providing detailed insights into cellular dynamics and disease mechanisms (Y. Lei et al., 2021). Recent advancements in scRNA-Seq technology have improved its throughput and accuracy, providing deeper insights into clinical mechanisms and potential applications. ScRNA-Seq is essential for studying embryonic development and organ formation by tracking cell differentiation and identifying key genes. It maps gene expression in developing tissues like the brain, heart, and liver, helping understand organ development and disease mechanisms. In cancer research, scRNA-Seq reveals tumor cell diversity, uncovers tumorigenesis mechanisms, identifies new biomarkers, and informs treatment strategies by mapping the tumor microenvironment and understanding drug resistance (Liao et al., 2023; S. Wang et al., 2023).

## 2.4. Genome Annotation

The journey of sequencing and understanding the human genome saw remarkable progress, starting slow but accelerating rapidly. Before the actual sequencing, scientists identified many aspects of the genome through clever experiments. When DNA sequencing became feasible, these theoretical concepts found real-world applications, connecting genetic sequences to observable traits, and forming a comprehensive annotation (Zerbino, Frankish, & Flicek, 2020). Genome annotation is a crucial process in genomics that involves identifying and assigning functional elements and characteristics to DNA sequences within a genome. This annotation is vital for understanding the genetic makeup and functionality of organisms, providing insights into gene structure, regulatory elements, coding sequences, and non-coding RNAs (Ejigu & Jung, 2020). Annotation pipelines and databases are essential for integrating experimental data, computational predictions, and comparative genomics analyses to accurately annotate genomic features (Ejigu & Jung, 2020). Different types of genome annotations include structural annotation, which identifies gene structures and regulatory elements, and functional annotation, which assigns biological functions to genes and other genomic elements. Comparative annotation entails comparing genomes across species to identify evolutionarily conserved elements and infer functional significance. Despite advancements in annotation methodologies, challenges remain in accurately predicting gene structures in complex genomes and annotating non-coding regions (Ejigu & Jung, 2020). As genomic data continue to grow in volume and complexity, future perspectives in genome annotation will likely involve developing more sophisticated computational tools, integrating multi-omics data, and collaborating to improve annotation accuracy and comprehensiveness across diverse organisms and ecosystems (Bianconi, Aschbacher, & Pagani, 2023).

The non-coding regions constitute a substantial proportion of the genome, presenting a unique challenge due to their lack of direct protein-coding potential. Despite this, they harbor crucial regulatory elements such as promoters, enhancers, and insulators, pivotal in modulating gene expression dynamics. However, elucidating the functionality of these elements proves intricate given their non-protein-coding nature, resulting in indirect and context-dependent effects. Compounding this complexity is the dynamic interplay between these regulatory elements and a plethora of transcription factors and chromatin modifiers, whose interactions vary across different cellular contexts and developmental stages, underscoring the nuanced nature of their identification and annotation(Micheel, Safrastyan, & Wollny, 2021; Necsulea, 2020; Zerbino et al., 2020).The study of long non-coding RNAs (lncRNAs) has been marked by uncertainty and debate since its inception, largely due to their misleading name and the initial belief that RNA primarily acted as a protein intermediary. The perception of RNA evolved significantly with the discovery of regulatory molecules like siRNAs and miRNAs in the early 2000s. Despite earlier observations of long regulatory RNAs, they were often viewed as exceptions. However, transcriptomic analyses later unveiled extensive transcription of non-coding RNAs throughout genomes, challenging existing notions of genetic information. Initially dismissed as mere background noise, lncRNAs are now acknowledged for their dynamic expression patterns and biological relevance, though many still lack comprehensive annotations. This prevalence of lncRNAs necessitates a reexamination of genomic processes in complex organisms (Fang et al., 2018; Mattick et al., 2023; Uszczynska-Ratajczak, Lagarde, Frankish, Guigó, & Johnson, 2018). To tackle these challenges, we need better computer tools, combining different types of biological data, and more teamwork among researchers. This will help us make more accurate and comprehensive annotations of genomes in various species.

## 2.5. Applications of Genome Sequencing

Genome sequencing has transformed various fields, including human genomics, playing a key role in personalized medicine, disease diagnosis, and pharmacogenomics. By sequencing an individual's genome, clinicians can identify genetic variations associated with diseases, predict disease risk, and customize treatment strategies (Brittain et al., 2017). In agriculture, genome sequencing aids in crop improvement and livestock breeding by pinpointing genes linked to desirable traits like yield, disease resistance, and nutritional content, thereby boosting agricultural productivity and sustainability (Martina et al., 2024). In microbial genomics, genome sequencing allows for infectious disease surveillance and monitoring of antibiotic resistance by identifying and characterizing microbial pathogens and their genetic determinants (Bianconi et al., 2023; Mousavi Sagharchi & Mahmoudi Nasab, 2023). In evolutionary biology, genome sequencing offers insights into phylogenetics and comparative genomics, illuminating evolutionary relationships among species and the genetic mechanisms of adaptation to diverse environments (Alföldi & Lindblad-Toh, 2013). Overall, genome sequencing serves as a potent tool for comprehending biological systems, fostering innovation across diverse domains, and tackling critical challenges in human health, agriculture, microbial ecology, and evolutionary biology (E. D. Green et al., 2020).

## 2.6. Future Directions and Challenges

Future directions in genome sequencing are marked by emerging technologies and trends aimed at enhancing efficiency, accuracy, and accessibility. Advancements in sequencing technologies, such as the continued development of TGS platforms and the emergence of novel sequencing methodologies, promise to further improve read lengths, throughput, and cost-effectiveness. Additionally, there is a growing focus on multi-omics integration, combining genomic data with other omics datasets such as transcriptomics, epigenomics, and proteomics, to provide a more comprehensive understanding of biological systems (Schwarz et al., 2019). However, along with these opportunities come challenges in data analysis, interpretation, and storage. The exponential growth of genomic data poses significant computational and storage challenges, requiring scalable and efficient algorithms and infrastructure for data management and analysis. Moreover, ethical considerations and societal implications surrounding genome sequencing, including issues related to privacy, data security, and equitable access to genomic information, must be carefully addressed. Strategies for overcoming these challenges and advancing the field include collaborative efforts between researchers, clinicians, policymakers, and ethicists to develop robust frameworks for data sharing, privacy protection, and ethical use of genomic information. Additionally, investment in education and public engagement is essential to ensure broader awareness and understanding of the implications of genome sequencing, fostering informed decision-making and responsible implementation of genomic technologies in various domains. Through concerted efforts and innovative approaches, genome sequencing holds the promise of unlocking new insights into biology, medicine, agriculture, and beyond, paving the way for transformative advancements in science and society (Burton, Adams, Bunton, & Schröder-Bäck, 2008; Eric D Green et al., 2020; Juengst, McGowan, Fishman, & Settersten Jr, 2016; Meagher & Lee, 2016). Advancements in artificial intelligence (AI) and machine learning are shaping the future of genome annotation, promising more efficient and accurate analysis. AI aids in variant classification, genotype-to-phenotype mapping, and clinical diagnostics, revolutionizing genetic analysis and disease prediction. Specific applications include enhanced genomic variant interpretation, polygenic risk score calculation, drug discovery, and single-cell genomics analysis. However, ethical and regulatory considerations, such as data privacy and responsible use, must be addressed as AI-driven genomics applications become more prevalent (Z. Chen, Ain, Zhao, & Zhang, 2024; Lin & Ngiam, 2023). While deep learning excels in handling complex data, it's not always superior to machine learning, especially with small datasets. Hardware support and high parallelism requirements may limit deep learning's applicability. Integrating natural language processing into genomics poses interpretability challenges. While deep learning can extract features effectively, its opaque nature hinders biological interpretability. Currently, inferring features and evaluating model weights based on input-output relationships are common approaches. Choosing the right deep learning model, such as CNNs for capturing omics data attributes, RNNs for sequence contextual information, and autoencoders for revealing hidden data distributions, requires careful consideration based on research objectives (Z. Chen et al., 2024).

## 3. ANALYTIC TOOLS

Genome sequencing has transformed our comprehension of life at the molecular level, exposing the complex genetic codes that regulate organisms' biology. Yet, the vast data from sequencing experiments present a challenge: how to extract meaningful insights effectively. Advanced analytic tools are essential in genomic research, enabling scientists to decode the mysteries within the genome. From online databases with extensive genomic repositories to specialized software for data analysis and statistical inference, each tool contributes to genomic research.

### 3.1. Online Databases for Genomic Data Collection

Online databases are essential for modern biological research, providing researchers with a wide range of genomic information from various organisms. These databases store genomic sequences, annotations, and metadata, while also offering tools for data analysis and interpretation (Vinatzer et al., 2019). Here are some notable online databases dedicated to genomic data:

NCBI GenBank: This repository, established by the NCBI contains nucleotide sequence data from a diverse range of organisms, with detailed annotations and metadata for each sequence (Benson, Karsch-Mizrachi, Lipman, Ostell, & Sayers, 2010).

ENSEMBL: A comprehensive genome annotation database that integrates data from multiple sources, including genome assemblies, gene annotations, and variation data. It provides an intuitive interface for browsing and analyzing genomic data, as well as tools for comparative genomics and functional annotation (H. Jung et al., 2020).

UCSC Genome Browser: Known for visualizing and analyzing genomic data, this browser offers interactive access to genome assemblies, gene annotations, regulatory elements, and other genomic features across various organisms. It provides a wide range of visualization options and tools for analyzing genomic tracks (Karolchik, Hinrichs, & Kent, 2011).

European Nucleotide Archive (ENA): This repository stores nucleotide sequence data, including raw sequencing reads, assembled genomes, and transcriptomic data. It supports data generated by different sequencing technologies and offers tools for data submission, retrieval, and analysis (Leinonen et al., 2011).

Genomic Data Commons (GDC): Established by the National Cancer Institute (NCI), the GDC is a data portal that provides access to genomic and clinical data from cancer studies. It contains a wealth of genomic data, such as somatic mutations, gene expression profiles, and clinical metadata, along with tools for data exploration and analysis (Jensen, Ferretti, Grossman, & Staudt, 2017).

Exome Aggregation Consortium (ExAC): This repository aggregates exome sequencing data from large-scale studies of human genetic variation, offering allele frequency data for genetic variants across diverse human populations. Researchers can use this resource to assess the frequency and distribution of genetic variation in the genome (Karczewski et al., 2017).

*Table 1. List of related databases to genome data extraction*

| Database Name | Description |
|---|---|
| NCBI GenBank | An extensive repository of DNA sequences and annotations, maintained by the NCBI. |
| ENSEMBL | Sequences and annotations for vertebrate species, providing access to gene annotations, comparative genomics data, and functional annotations. |
| UCSC Genome Browser | A comprehensive database of genomic A web-based tool for visualizing and analyzing genomic data, offering a user-friendly interface for exploring genome assemblies, gene annotations, and epigenomic data. |
| ENA | A centralized repository for storing and sharing nucleotide sequence data, providing access to raw sequencing data, assembled genomes, and metadata. |
| GDC | A data portal for accessing genomic data from cancer studies, providing access to genomic profiles, clinical data, and analysis tools for cancer research. |
| ExAC | A database containing exome sequencing data from over 60,000 individuals, enabling the exploration of genetic variation in human populations. |

## 3.2. Data Structures and File Formats

Data structures and file formats are crucial elements in the management and analysis of genomic data, enabling the storage, retrieval, and manipulation of genetic information. Two commonly used formats in genome sequencing are the FASTA and FASTQ formats. FASTA format is a widely adopted text-based format for representing nucleotide or protein sequences and their metadata. Each sequence in FASTA is represented by a header line starting with ">", followed by lines containing the sequence data. The header typically includes information like the sequence identifier or description. The sequence data can span multiple lines, each limited in length for readability. The FASTQ format is a text-based format used to store sequencing reads from technologies like Illumina sequencing (Bao et al., 2014).

### 3.2.1. Applications and Usage in Genome Sequencing

The FASTA and FASTQ formats are fundamental for storing and exchanging genomic data in bioinformatics. FASTA files are commonly used to store reference genomes, gene sequences, and other biological sequences, making them essential for genome annotation, sequence alignment, and comparative genomics analysis (Bao et al., 2014). On the other hand, FASTQ files are indispensable for storing sequencing reads generated from high-throughput sequencing experiments. These files are used in various applications, including read mapping, variant calling, transcriptome analysis, and metagenomics studies (Bao et al., 2014).

### 3.2.2. Advantages and Limitations

The FASTA format's simplicity and versatility make it widely compatible with bioinformatics tools and databases, facilitating seamless data exchange and analysis. However, its lack of standardized metadata fields can lead to inconsistencies in data interpretation and integration. In contrast, the FASTQ format's inclusion of quality scores enables accurate assessment of sequencing read quality and facilitates downstream analysis, such as read trimming and filtering. Nevertheless, the verbosity of the FASTQ format can result in larger file sizes, posing challenges for data storage and processing, particularly with large-scale sequencing experiments (Mrozek, Stępień, Grzesik, & Małysiak-Mrozek, 2021). Additionally, both formats may exhibit limitations in representing complex genomic features, such as structural variations or epigenetic modifications, necessitating supplementary data formats or specialized tools for comprehensive analysis (H. Satam et al., 2023). The limitations of using FASTA and FASTQ formats in bioinformatics stem from their design, which, while useful in some ways, presents challenges in others. For FASTA, the lack of standardized metadata leads to inconsistencies in data interpretation and integration across different tools and databases, potentially causing discrepancies in analysis and hindering research reproducibility. Additionally, this lack of metadata complicates software integration and efficient data retrieval, particularly in large-scale genomic studies. For FASTQ, the verbosity results in large file sizes, affecting workflow efficiency and accuracy (Hoogstrate, Jenster, & Werken, 2021; Nielsen et al., 2023; Privitera, Alaimo, Ferro, & Pulvirenti, 2022). The large file sizes of FASTQ files, due to their inclusion of quality scores for each base, pose significant challenges. These include difficulties in data storage and processing, especially in large-scale sequencing experiments, making data management cumbersome. Additionally, the size of FASTQ files can hinder certain bioinformatics analyses that require loading entire datasets into memory, potentially slowing down research and increasing computational costs (Du et al., 2021; Kryukov, Ueda, Nakagawa, & Imanishi, 2020; Telatin, Fariselli, & Birolo, 2021). To address these limitations, the bioinformatics community is developing new formats and strategies. Compressed formats like CRAM and tools such as the FASTA File System (FASTAFS) are designed to balance efficient data storage with the need for detailed sequence and metadata information. CRAM, for instance, reduces file sizes by compressing the data, while FASTAFS enhances data retrieval and integration by structuring the metadata more effectively. These innovations aim to overcome the challenges posed by the verbosity and lack of standardization in the current FASTA and FASTQ formats (Hoogstrate et al., 2021). By improving storage efficiency and enabling better data access and analysis, these new approaches are paving the way for more effective and streamlined genomic research.

## 3.3. Statistics in Genomic Data Analysis

Statistics plays a crucial role in genomic data analysis, enabling researchers to extract meaningful insights from complex biological datasets. Here, we explore various statistical methods commonly employed in genomic research:

Descriptive Statistics: Descriptive statistics involve summarizing and describing the characteristics of genomic data, including measures of central tendency (e.g., mean, median, mode), dispersion (e.g., standard deviation, variance), and distribution (e.g., histograms, box plots). Descriptive statistics provide a comprehensive overview of genomic data, aiding in data exploration and visualization (Cooksey, 2020).

Inferential Statistics: Inferential statistics involves making inferences or predictions about a population based on a sample of data. In genomic research, inferential statistics are used to test hypotheses, assess relationships between variables, and make predictions about biological phenomena. Common inferential techniques include regression analysis, analysis of variance (ANOVA), and correlation analysis (Geman, Ochs, Price, Tomasetti, & Younes, 2015).

Regression Analysis: Regression analysis is a statistical method used to model the relationship between one or more independent variables (predictors) and a dependent variable (outcome). In genomic research, regression analysis is employed to identify associations between genetic variants and phenotypic traits, such as disease susceptibility or drug response. Linear regression, logistic regression, and

Cox regression are among the regression techniques commonly used in genomic studies (Alamin, Sultana, Xu, & Mollah, 2020).

Hypothesis Testing: Hypothesis testing is a statistical method used to evaluate the validity of a hypothesis based on observed data. In genomic research, hypothesis testing is employed to assess whether observed differences in genetic variation or gene expression are statistically significant. Common hypothesis tests include t-tests, chi-square tests, and Fisher's exact tests, depending on the nature of the data and the research question (Emmert-Streib & Dehmer, 2019).

Applications in Genomic Research: Statistics finds extensive applications in genomic research across various domains, including genome-wide association studies (GWAS), gene expression analysis, and pathway analysis. GWAS involves identifying genetic variants associated with complex traits or diseases by analyzing genotype-phenotype associations in large populations (Adam, Samtal, Brandenburg, Falola, & Adebiyi, 2021). Gene expression analysis assesses changes in gene expression levels under different experimental conditions, such as disease states or drug treatments, using statistical techniques such as differential gene expression analysis. Pathway analysis involves identifying biological pathways or networks enriched with genes associated with a specific trait or characteristic or disease, using statistical methods to assess pathway-level significance (Rosati et al., 2024).

## 3.4. Introduction to R for Genomic Data Analysis

R is a powerful open-source programming language and environment for statistical computing and graphics. It provides a wide range of statistical and graphical techniques and is widely used in genomic data analysis due to its flexibility, extensibility, and rich ecosystem of packages. R features a syntax that is conducive to data manipulation and analysis, making it well-suited for handling large genomic datasets (Schatz et al., 2022).

Bioconductor: Bioconductor is a collection of R packages specifically tailored for the analysis and comprehension of genomic data. These packages cover a broad spectrum of genomic applications, including sequence analysis, microarray analysis, NGS data processing, and pathway analysis. Bioconductor provides a unified framework for accessing and analyzing genomic data, facilitating reproducible research and collaboration within the scientific community (Huber et al., 2015).

One notable tool is GeneTonic, a R/Bioconductor package that streamlines RNA-seq data interpretation, integrating quantification, differential expression, and functional enrichment into a single workflow. GeneTonic's interactive Shiny application and RMarkdown reports ensure reproducibility and ease of use. R and Bioconductor's modularity, interoperability, and open-source nature foster transparency and innovation, making them ideal for clinical bioinformatics. These tools support the development of validated, standardized pipelines, crucial for advancing personalized medicine and patient care (Sepulveda, 2020).

Interpreting RNA sequencing (RNA-seq) data is often challenging because essential information is scattered across various formats. The GeneTonic software package simplifies this process by streamlining the interpretation of functional enrichment analyses. It integrates expression values in a differential expression context and offers interactive visualizations through R and Shiny. Compatible with Bioconductor workflows, GeneTonic supports many tools for pathway analysis, allowing users to navigate data, bookmark features, and create customized HTML reports for enhanced interactivity and reproducibility. GeneTonic is an R package available through Bioconductor under the MIT license. It aims to make complex RNA-seq data more accessible by offering both broad overviews and detailed inspections of genes and their relationships. This tool bridges the gap between different software solutions and file formats, providing user-friendly summaries and visualizations. GeneTonic helps researchers generate new insights and hypotheses efficiently. The package can be found on Bioconductor, with a public instance available for demonstration purposes (Marini, Ludt, Linke, & Strauch, 2021).

Data Visualization in R: R offers extensive capabilities for data visualization, allowing researchers to create high-quality graphics for exploratory data analysis and presentation. The "ggplot2" package, for instance, provides a flexible and intuitive system for creating elegant and customizable plots. Other packages, such as Heatmaply and Complex Heatmap, enable the visualization of complex genomic data, such as gene expression profiles or genomic variation data, in the form of heatmaps and hierarchical clustering dendrograms (Huber et al., 2015).

Statistical Analysis with R: R provides a comprehensive suite of statistical tools and techniques for analyzing genomic data. From basic descriptive statistics to advanced modeling and hypothesis testing, R facilitates a wide range of statistical analyses. Packages such as limma and DESeq2 are commonly used for differential gene expression analysis, while GWAS and QTL analysis can be performed using packages such as GenABEL and qvalue. Additionally, R offers support for survival analysis, regression modeling, and machine learning algorithms, allowing researchers to address diverse research questions in genomic analysis (Ritchie et al., 2015).

Integration with Other Tools and Databases: One of R's strengths is its seamless integration with other tools and databases commonly used in genomic research. R packages often provide interfaces for accessing and querying public databases, such as NCBI's GenBank and the UCSC Genome Browser, allowing researchers to retrieve genomic data directly within their R environment. Furthermore, R can be integrated with other programming languages and software tools, enabling interoperability and enhancing the analytical capabilities of genomic researchers (Hershberg et al., 2021).

## 3.5. Applications of Analytic Tools in Genome Sequencing

Analytic tools in genome sequencing play a pivotal role in human genomics, enabling personalized medicine, disease diagnosis, and pharmacogenomics. These tools are used to identify genetic variants associated with diseases, predict disease risk, and guide treatment strategies. For example, GWAS utilizes statistical techniques to identify genetic variants linked to common diseases such as diabetes, cancer, and cardiovascular diseases (Esplin, Oei, & Snyder, 2014). Analytic tools also facilitate the interpretation of genomic data in clinical settings, allowing clinicians to tailor treatments based on an individual's genetic makeup, ultimately improving patient outcomes and healthcare delivery (Hassan et al., 2022).

Agricultural Genomics: In agricultural genomics, analytic tools are instrumental in crop improvement, livestock breeding, and trait selection. Genomic selection, for instance, leverages statistical models to predict the performance of crops or livestock based on their genetic profiles, enabling breeders to make informed decisions about which individuals to select for breeding (Sinha et al., 2023). Additionally, tools such as GWAS and transcriptomic analysis aid in identifying genes associated with desirable traits such as yield, disease resistance, and nutritional content. By integrating genomic data with phenotypic information, researchers can accelerate the breeding process and develop improved crop varieties and livestock breeds that are more resilient, productive, and sustainable (Sahito et al., 2024).

Microbial Genomics: Analytical tools are vital in microbial genomics, allowing researchers to explore microbial diversity, evolution, and pathogenesis. Metagenomic analysis, for instance, employs bioinformatics tools to analyze DNA sequences obtained directly from environmental samples, providing insights into the composition and functional potential of microbial communities. In infectious disease surveillance, genomic epidemiology tools are used to trace pathogen transmission, detect outbreaks, and monitor the development of antimicrobial resistance (Nam, Do, Loan Trinh, & Lee, 2023). Additionally, comparative genomics techniques enable the identification of genomic features related to pathogenicity, such as virulence factors and antibiotic resistance genes. These advancements aid in the development of diagnostic methods and therapeutic interventions for infectious diseases (Bianconi et al., 2023). The application of analytical tools in microbial genomics is thus essential for understanding and managing microbial populations, tracking disease outbreaks, and tackling the global challenge of antibiotic resistance.

Evolutionary Genomics: Analytic tools play a crucial role in evolutionary genomics, enabling researchers to investigate the genetic underpinnings of evolutionary processes, speciation, and adaptation. Comparative genomics methods, including genome alignment and phylogenetic analysis, allow scientists to pinpoint genomic similarities and differences among species, deduce evolutionary relationships, and recreate ancestral genomes. Population genomics techniques, such as demographic modeling and selection scans, provide insights into population dynamics, genetic diversity, and adaptive evolution (Congrains et al., 2023). By integrating genomic data with ecological and paleontological information, evolutionary biologists can uncover the mechanisms driving genetic divergence and adaptation across a wide range of organisms, from microbes to plants and animals (Spyrou, Bos, Herbig, & Krause, 2019). These tools help us understand the complex interplay of genes, environment, and natural selection in shaping the evolution of life on Earth.

### 3.5.1. Case Studies and Examples

Numerous case studies exemplify the diverse applications of analytical tools in genome sequencing across various fields. A notable example is the 1000 Genomes Project, which employed advanced bioinformatics tools to analyze genetic variation in human populations. This project generated a comprehensive resource of human genetic diversity, enabling more effective genetic studies of complex diseases. In agriculture, genomic selection and marker-assisted breeding techniques have been instrumental in developing high-yielding and disease-resistant crop varieties. The Green Revolution in rice and wheat production is a testament to the power of these approaches (H. Satam et al., 2023). In microbial genomics, real-time sequencing and bioinformatics analysis played a crucial role in tracking the spread of pathogens during outbreaks, such as Ebola and Zika, and informing public health interventions. Similarly, in evolutionary genomics, researchers have harnessed comparative genomics and population genetics techniques to investigate the genetic basis of species diversification. Studies of Darwin's finches have provided insights into adaptive radiation and the influence of environmental changes on the evolution of these birds. These case studies highlight the transformative impact of analytical tools in genome sequencing on scientific research, agriculture, public health, and our understanding of biodiversity and evolution. (Gardy & Loman, 2018).

## 3.6. Emerging Trends and Future Directions

The realm of genome sequencing is experiencing significant advancements in sequencing technologies, driving innovation and broadening the horizons of genomic research. TGS platforms, such as PacBio and Oxford Nanopore, offer long-read sequencing capabilities and real-time sequencing without requiring amplification or fragmentation (H. Satam et al., 2023). These innovations help overcome challenges associated with short-read sequencing technologies and enable the generation of more continuous genome assemblies while facilitating the identification of structural variants and complex genomic rearrangements. Furthermore, advancements in sequencing accuracy, throughput, and cost-effectiveness are ushering in large-scale genomic initiatives and personalized medicine applications. These improvements are making genomic information more accessible to the broader community (Ben Khedher, Ghedira, Rolain, Ruimy, & Croce, 2022).

Integration of multi-omics Data: A rising approach in genomic research involves the integration of multi-omics data, which includes genomics, transcriptomics, proteomics, and metabolomics information. This holistic strategy allows researchers to gain a deeper understanding of the complex interactions within biological systems (Roychowdhury et al., 2023). Integrative multi-omics methods enable the exploration of intricate molecular relationships and regulatory networks underlying biological processes, disease mechanisms, and drug responses. By combining data from various omics layers, researchers can identify biomarkers, therapeutic targets, and pathways associated with diseases. This facilitates the development of personalized precision medicine strategies for individual patients. Additionally, multi-omics integration improves our capacity to understand gene-environment interactions, microbial-host interactions, and ecosystem dynamics, providing valuable insights into a wide range of biological phenomena across different scales (R. C. Wang & Wang, 2023).

Challenges and Opportunities: Despite advances in genome sequencing, various challenges remain, requiring dedicated efforts to overcome them. Computational and analytical hurdles associated with processing and analyzing large genomic datasets necessitate scalable algorithms, reliable infrastructure, and skilled bioinformatics personnel (Shi & Wang, 2019). Moreover, discrepancies in data access, data sharing protocols, and data quality standards pose obstacles to collaborative research and data-driven discovery. Ethical considerations involving data privacy, informed consent, and equitable access to genomic data must be addressed to ensure responsible and ethical use of genomic technologies (Borry et al., 2018). However, these challenges also create opportunities for innovation and collaboration. Developments in artificial intelligence, machine learning, and cloud computing show potential in tackling computational challenges and expediting genomic data analysis. International partnerships, data-sharing initiatives, and community-driven efforts to establish data standards and best practices enhance transparency, reproducibility, and data interoperability, furthering genomic research and its applications (Martínez-García & Hernández-Lemus, 2021).

Ethical Considerations and Societal Implications: Integrating ethical considerations into genomic projects requires balancing technological advances with protecting individual rights, promoting equity, and aligning with societal values. Key challenges include ensuring privacy and confidentiality, obtaining informed consent, defining data ownership, and addressing disparities in access to genomic technologies, particularly in low-income regions. Additionally, integrating genomic data into healthcare systems poses storage and interpretation challenges, raising concerns about patient privacy and the clinical significance of genetic findings (McGuire, Caulfield, & Cho, 2008; Rosenquist et al., 2022).

Addressing these challenges necessitates engaging a broad range of stakeholders, including researchers, clinicians, policymakers, and the public, to align genomic projects with societal values and build trust (De Vries et al., 2011; McGuire et al., 2008). Interdisciplinary collaboration is essential for developing comprehensive policies on privacy, consent, and data management. Conducting empirical studies can guide policy development by understanding the effects of data sharing and determining acceptable uses of genomic data. Moreover, supporting researchers in low-income countries through capacity-building collaborations ensures ethical data sharing and mutual benefits. A proactive and collaborative approach will help navigate the ethical landscape, ensuring genomic projects contribute positively to society while upholding individual rights and promoting equity (McGuire et al., 2008).

## 4. APPLYING GENOMICS

Omics is the study of all biological molecules in an organism, aiming to understand their characteristics, quantify their presence, and explore how they interact and impact the organism as a whole. Genomics, focusing on the entire genome, is a key application of omics in medicine, with notable projects such as HGP leading the way (C. Chen et al., 2023). Genetics looks at how traits are passed down and inherited, while genomics takes a wider view by studying all the genetic material in cells. Genomics also looks at how genes and the environment interact, helping us understand how biological traits develop (Kessler, 2018).

### 4.1. Sub-Disciplines of Genomics

Within the field of genomics, there are several sub-disciplines, including functional genomics, structural genomics, pharmacogenomics, comparative genomics, metagenomics, epigenomics, and more. These sub-disciplines utilize bioinformatics and computational tools to delve into the features and traits of genomes. The introduction of NGS platforms has revolutionized genomics, turning it from a mere scientific discipline into a widely adopted technology in laboratories worldwide. Genomics now plays a crucial role in various fields, such as medicine, biotechnology, and agriculture (Saravanan et al., 2022).

## 4.2. Genetic Tests and Genome-Wide Association Studies (GWAS)

In the field of genomics, numerous innovative genetic tests have been developed to aid in the identification of both polygenic and monogenic markers. These tests encompass various approaches, such as the analysis of familial genetic linkages, candidate gene investigations (Cano-Gamez & Trynka, 2020). GWAS, for instance, involves the examination of a vast array of genetic variants in both affected and unaffected individuals from a specific population. This comprehensive analysis aims to unveil the genetic variants that are associated with a particular disease or condition. The introduction of high-density genotyping technologies and GWAS has significantly advanced genomic medicine. These breakthroughs have enabled the detection of SNPs by analyzing hundreds of thousands of genetic markers in numerous individuals. As a result, GWAS has made substantial contributions to understanding the genetic basis of various diseases and conditions (Cano-Gamez & Trynka, 2020; McGrath & Ghersi, 2016). The usefulness of these genetic variants in a clinical setting relies on the extent to which they can provide additional information regarding prognosis, diagnosis, or aid in managing diseases. There are various types of tests available, such as carrier, predictive, diagnostic, and susceptibility tests. Diagnostic tests are used to confirm a suspected diagnosis based on observable symptoms. Carrier testing is a process that helps identify individuals who possess one copy of a gene mutation, which may be transmitted to their offspring. This information can inform family planning decisions and assess the risk of passing on genetic disorders. Predictive and susceptibility testing identifies mutations that elevate the risk of developing genetic-based disorders. These tests can assist individuals in making informed decisions about their lifestyle choices and medical treatment (Oliveri, Ferrari, Manfrinati, & Pravettoni, 2018).

## 4.3. Applications of Genomics in Medicine and Public Health

The use of genomic information presents a chance for tailored treatment and prevention in both clinical and public health settings (Galasso, 2024). Traditionally, these efforts have concentrated on common complex diseases like cancer, heart disease, and neurodegenerative conditions, as well as less common inherited diseases. This includes activities such as risk assessment, diagnostic sequencing, and pharmacogenomics (C. Chen et al., 2023; Oliveri et al., 2018; Strianese et al., 2020). A significant goal in the field of medicine is to ensure tailored treatment plans that start with personalized approaches. In the era of genomics, we are currently experiencing a thrilling phase in medicine where the integration of bioinformatics, genomics, and advanced molecular techniques holds immense promise in enhancing our comprehension of the genetic foundations of various diseases. This enhanced understanding opens up new avenues for redefining disease mechanisms at a more detailed level, which in turn enables us to develop more precise therapies (Strianese et al., 2020). For instance, we can now focus on preventing the onset of diseases, facilitating early detection, and customizing treatments based on individual patient characteristics (Tuena, Semonella, Fernández-Álvarez, Colombo, & Cipresso, 2020).

However, there is now a rising interest in applying genomics to address infectious diseases and epidemics, which are significant global public health challenges (C. Chen et al., 2023; Kwok, Mentzer, & Knight, 2021; Solís-Moruno, Batlle-Masó, Bonet, Aróstegui, & Casals, 2023). The quick and extensive sequencing of pathogen genomes is now being used extensively for managing disease outbreaks. This approach offers more reliable evidence for tracing the source and contacts of infections than was previously available. A notable example of this application was during the Ebola outbreak in West Africa (Kinganda-Lusamaki et al., 2021; Mbala-Kingebeni et al., 2019).

## 4.4. Genomics and the COVID-19 Pandemic

Recently, the global impact of the Coronavirus disease 2019 (COVID-19) was swift and devastating, making it one of the most destructive outbreaks in recorded human history. As the first pandemic in the post-genomic era, advancements in genomics have played a crucial role in enhancing our scientific understanding and public health approach to COVID-19 response. Researchers worldwide have utilized genomic technologies to gain insights into the biology of SARS-CoV-2, including its origins, genomic diversity, and evolution. The availability of global genomic resources has greatly facilitated investigations into the COVID-19 pandemic (Uddin et al., 2020). This pandemic has marked the advent of a new era in genomic surveillance, where scientists are actively monitoring real-time changes in the SARS-CoV-2 genome at both national and international levels. Access to genomics and proteomics information has accelerated the development of molecular diagnostics and therapeutics. The progress of NGS has enabled rapid and cost-effective production of complete viral genomes. Genomic sequencing plays a critical role in continuously monitoring the evolutionary trajectory of the SARS-CoV-2 genome. Recognizing its importance, the World Health Organization (WHO) has encouraged nations to expedite genome sequencing efforts and share the resulting data and findings in a coordinated manner through publicly accessible databases. To streamline sequencing operations, several initiatives and consortia have been established in different countries (Saravanan et al., 2022). SARS-CoV-2, like other viruses, undergoes changes to adapt to its surroundings. Most of these mutations are neutral and have minimal impact on the virus's functionality. However, certain mutations can be significant, particularly when they affect crucial components like the spike glycoprotein, which allows the virus to enter host cells and initiate infection (Zhang et al., 2020). Genomic analyses suggest that certain changes may confer the virus with a selective advantage, enhancing its fitness by, for instance, developing resistance to antiviral drugs and evading the immune system (Harvey et al., 2021; Saravanan et al., 2022).

## 4.5. Genetic Epidemiology and Population Health

Other applications of this technology include accurately diagnosing microbial infections, mapping out the spread of infections, studying the genetic makeup of drug-resistant strains, and identifying potential targets for new treatments and vaccines. Research indicates that not only do genetic factors of pathogens play a role, but also the genetic factors of hosts, and the interactions among vectors, hosts, and pathogens influence infection rates, susceptibility to disease, immune responses, treatment effectiveness, disease progression, and response to preventive measures (Nguyen, 2024; Rahman & Schellhorn, 2023). Furthermore, various countries have put forth significant efforts to explore and assess vaccine safety through the concept of vaccinomics. The credibility of vaccinomics has been reinforced by a review in Scientific American journal, which hailed it as one of the most groundbreaking scientific ideas of the decade. Similarly, The Scientist recognized vaccinomics as one of the most prominent fields in omics. Additionally, scientists working in this area have integrated vaccinomics principles into different aspects of vaccine biology, such as systems or bioinformatics (Aiman et al., 2023; Poland et al., 2013; Rosario-Cruz, Domínguez-García, López-Silva, & Rosario-Domínguez, 2023). Genomic research plays a crucial role in enhancing our comprehension of how infectious diseases develop and how the immune system responds to them. This valuable knowledge holds the potential to guide us in formulating effective strategies for developing vaccines and devising treatments that can combat these diseases more efficiently in the future (Eckhardt, Hultquist, Kaake, Hüttenhain, & Krogan, 2020; Excler, Saville, Berkley, & Kim, 2021).

## 4.6. Genomics in Agriculture

In agriculture, genomics has played a crucial role in crop improvement and livestock breeding. By sequencing the genomes of crop plants and livestock species, researchers can identify genes that are associated with desirable traits such as yield, disease resistance, and nutritional content. This information can then be used to develop new varieties or breeds with improved characteristics, leading to more sustainable and productive agricultural systems. On the other hand, plant viruses are commonly recognized for causing diseases, but their role extends beyond that. Extensive research has revealed various possibilities regarding the function of these pathogens. Geminiviruses and other viruses offer significant potential in genome editing for managing virus-induced agricultural problems. Modified viruses have successfully transformed crops for example chili pepper, cassava, and, wheat enhancing host genomes. Rapid assays, such as transient Geminivirus-induced gene silencing (GIGS) procedures, enable the discovery of new genes and their characteristics. Gemini-vector design has played a pivotal role in plant improvement, finding applications in ncDNA expression, vaccine production, gene silencing, and clustered regularly interspaced palindromic repeats (CRISPR)/Cas9 system construction. The biopharmaceutical industry recognizes Gemini-vectors' potential for vaccine development. Plant viruses pose no threat to human health through infected food consumption. Geminiviruses can serve as a model for shorter infiltration strategies, replacing lengthy transformation procedures. They can deliver small RNAs to modulate regulatory processes and enhance desirable traits in plants. Additionally, Gemini vectors facilitate the overexpression of small peptides with various applications. Understanding Gemini vectors is crucial for improving crop health and bio-pharmaceutical applications, benefiting middle- and low-income countries (Bhattacharjee & Hallan, 2022; Olivares et al., 2021).

## 4.7. Harnessing the Power of Yeast Genomics for Bioproduction

Numerous natural compounds serve various purposes in pharmaceuticals, flavors, nutraceuticals, and fragrances. However, their extraction from plants is often challenging and yields are low, while organisms used for synthesis face limitations in industrial-scale production. Microbial recombinant protein synthesis has gained attention due to its speed, cost-effectiveness, and ease of gene manipulation (Madhavan et al., 2019). Baker's yeast, or *Saccharomyces cerevisiae*, is a well-researched and often-used host for gene editing and protein synthesis. Furthermore, other yeast species have surfaced as replacement hosts, including *Pichia pastoris, Hansenula polymorpha, Kluyveromyces lactis, Yarrowia lipolytica,* and *Schizosaccharomyces pombe*. The advent of functional genomics, complete genome sequences, and systems biology has provided valuable insights into yeast physiology. These innovative technologies and approaches contribute to addressing production challenges and optimizing yeast-based industrial processes (Madhavan et al., 2021). Also, synthetic biology and CRISPR-based genome manipulation techniques have revolutionized yeast cell factory development. Yeast has long been harnessed for synthesizing a diverse array of compounds, encompassing therapeutic proteins like monoclonal antibodies and anti-tumor necrosis factor (TNF) agents for cancer therapy. These applications have made substantial contributions to the thriving multibillion-dollar biopharmaceutical sector. It exhibits high productivity, low contamination risks, and tolerance to harsh conditions. Moreover, yeast efficiently processes recombinant proteins post-translationally (Madhavan et al., 2021; Thak, Yoo, Moon, & Kang, 2020).

## 5. CONCLUSION

The rise of high-throughput technologies has ushered us into a new era where we're deciphering the complexities of biological systems faster than ever before. These cutting-edge methods, known collectively as high-throughput, have completely changed how we approach biological research by allowing us to generate massive datasets across various omics sciences. In this chapter, we've explored the profound impact of high-throughput technologies, the crucial role of omics sciences in deepening our understanding of biological processes, the indispensable role of bioinformatic tools in handling genomic data, the significant implications of genomics in treating diseases, and the wide-ranging applications of genomics across different organisms and fields. High-throughput technologies encompass a range of powerful methods aimed at rapidly acquiring biological data on a large scale. From NGS to microarray analysis, these tools have transformed how researchers study complex biological phenomena, giving us unparalleled insights into the inner workings of life. While these technologies generate a wealth of data, their sheer volume presents a duality of opportunities and challenges. As a result, there is an increasing demand for groundbreaking computational and analytical techniques to effectively interpret and harness this vast information. The omics sciences, including genomics, transcriptomics, proteomics, metabolomics, and epigenomics, are essential for unraveling the molecular foundations of biological processes. By providing comprehensive profiles of various molecular components within biological systems, omics sciences offer holistic insights into cellular function, disease mechanisms, and organismal physiology. Leveraging high-throughput technologies, omics studies have uncovered intricate regulatory networks, identified disease biomarkers, and shed light on the molecular basis of complex traits and disorders. Bioinformatic tools are the backbone of genomic research, enabling the analysis, interpretation, and integration of large-scale genomic datasets. From aligning sequences to deciphering gene regulation networks, these tools empower researchers to extract valuable information from genomic data, helping us understand gene function, annotate genetic variants, and explore evolutionary dynamics. Genomics, in particular, has revolutionized our understanding of the genetic basis of disease and personalized medicine. By decoding the genome, genomics has enabled us to identify genetic variants associated with disease susceptibility, drug response, and treatment outcomes. This knowledge is paving the way for precision medicine, where therapies can be tailored to individual patients based on their genetic make-up. Beyond human health, genomics has applications across diverse organisms and fields, from agriculture and environmental conservation to evolutionary biology and synthetic biology. In agriculture, genomic technologies offer opportunities for crop improvement and disease-resistance breeding. In environmental science, genomics provides insights into microbial communities and ecosystem dynamics. In evolutionary biology, genomics helps us understand the genetic basis of adaptation and speciation. In synthetic biology, genomics serves as a blueprint for designing novel biological systems and addressing global challenges. In brief, the genomic revolution is propelled by high-throughput technologies, omics sciences, bioinformatic tools, and genomics, which together transform biology and its applications. As we continue to use genomics to unlock life's mysteries, interdisciplinary collaboration, technological innovation, and ethical considerations will be critical to realizing the full potential of genomics in influencing science and medicine.

## REFERENCES


Abdul Aziz, M., Brini, F., Rouached, H., & Masmoudi, K. (2022). Genetically engineered crops for sustainably enhanced food production systems. DOI: 10.3389/fpls.2022.1027828

Adam, Y., Samtal, C., Brandenburg, J. T., Falola, O., & Adebiyi, E. (2021). Performing post-genome-wide association study analysis: Overview, challenges and recommendations. *F1000 Research*, 10, 1002. DOI: 10.12688/f1000research.53962.1 PMID: 35222990

Aiman, S., Ahmad, A., Khan, A., Ali, Y., Malik, A., Alkholief, M., & Jalil, F. (2023). Vaccinomics-aided next-generation novel multi-epitope-based vaccine engineering against multidrug resistant Shigella Sonnei: Immunoinformatics and chemoinformatics approaches. *PLoS One*, 18(11), e0289773. DOI: 10.1371/journal.pone.0289773 PMID: 37992050

Akaçin, İ., Ersoy, Ş., Doluca, O., & Güngörmüşler, M. (2022). Comparing the significance of the utilization of next generation and third generation sequencing technologies in microbial metagenomics. *Microbiological Research*, 264, 127154. DOI: 10.1016/j.micres.2022.127154 PMID: 35961096

Alamin, M., Sultana, M. H., Xu, H., & Mollah, M. N. H. (2020). Robustification of Linear Regression and Its Application in Genome-Wide Association Studies. *Frontiers in Genetics*, 11, 549. DOI: 10.3389/fgene.2020.00549 PMID: 32582288

Alex, A., Brundha, M., & Prathap, L. (2020). Sanger sequencing and its recent advances-a review. *PalArch's Journal of Archaeology of Egypt/Egyptology, 17*(7), 698-705.

Alföldi, J., & Lindblad-Toh, K. (2013). Comparative genomics as a tool to understand evolution and disease. *Genome Research*, 23(7), 1063–1068. DOI: 10.1101/gr.157503.113 PMID: 23817047

Athanasopoulou, K., Boti, M. A., Adamopoulos, P. G., Skourou, P. C., & Scorilas, A. (2021). Third-Generation Sequencing: The Spearhead towards the Radical Transformation of Modern Genomics. *Life (Basel, Switzerland)*, 12(1), 30. Advance online publication. DOI: 10.3390/life12010030 PMID: 35054423

Bao, R., Huang, L., Andrade, J., Tan, W., Kibbe, W. A., Jiang, H., & Feng, G. (2014). Review of current methods, applications, and data management for the bioinformatics analysis of whole exome sequencing. *Cancer Informatics*, 13(Suppl 2), 67–82. DOI: 10.4137/CIN.S13779 PMID: 25288881

Ben Khedher, M., Ghedira, K., Rolain, J. M., Ruimy, R., & Croce, O. (2022). Application and Challenge of 3rd Generation Sequencing for Clinical Bacterial Studies. *International Journal of Molecular Sciences*, 23(3), 1395. Advance online publication. DOI: 10.3390/ijms23031395 PMID: 35163319

Benson, D. A., Karsch-Mizrachi, I., Lipman, D. J., Ostell, J., & Sayers, E. W. (2010). GenBank. *Nucleic Acids Research*, 38(suppl_1), D46–D51. DOI: 10.1093/nar/gkp1024 PMID: 19910366

Bhattacharjee, B., & Hallan, V. (2022). Geminivirus-derived vectors as tools for functional genomics. *Frontiers in Microbiology*, 13, 799345. DOI: 10.3389/fmicb.2022.799345 PMID: 35432267

Bianconi, I., Aschbacher, R., & Pagani, E. (2023). Current Uses and Future Perspectives of Genomic Technologies in Clinical Microbiology. *Antibiotics (Basel, Switzerland)*, 12(11), 1580. Advance online publication. DOI: 10.3390/antibiotics12111580 PMID: 37998782

Bornstein, K., Gryan, G., Chang, E. S., Marchler-Bauer, A., & Schneider, V. A. (2023). The NIH Comparative Genomics Resource: Addressing the promises and challenges of comparative genomics on human health. *BMC Genomics*, 24(1), 575. DOI: 10.1186/s12864-023-09643-4 PMID: 37759191

Borry, P., Bentzen, H. B., Budin-Ljøsne, I., Cornel, M. C., Howard, H. C., Feeney, O., Jackson, L., Mascalzoni, D., Mendes, Á., Peterlin, B., Riso, B., Shabani, M., Skirton, H., Sterckx, S., Vears, D., Wjst, M., & Felzmann, H. (2018). The challenges of the expanded availability of genomic information: An agenda-setting paper. *Journal of Community Genetics*, 9(2), 103–116. DOI: 10.1007/s12687-017-0331-7 PMID: 28952070

Brittain, H. K., Scott, R., & Thomas, E. (2017). The rise of the genome and personalised medicine. *Clinical Medicine (London, England)*, 17(6), 545–551. DOI: 10.7861/clinmedicine.17-6-545 PMID: 29196356

Burton, H., Adams, M., Bunton, R., & Schröder-Bäck, P. (2008). Developing stakeholder involvement for introducing public health genomics into public policy. *Public Health Genomics*, 12(1), 11–19. DOI: 10.1159/000153426 PMID: 19023186

Cano-Gamez, E., & Trynka, G. (2020). From GWAS to function: Using functional genomics to identify the mechanisms underlying complex diseases. *Frontiers in Genetics*, 11, 505357. DOI: 10.3389/fgene.2020.00424 PMID: 32477401

Chan, E. Y. (2005). Advances in sequencing technology. *Mutation Research. Fundamental and Molecular Mechanisms of Mutagenesis*, 573(1-2), 13–40. DOI: 10.1016/j.mrfmmm.2005.01.004 PMID: 15829235

Chen, C., Wang, J., Pan, D., Wang, X., Xu, Y., Yan, J., Wang, L., Yang, X., Yang, M., & Liu, G. P. (2023). Applications of multi-omics analysis in human diseases. *MedComm*, 4(4), e315. DOI: 10.1002/mco2.315 PMID: 37533767

Chen, Y., Song, J., Ruan, Q., Zeng, X., Wu, L., Cai, L., Wang, X., & Yang, C. (2021). Single-cell sequencing methodologies: From transcriptome to multi-dimensional measurement. *Small Methods*, 5(6), 2100111. DOI: 10.1002/smtd.202100111 PMID: 34927917

Chen, Z., Ain, N., Zhao, Q., & Zhang, X. (2024). From tradition to innovation: Conventional and deep learning frameworks in genome annotation. *Briefings in Bioinformatics*, 25(3), bbae138. DOI: 10.1093/bib/bbae138 PMID: 38581418

Cheng, C., Fei, Z., & Xiao, P. (2023). Methods to improve the accuracy of next-generation sequencing. *Frontiers in Bioengineering and Biotechnology*, 11, 982111. DOI: 10.3389/fbioe.2023.982111 PMID: 36741756

Chiu, C. Y., & Miller, S. A. (2019). Clinical metagenomics. *Nature Reviews. Genetics*, 20(6), 341–355. DOI: 10.1038/s41576-019-0113-7 PMID: 30918369

Collins, F. S., & Fink, L. (1995). The Human Genome Project. *Alcohol Health and Research World*, 19(3), 190–195. PMID: 31798046

Congrains, C., Dupuis, J. R., Rodriguez, E. J., Norrbom, A. L., Steck, G., Sutton, B., Nolazco, N., de Brito, R. A., & Geib, S. M. (2023). Phylogenomic analysis provides diagnostic tools for the identification of Anastrepha fraterculus (Diptera: Tephritidae) species complex. *Evolutionary Applications*, 16(9), 1598–1618. DOI: 10.1111/eva.13589 PMID: 37752958

Cooksey, R. W. (2020). Descriptive Statistics for Summarising Data. In *Illustrating Statistical Procedures: Finding Meaning in Quantitative Data* (pp. 61–139). Springer Singapore. DOI: 10.1007/978-981-15-2537-7_5

Crossley, B. M., Bai, J., Glaser, A., Maes, R., Porter, E., Killian, M. L., Clement, T., & Toohey-Kurth, K. (2020). Guidelines for Sanger sequencing and molecular assay monitoring. *Journal of Veterinary Diagnostic Investigation*, 32(6), 767–775. DOI: 10.1177/1040638720905833 PMID: 32070230

De Vries, J., Bull, S. J., Doumbo, O., Ibrahim, M., Mercereau-Puijalon, O., Kwiatkowski, D., & Parker, M. (2011). Ethical issues in human genomics research in developing countries. *BMC Medical Ethics*, 12(1), 1–10. DOI: 10.1186/1472-6939-12-5 PMID: 21418562

Du, L., Liu, Q., Fan, Z., Tang, J., Zhang, X., Price, M., Yue, B., & Zhao, K. (2021). Pyfastx: A robust Python package for fast random access to sequences from plain and gzipped FASTA/Q files. *Briefings in Bioinformatics*, 22(4), bbaa368. DOI: 10.1093/bib/bbaa368 PMID: 33341884

Durland, J., & Ahmadian-Moghadam, H. (2024). Genetics, Mutagenesis. In *StatPearls*. StatPearls Publishing.

Eckhardt, M., Hultquist, J. F., Kaake, R. M., Hüttenhain, R., & Krogan, N. J. (2020). A systems approach to infectious disease. *Nature Reviews. Genetics*, 21(6), 339–354. DOI: 10.1038/s41576-020-0212-5 PMID: 32060427

Ejigu, G. F., & Jung, J. (2020). Review on the Computational Genome Annotation of Sequences Obtained by Next-Generation Sequencing. *Biology (Basel)*, 9(9), 295. Advance online publication. DOI: 10.3390/biology9090295 PMID: 32962098

Emmert-Streib, F., & Dehmer, M. (2019). Understanding Statistical Hypothesis Testing: The Logic of Statistical Inference. Retrieved from https://www.mdpi.com/2504-4990/1/3/54

Esplin, E. D., Oei, L., & Snyder, M. P. (2014). Personalized sequencing and the future of medicine: Discovery, diagnosis and defeat of disease. *Pharmacogenomics*, 15(14), 1771–1790. DOI: 10.2217/pgs.14.117 PMID: 25493570

Excler, J.-L., Saville, M., Berkley, S., & Kim, J. H. (2021). Vaccine development for emerging infectious diseases. *Nature Medicine*, 27(4), 591–600. DOI: 10.1038/s41591-021-01301-0 PMID: 33846611

Fang, S., Zhang, L., Guo, J., Niu, Y., Wu, Y., Li, H., & Sun, X. (2018). NONCODEV5: A comprehensive annotation database for long non-coding RNAs. *Nucleic Acids Research*, 46(D1), D308–D314. DOI: 10.1093/nar/gkx1107 PMID: 29140524

Feng, L. B., Grosse, S. D., Green, R. F., Fink, A. K., & Sawicki, G. S. (2018). Precision Medicine In Action: The Impact Of Ivacaftor On Cystic Fibrosis-Related Hospitalizations. *Health Affairs (Project Hope)*, 37(5), 773–779. DOI: 10.1377/hlthaff.2017.1554 PMID: 29733727

Flynn, E. D., & Lappalainen, T. (2022). Functional characterization of genetic variant effects on expression. *Annual Review of Biomedical Data Science*, 5(1), 119–139. DOI: 10.1146/annurev-biodatasci-122120-010010 PMID: 35483347

Galasso, I. (2024). Precision medicine for whom? public health outputs from "Genomics England" and "All of Us" to make up for upstream and downstream exclusion. *The American Journal of Bioethics*, 24(3), 71–85. DOI: 10.1080/15265161.2023.2180108 PMID: 36876959

Gao, C. (2021). Genome engineering for crop improvement and future agriculture. *Cell*, 184(6), 1621–1635. DOI: 10.1016/j.cell.2021.01.005 PMID: 33581057

Gardy, J. L., & Loman, N. J. (2018). Towards a genomics-informed, real-time, global pathogen surveillance system. *Nature Reviews. Genetics*, 19(1), 9–20. DOI: 10.1038/nrg.2017.88 PMID: 29129921

Geman, D., Ochs, M., Price, N. D., Tomasetti, C., & Younes, L. (2015). An argument for mechanism-based statistical inference in cancer. *Human Genetics*, 134(5), 479–495. DOI: 10.1007/s00439-014-1501-x PMID: 25381197

Ghurye, J., & Pop, M. (2019). Modern technologies and algorithms for scaffolding assembled genomes. *PLoS Computational Biology*, 15(6), e1006994. DOI: 10.1371/journal.pcbi.1006994 PMID: 31166948

Giani, A. M., Gallo, G. R., Gianfranceschi, L., & Formenti, G. (2020). Long walk to genomics: History and current approaches to genome sequencing and assembly. *Computational and Structural Biotechnology Journal*, 18, 9–19. DOI: 10.1016/j.csbj.2019.11.002 PMID: 31890139

Green, E. D., Gunter, C., Biesecker, L. G., Di Francesco, V., Easter, C. L., Feingold, E. A., Felsenfeld, A. L., Kaufman, D. J., Ostrander, E. A., Pavan, W. J., Phillippy, A. M., Wise, A. L., Dayal, J. G., Kish, B. J., Mandich, A., Wellington, C. R., Wetterstrand, K. A., Bates, S. A., Leja, D., & Manolio, T. A. (2020). Strategic vision for improving human health at The Forefront of Genomics. *Nature*, 586(7831), 683–692. DOI: 10.1038/s41586-020-2817-4 PMID: 33116284

Green, E. D., Gunter, C., Biesecker, L. G., Di Francesco, V., Easter, C. L., Feingold, E. A., & Pavan, W. J. (2020). Strategic vision for improving human health at The Forefront of Genomics. *Nature*, 586(7831), 683–692. DOI: 10.1038/s41586-020-2817-4 PMID: 33116284

Gupta, N., & Verma, V. K. (2019). Next-Generation Sequencing and Its Application: Empowering in Public Health Beyond Reality. In Arora, P. K. (Ed.), *Microbial Technology for the Welfare of Society* (pp. 313–341). Springer Singapore. DOI: 10.1007/978-981-13-8844-6_15

Gupta, P. D. (2015). Pharmacogenetics, pharmacogenomics and ayurgenomics for personalized medicine: A paradigm shift. *Indian Journal of Pharmaceutical Sciences*, 77(2), 135–141. DOI: 10.4103/0250-474X.156543 PMID: 26009644

Harvey, W. T., Carabelli, A. M., Jackson, B., Gupta, R. K., Thomson, E. C., Harrison, E. M., & Peacock, S. J. (2021). SARS-CoV-2 variants, spike mutations and immune escape. *Nature Reviews. Microbiology*, 19(7), 409–424. DOI: 10.1038/s41579-021-00573-0 PMID: 34075212

Hasin, Y., Seldin, M., & Lusis, A. (2017). Multi-omics approaches to disease. *Genome Biology*, 18(1), 83. DOI: 10.1186/s13059-017-1215-1 PMID: 28476144

Hasin, Y., Seldin, M., & Lusis, A. (2017). Multi-omics approaches to disease. *Genome Biology*, 18(1), 1–15. DOI: 10.1186/s13059-017-1215-1 PMID: 28476144

Hassan, M., Awan, F. M., Naz, A., deAndrés-Galiana, E. J., Alvarez, O., Cernea, A., Fernández-Brillet, L., Fernández-Martínez, J. L., & Kloczkowski, A. (2022). Innovations in Genomics and Big Data Analytics for Personalized Medicine and Health Care: A Review. *International Journal of Molecular Sciences*, 23(9), 4645. Advance online publication. DOI: 10.3390/ijms23094645 PMID: 35563034

Heather, J. M., & Chain, B. (2016). The sequence of sequencers: The history of sequencing DNA. *Genomics*, 107(1), 1–8. DOI: 10.1016/j.ygeno.2015.11.003 PMID: 26554401

Hershberg, E. A., Stevens, G., Diesh, C., Xie, P., De Jesus Martinez, T., Buels, R., Stein, L., & Holmes, I. (2021). JBrowseR: An R interface to the JBrowse 2 genome browser. *Bioinformatics (Oxford, England)*, 37(21), 3914–3915. DOI: 10.1093/bioinformatics/btab459 PMID: 34196689

Hood, L., & Rowen, L. (2013). The Human Genome Project: Big science transforms biology and medicine. *Genome Medicine*, 5(9), 79. DOI: 10.1186/gm483 PMID: 24040834

Hoogstrate, Y., Jenster, G. W., & Werken, H. J. (2021). FASTAFS: File system virtualisation of random access compressed FASTA files. *BMC Bioinformatics*, 22(1), 1–12. DOI: 10.1186/s12859-021-04455-3 PMID: 34724897

Huang, B., Kong, L., Wang, C., Ju, F., Zhang, Q., Zhu, J., Gong, T., Zhang, H., Yu, C., Zheng, W.-M., & Bu, D. (2023). Protein Structure Prediction: Challenges, Advances, and the Shift of Research Paradigms. *Genomics, Proteomics & Bioinformatics*, 21(5), 913–925. DOI: 10.1016/j.gpb.2022.11.014 PMID: 37001856

Huber, W., Carey, V. J., Gentleman, R., Anders, S., Carlson, M., Carvalho, B. S., Bravo, H. C., Davis, S., Gatto, L., Girke, T., Gottardo, R., Hahne, F., Hansen, K. D., Irizarry, R. A., Lawrence, M., Love, M. I., MacDonald, J., Obenchain, V., Oleś, A. K., & Morgan, M. (2015). Orchestrating high-throughput genomic analysis with Bioconductor. *Nature Methods*, 12(2), 115–121. DOI: 10.1038/nmeth.3252 PMID: 25633503

Hutchins, J. R. (2014). What's that gene (or protein)? Online resources for exploring functions of genes, transcripts, and proteins. *Molecular Biology of the Cell*, 25(8), 1187–1201. DOI: 10.1091/mbc.e13-10-0602 PMID: 24723265

Jensen, M. A., Ferretti, V., Grossman, R. L., & Staudt, L. M. (2017). The NCI Genomic Data Commons as an engine for precision medicine. *Blood*, 130(4), 453–459. DOI: 10.1182/blood-2017-03-735654 PMID: 28600341

Johnson, J. A., & Cavallari, L. H. (2015). Warfarin pharmacogenetics. *Trends in Cardiovascular Medicine*, 25(1), 33–41. DOI: 10.1016/j.tcm.2014.09.001 PMID: 25282448

Jovic, D., Liang, X., Zeng, H., Lin, L., Xu, F., & Luo, Y. (2022). Single-cell RNA sequencing technologies and applications: A brief overview. *Clinical and Translational Medicine*, 12(3), e694. DOI: 10.1002/ctm2.694 PMID: 35352511

Jowett, T. (1999). Analysis of protein and gene expression. *Methods in Cell Biology*, 59, 63–85. DOI: 10.1016/S0091-679X(08)61821-X PMID: 9891356

Juengst, E., McGowan, M. L., Fishman, J. R., & Settersten, R. A.Jr. (2016). From "personalized" to "precision" medicine: The ethical and social implications of rhetorical reform in genomic medicine. *The Hastings Center Report*, 46(5), 21–33. DOI: 10.1002/hast.614 PMID: 27649826

Jung, H., Ventura, T., Chung, J. S., Kim, W.-J., Nam, B.-H., Kong, H. J., Kim, Y.-O., Jeon, M.-S., & Eyun, S.-i. (2020). Twelve quick steps for genome assembly and annotation in the classroom. *PLoS Computational Biology*, 16(11), e1008325. DOI: 10.1371/journal.pcbi.1008325 PMID: 33180771

Jung, H., Ventura, T., Chung, J. S., Kim, W. J., Nam, B. H., Kong, H. J., Kim, Y.-O., Jeon, M.-S., & Eyun, S. I. (2020). Twelve quick steps for genome assembly and annotation in the classroom. *PLoS Computational Biology*, 16(11), e1008325. DOI: 10.1371/journal.pcbi.1008325 PMID: 33180771

Karczewski, K. J., Weisburd, B., Thomas, B., Solomonson, M., Ruderfer, D. M., Kavanagh, D., Hamamsy, T., Lek, M., Samocha, K. E., Cummings, B. B., Birnbaum, D., Daly, M. J., & MacArthur, D. G. (2017). The ExAC browser: Displaying reference data information from over 60 000 exomes. *Nucleic Acids Research*, 45(D1), D840–d845. DOI: 10.1093/nar/gkw971 PMID: 27899611

Karolchik, D., Hinrichs, A. S., & Kent, W. J. (2011). The UCSC Genome Browser. *Curr Protoc Hum Genet*. DOI: 10.1002/0471142905.hg1806s71

Kessler, C. (2018). Genomics and Precision Medicine: Implications for Critical Care. *AACN Advanced Critical Care*, 29(1), 28–35. DOI: 10.4037/aacnacc2018521 PMID: 29496711

Kinganda-Lusamaki, E., Black, A., Mukadi, D. B., Hadfield, J., Mbala-Kingebeni, P., Pratt, C. B., & Bisento, N. (2021). Integration of genomic sequencing into the response to the Ebola virus outbreak in Nord Kivu, Democratic Republic of the Congo. *Nature Medicine*, 27(4), 710–716. DOI: 10.1038/s41591-021-01302-z PMID: 33846610

Koboldt, D. C., Steinberg, K. M., Larson, D. E., Wilson, R. K., & Mardis, E. R. (2013). The next-generation sequencing revolution and its impact on genomics. *Cell*, 155(1), 27–38. DOI: 10.1016/j.cell.2013.09.006 PMID: 24074859

Kotha, K., & Clancy, J. P. (2013). Ivacaftor treatment of cystic fibrosis patients with the G551D mutation: A review of the evidence. *Therapeutic Advances in Respiratory Disease*, 7(5), 288–296. DOI: 10.1177/1753465813502115 PMID: 24004658

Kryukov, K., Ueda, M. T., Nakagawa, S., & Imanishi, T. (2020). Sequence Compression Benchmark (SCB) database—A comprehensive evaluation of reference-free compressors for FASTA-formatted sequences. *GigaScience*, 9(7), giaa072. Advance online publication. DOI: 10.1093/gigascience/giaa072 PMID: 32627830

Kwok, A. J., Mentzer, A., & Knight, J. C. (2021). Host genetics and infectious disease: New tools, insights and translational opportunities. *Nature Reviews. Genetics*, 22(3), 137–153. DOI: 10.1038/s41576-020-00297-6 PMID: 33277640

Lamond, A. I. (1993). The spliceosome. *BioEssays*, 15(9), 595–603. DOI: 10.1002/bies.950150905 PMID: 8240312

Lee, T. I., & Young, R. A. (2000). Transcription of eukaryotic protein-coding genes. *Annual Review of Genetics*, 34(1), 77–137. DOI: 10.1146/annurev.genet.34.1.77 PMID: 11092823

Lei, Y., Meng, Y., Guo, X., Ning, K., Bian, Y., Li, L., & Dong, Y. (2022). Overview of structural variation calling: Simulation, identification, and visualization. *Computers in Biology and Medicine*, 145, 105534. DOI: 10.1016/j.compbiomed.2022.105534 PMID: 35585730

Lei, Y., Tang, R., Xu, J., Wang, W., Zhang, B., Liu, J., Yu, X., & Shi, S. (2021). Applications of single-cell sequencing in cancer research: Progress and perspectives. *Journal of Hematology & Oncology*, 14(1), 91. DOI: 10.1186/s13045-021-01105-2 PMID: 34108022

Leinonen, R., Akhtar, R., Birney, E., Bower, L., Cerdeno-Tárraga, A., Cheng, Y., Cleland, I., Faruque, N., Goodgame, N., Gibson, R., Hoad, G., Jang, M., Pakseresht, N., Plaister, S., Radhakrishnan, R., Reddy, K., Sobhany, S., Ten Hoopen, P., Vaughan, R., & Cochrane, G. (2011). The European Nucleotide Archive. *Nucleic Acids Research*, 39(Database issue), D28–D31. DOI: 10.1093/nar/gkq967 PMID: 20972220

Lesk, A. M. (2017). *Introduction to genomics*. Oxford University Press. DOI: 10.1093/hesc/9780198754831.001.0001

Li, J., & Liu, C. (2019). Coding or Noncoding, the Converging Concepts of RNAs. *Frontiers in Genetics*, 10, 496. DOI: 10.3389/fgene.2019.00496 PMID: 31178900

Liao, Y., Liu, Z., Zhang, Y., Lu, P., Wen, L., & Tang, F. (2023). High-throughput and high-sensitivity full-length single-cell RNA-seq analysis on third-generation sequencing platform. *Cell Discovery*, 9(1), 5. DOI: 10.1038/s41421-022-00500-4 PMID: 36631434

Lin, J., & Ngiam, K. Y. (2023). How data science and AI-based technologies impact genomics. *Singapore Medical Journal*, 64(1), 59–66. DOI: 10.4103/singaporemedj.SMJ-2021-438 PMID: 36722518

Logsdon, G. A., Vollger, M. R., & Eichler, E. E. (2020). Long-read human genome sequencing and its applications. *Nature Reviews. Genetics*, 21(10), 597–614. DOI: 10.1038/s41576-020-0236-x PMID: 32504078

Madhavan, A., Arun, K., Sindhu, R., Krishnamoorthy, J., Reshmy, R., Sirohi, R., Pugazhendi, A., Awasthi, M. K., Szakacs, G., & Binod, P. (2021). Customized yeast cell factories for biopharmaceuticals: From cell engineering to process scale up. *Microbial Cell Factories*, 20(1), 124. DOI: 10.1186/s12934-021-01617-z PMID: 34193127

Madhavan, A., Arun, K. B., Sindhu, R., Binod, P., Kim, S. H., & Pandey, A. (2019). Tailoring of microbes for the production of high value plant-derived compounds: From pathway engineering to fermentative production. *Biochimica et Biophysica Acta (BBA)-. Biochimica et Biophysica Acta. Proteins and Proteomics*, 1867(11), 140262. DOI: 10.1016/j.bbapap.2019.140262 PMID: 31404685

Marini, F., Ludt, A., Linke, J., & Strauch, K. (2021). GeneTonic: An R/Bioconductor package for streamlining the interpretation of RNA-seq data. *BMC Bioinformatics*, 22(1), 610. DOI: 10.1186/s12859-021-04461-5 PMID: 34949163

Martina, M., De Rosa, V., Magon, G., Acquadro, A., Barchi, L., Barcaccia, G., De Paoli, E., Vannozzi, A., & Portis, E. (2024). Revitalizing agriculture: Next-generation genotyping and -omics technologies enabling molecular prediction of resilient traits in the Solanaceae family. *Frontiers in Plant Science*, 15, 1278760. DOI: 10.3389/fpls.2024.1278760 PMID: 38375087

Martínez-García, M., & Hernández-Lemus, E. (2021). Data Integration Challenges for Machine Learning in Precision Medicine. *Frontiers in Medicine*, 8, 784455. DOI: 10.3389/fmed.2021.784455 PMID: 35145977

Mattick, J. S., Amaral, P. P., Carninci, P., Carpenter, S., Chang, H. Y., Chen, L.-L., & Fitzgerald, K. A. (2023). Long non-coding RNAs: Definitions, functions, challenges and recommendations. *Nature Reviews. Molecular Cell Biology*, 24(6), 430–447. DOI: 10.1038/s41580-022-00566-8 PMID: 36596869

Mbala-Kingebeni, P., Aziza, A., Di Paola, N., Wiley, M. R., Makiala-Mandanda, S., Caviness, K., & Prieto, K. (2019). Medical countermeasures during the 2018 Ebola virus disease outbreak in the North Kivu and Ituri Provinces of the Democratic Republic of the Congo: A rapid genomic assessment. *The Lancet. Infectious Diseases*, 19(6), 648–657. DOI: 10.1016/S1473-3099(19)30118-5 PMID: 31000464

McGrath, S., & Ghersi, D. (2016). Building towards precision medicine: Empowering medical professionals for the next revolution. *BMC Medical Genomics*, 9(1), 1–6. DOI: 10.1186/s12920-016-0183-8 PMID: 27160306

McGuire, A. L., Caulfield, T., & Cho, M. K. (2008). Research ethics and the challenge of whole-genome sequencing. *Nature Reviews. Genetics*, 9(2), 152–156. DOI: 10.1038/nrg2302 PMID: 18087293

Meagher, K. M., & Lee, L. M. (2016). Integrating public health and deliberative public bioethics: Lessons from the human genome project ethical, legal, and social implications program. *Public Health Reports*, 131(1), 44–51. DOI: 10.1177/003335491613100110 PMID: 26843669

Micheel, J., Safrastyan, A., & Wollny, D. (2021). Advances in non-coding RNA sequencing. *Non-Coding RNA*, 7(4), 70. DOI: 10.3390/ncrna7040070 PMID: 34842804

Morange, M. (2006). The protein side of the central dogma: Permanence and change. *History and Philosophy of the Life Sciences*, 513–524. PMID: 18351049

Mousavi Sagharchi, S., & Mahmoudi Nasab, A. (2023). *Investigation of Rifampicin and Isoniazid Resistance Mutated Genes in Mycobacterium tuberculosis Isolated from Patients.* Paper presented at the ICMAAA 2023: International Conference on Mechanisms of Action of Antimicrobial Agents, Venice, Italy.

Mrozek, D., Stępień, K., Grzesik, P., & Małysiak-Mrozek, B. (2021). A Large-Scale and Serverless Computational Approach for Improving Quality of NGS Data Supporting Big Multi-Omics Data Analyses. *Frontiers in Genetics*, 12, 699280. DOI: 10.3389/fgene.2021.699280 PMID: 34326863

Nam, N. N., Do, H. D. K., Loan Trinh, K. T., & Lee, N. Y. (2023). Metagenomics: An Effective Approach for Exploring Microbial Diversity and Functions. *Foods*, 12(11), 2140. Advance online publication. DOI: 10.3390/foods12112140 PMID: 37297385

Necsulea, A. (2020). Phylogenomics and genome annotation.

Nguyen, N. H. (2024). Genetics and Genomics of Infectious Diseases in Key Aquaculture Species. *Biology (Basel)*, 13(1), 29. DOI: 10.3390/biology13010029 PMID: 38248460

Nielsen, T. K., Forero-Junco, L. M., Kot, W., Moineau, S., Hansen, L. H., & Riber, L. (2023). Detection of nucleotide modifications in bacteria and bacteriophages: Strengths and limitations of current technologies and software. *Molecular Ecology*, 32(6), 1236–1247. DOI: 10.1111/mec.16679 PMID: 36052951

Nusrat, S., Harbig, T., & Gehlenborg, N. (2019). Tasks, Techniques, and Tools for Genomic Data Visualization. *Computer Graphics Forum*, 38(3), 781–805. DOI: 10.1111/cgf.13727 PMID: 31768085

Olivares, F., Loyola, R., Olmedo, B., Miccono, M. Á., Aguirre, C., Vergara, R., & Mora, R. (2021). CRISPR/Cas9 targeted editing of genes associated with fungal susceptibility in Vitis vinifera L. Cv. Thompson Seedless using Geminivirus-derived replicons. *Frontiers in Plant Science*, 12, 791030. DOI: 10.3389/fpls.2021.791030 PMID: 35003180

Oliveri, S., Ferrari, F., Manfrinati, A., & Pravettoni, G. (2018). A systematic review of the psychological implications of genetic testing: A comparative analysis among cardiovascular, neurodegenerative and cancer diseases. *Frontiers in Genetics*, 9, 322074. DOI: 10.3389/fgene.2018.00624 PMID: 30619456

Pereira, F. (2019). Metagenomics: A gateway to drug discovery. In *Advances in Biological Science Research* (pp. 453-468). Elsevier.

Poland, G. A., Kennedy, R. B., McKinney, B. A., Ovsyannikova, I. G., Lambert, N. D., Jacobson, R. M., & Oberg, A. L. (2013). *Vaccinomics, adversomics, and the immune response network theory: individualized vaccinology in the 21st century.* Paper presented at the Seminars in immunology. DOI: 10.1016/j.smim.2013.04.007

Privitera, G. F., Alaimo, S., Ferro, A., & Pulvirenti, A. (2022). Virus finding tools: Current solutions and limitations. *Briefings in Bioinformatics*, 23(4), bbac235. DOI: 10.1093/bib/bbac235 PMID: 35753694

Rahman, M., & Schellhorn, H. E. (2023). Metabolomics of infectious diseases in the era of personalized medicine. *Frontiers in Molecular Biosciences*, 10, 1120376. DOI: 10.3389/fmolb.2023.1120376 PMID: 37275959

Rebets, Y., Kormanec, J., Luzhetskyy, A., Bernaerts, K., & Anné, J. (2017). Cloning and expression of metagenomic DNA in Streptomyces lividans and subsequent fermentation for optimized production. *Metagenomics: Methods and Protocols*, 99-144.

Ritchie, M. E., Phipson, B., Wu, D., Hu, Y., Law, C. W., Shi, W., & Smyth, G. K. (2015). limma powers differential expression analyses for RNA-sequencing and microarray studies. *Nucleic Acids Research*, 43(7), e47. DOI: 10.1093/nar/gkv007 PMID: 25605792

Rosario-Cruz, R., Domínguez-García, D. I., López-Silva, S., & Rosario-Domínguez, F. (2023). Immunoinformatics and tick vaccinology. *Exploration of Immunology*, 3(1), 1–16. DOI: 10.37349/ei.2023.00085

Rosati, D., Palmieri, M., Brunelli, G., Morrione, A., Iannelli, F., Frullanti, E., & Giordano, A. (2024). Differential gene expression analysis pipelines and bioinformatic tools for the identification of specific biomarkers: A review. *Computational and Structural Biotechnology Journal*, 23, 1154–1168. DOI: 10.1016/j.csbj.2024.02.018 PMID: 38510977

Rosenquist, R., Cuppen, E., Buettner, R., Caldas, C., Dreau, H., Elemento, O., . . . Jobanputra, V. (2022). *Clinical utility of whole-genome sequencing in precision oncology.* Paper presented at the Seminars in Cancer Biology. DOI: 10.1016/j.semcancer.2021.06.018

Roychowdhury, R., Das, S. P., Gupta, A., Parihar, P., Chandrasekhar, K., Sarker, U., Kumar, A., Ramrao, D. P., & Sudhakar, C. (2023). Multi-Omics Pipeline and Omics-Integration Approach to Decipher Plant's Abiotic Stress Tolerance Responses. *Genes*, 14(6), 1281. Advance online publication. DOI: 10.3390/genes14061281 PMID: 37372461

Sahito, J. H., Zhang, H., Gishkori, Z. G. N., Ma, C., Wang, Z., Ding, D., Zhang, X., & Tang, J. (2024). Advancements and Prospects of Genome-Wide Association Studies (GWAS) in Maize. *International Journal of Molecular Sciences*, 25(3), 1918. Advance online publication. DOI: 10.3390/ijms25031918 PMID: 38339196

Saravanan, K., Panigrahi, M., Kumar, H., Rajawat, D., Nayak, S. S., Bhushan, B., & Dutt, T. (2022). Role of genomics in combating COVID-19 pandemic. *Gene*, 823, 146387. DOI: 10.1016/j.gene.2022.146387 PMID: 35248659

Satam, H., Joshi, K., Mangrolia, U., Waghoo, S., Zaidi, G., Rawool, S., & Das, G. (2023). Next-generation sequencing technology: Current trends and advancements. *Biology (Basel)*, 12(7), 997. DOI: 10.3390/biology12070997 PMID: 37508427

Satam, H., Joshi, K., Mangrolia, U., Waghoo, S., Zaidi, G., Rawool, S., Thakare, R. P., Banday, S., Mishra, A. K., Das, G., & Malonia, S. K. (2023). Next-Generation Sequencing Technology: Current Trends and Advancements. *Biology (Basel)*, 12(7), 997. Advance online publication. DOI: 10.3390/biology12070997 PMID: 37508427

Schatz, M. C., Philippakis, A. A., Afgan, E., Banks, E., Carey, V. J., Carroll, R. J., & Wuichet, K. (2022). Inverting the model of genomics data sharing with the NHGRI Genomic Data Science Analysis, Visualization, and Informatics Lab-space. *Cell Genomics*, 2(1), 100085. Advance online publication. DOI: 10.1016/j.xgen.2021.100085 PMID: 35199087

Schilbert, H. M., Rempel, A., & Pucker, B. (2020). Comparison of Read Mapping and Variant Calling Tools for the Analysis of Plant NGS Data. *Plants*, 9(4), 439. Advance online publication. DOI: 10.3390/plants9040439 PMID: 32252268

Schwarz, U. I., Gulilat, M., & Kim, R. B. (2019). The Role of Next-Generation Sequencing in Pharmacogenetics and Pharmacogenomics. *Cold Spring Harbor Perspectives in Medicine*, 9(2), a033027. Advance online publication. DOI: 10.1101/cshperspect.a033027 PMID: 29844222

Sepulveda, J. L. (2020). Using R and Bioconductor in Clinical Genomics and Transcriptomics. *The Journal of Molecular Diagnostics*, 22(1), 3–20. DOI: 10.1016/j.jmoldx.2019.08.006 PMID: 31605800

Shi, L., & Wang, Z. (2019). Computational Strategies for Scalable Genomics Analysis. *Genes*, 10(12), 1017. Advance online publication. DOI: 10.3390/genes10121017 PMID: 31817630

Sinha, D., Maurya, A. K., Abdi, G., Majeed, M., Agarwal, R., Mukherjee, R., Ganguly, S., Aziz, R., Bhatia, M., Majgaonkar, A., Seal, S., Das, M., Banerjee, S., Chowdhury, S., Adeyemi, S. B., & Chen, J. T. (2023). Integrated Genomic Selection for Accelerating Breeding Programs of Climate-Smart Cereals. *Genes*, 14(7), 1484. Advance online publication. DOI: 10.3390/genes14071484 PMID: 37510388

Solís-Moruno, M., Batlle-Masó, L., Bonet, N., Aróstegui, J. I., & Casals, F. (2023). Somatic genetic variation in healthy tissue and non-cancer diseases. *European Journal of Human Genetics*, 31(1), 48–54. DOI: 10.1038/s41431-022-01213-8 PMID: 36289407

Spyrou, M. A., Bos, K. I., Herbig, A., & Krause, J. (2019). Ancient pathogen genomics as an emerging tool for infectious disease research. *Nature Reviews. Genetics*, 20(6), 323–340. DOI: 10.1038/s41576-019-0119-1 PMID: 30953039

Strianese, O., Rizzo, F., Ciccarelli, M., Galasso, G., D'Agostino, Y., Salvati, A., Del Giudice, C., Tesorio, P., & Rusciano, M. R. (2020). Precision and personalized medicine: How genomic approach improves the management of cardiovascular and neurodegenerative disease. *Genes*, 11(7), 747. DOI: 10.3390/genes11070747 PMID: 32640513

Subramanian, I., Verma, S., Kumar, S., Jere, A., & Anamika, K. (2020). Multi-omics data integration, interpretation, and its application. *Bioinformatics and Biology Insights*, 14. DOI: 10.1177/1177932219899051 PMID: 32076369

Telatin, A., Fariselli, P., & Birolo, G. (2021). SeqFu: A suite of utilities for the robust and reproducible manipulation of sequence files. *Bioengineering (Basel, Switzerland)*, 8(5), 59. DOI: 10.3390/bioengineering8050059 PMID: 34066939

Thak, E. J., Yoo, S. J., Moon, H. Y., & Kang, H. A. (2020). Yeast synthetic biology for designed cell factories producing secretory recombinant proteins. *FEMS Yeast Research*, 20(2), foaa009. DOI: 10.1093/femsyr/foaa009 PMID: 32009173

Thieffry, D., & Sarkar, S. (1998). Forty years under the central dogma. *Trends in Biochemical Sciences*, 23(8), 312–316. DOI: 10.1016/S0968-0004(98)01244-4 PMID: 9757833

Tuena, C., Semonella, M., Fernández-Álvarez, J., Colombo, D., & Cipresso, P. (2020). Predictive precision medicine: towards the computational challenge. *P5 eHealth: An Agenda for the Health Technologies of the Future*, 71-86.

Uddin, M., Mustafa, F., Rizvi, T. A., Loney, T., Al Suwaidi, H., Al-Marzouqi, A. H. H., & Stefanini, C. (2020). SARS-CoV-2/COVID-19: Viral genomics, epidemiology, vaccines, and therapeutic interventions. *Viruses*, 12(5), 526. DOI: 10.3390/v12050526 PMID: 32397688

Uszczynska-Ratajczak, B., Lagarde, J., Frankish, A., Guigó, R., & Johnson, R. (2018). Towards a complete map of the human long non-coding RNA transcriptome. *Nature Reviews. Genetics*, 19(9), 535–548. DOI: 10.1038/s41576-018-0017-y PMID: 29795125

Vinatzer, B. A., Heath, L. S., Almohri, H. M. J., Stulberg, M. J., Lowe, C., & Li, S. (2019). Cyberbiosecurity Challenges of Pathogen Genome Databases. *Frontiers in Bioengineering and Biotechnology*, 7, 106. DOI: 10.3389/fbioe.2019.00106 PMID: 31157218

Wang, R. C., & Wang, Z. (2023). Precision Medicine: Disease Subtyping and Tailored Treatment. *Cancers (Basel)*, 15(15), 3837. Advance online publication. DOI: 10.3390/cancers15153837 PMID: 37568653

Wang, S., Sun, S.-T., Zhang, X.-Y., Ding, H.-R., Yuan, Y., He, J.-J., Wang, M.-S., Yang, B., & Li, Y.-B. (2023). The evolution of single-cell RNA sequencing technology and application: Progress and perspectives. *International Journal of Molecular Sciences*, 24(3), 2943. DOI: 10.3390/ijms24032943 PMID: 36769267

Yang, C., Chowdhury, D., Zhang, Z., Cheung, W. K., Lu, A., Bian, Z., & Zhang, L. (2021). A review of computational tools for generating metagenome-assembled genomes from metagenomic sequencing data. *Computational and Structural Biotechnology Journal*, 19, 6301–6314. DOI: 10.1016/j.csbj.2021.11.028 PMID: 34900140

Zerbino, D. R., Frankish, A., & Flicek, P. (2020). Progress, challenges, and surprises in annotating the human genome. *Annual Review of Genomics and Human Genetics*, 21(1), 55–79. DOI: 10.1146/annurev-genom-121119-083418 PMID: 32421357

Zhang, L., Chen, F., Zeng, Z., Xu, M., Sun, F., Yang, L., & Hao, H. (2021). Advances in metagenomics and its application in environmental microorganisms. *Frontiers in Microbiology*, 12, 766364. DOI: 10.3389/fmicb.2021.766364 PMID: 34975791

Zhang, L., Jackson, C. B., Mou, H., Ojha, A., Peng, H., Quinlan, B. D., & Suthar, M. S. (2020). SARS-CoV-2 spike-protein D614G mutation increases virion spike density and infectivity. *Nature Communications*, 11(1), 6013. DOI: 10.1038/s41467-020-19808-4 PMID: 33243994

Zhao, J., Qin, B., Nikolay, R., Spahn, C. M. T., & Zhang, G. (2019). Translatomics: The Global View of Translation. *International Journal of Molecular Sciences*, 20(1), 212. Advance online publication. DOI: 10.3390/ijms20010212 PMID: 30626072

Zheng, L., Wang, H., Lin, J., Zhou, Y., Xiao, J., & Li, K. (2023). Population genomics provides insights into the genetic diversity and adaptation of the Pieris rapae in China. *PLoS One*, 18(11), e0294521. DOI: 10.1371/journal.pone.0294521 PMID: 37972203

# Chapter 3
# State of the Art of Immunoinformatics

**Eduarda Guimarães Sousa**
https://orcid.org/0000-0001-8326-3612
*Federal University of Minas Gerais, Brazil*

**Lucas Gabriel Rodrigues Gomes**
*Federal University of Minas Gerais, Brazil*

**Fernanda Diniz Prates**
*Federal University of Minas Gerais, Brazil*

**Talita Pereira Gomes**
*Federal University of Minas Gerais, Brazil*

**Gabriel Camargos Gomes**
*Federal University of Minas Gerais, Brazil*

**Janaíne Aparecida de Paula**
https://orcid.org/0000-0002-9932-2610
*Federal University of Minas Gerais, Brazil*

**Ana Lua de Oliveira Vinhal**
*Federal University of Minas Gerais, Brazil*

**Bernardo Buhr Alves Mendonça**
https://orcid.org/0009-0001-5779-8557
*Federal University of Minas Gerais, Brazil*

**Mariana Letícia Costa Pedrosa**
*Federal University of Minas Gerais, Brazil*

**Luiza Pereira Reis**
*Federal University of Minas Gerais, Brazil*

**Aline Ferreira Maciel de Oliveira**
*Federal University of Minas Gerais, Brazil*

**Marcus Vinicius Canário Viana**
*Federal University of Minas Gerais, Brazil*

**Arun Kumar Jaiswal**
*Federal University of Minas Gerais, Brazil*

DOI: 10.4018/979-8-3693-3192-7.ch003

**Siomar de Castro Soares**
https://orcid.org/0000-0001-7299-3724
*Federal University of Triângulo Mineiro, Brazil*

**Vasco Ariston de Carvalho Azevedo**
https://orcid.org/0000-0002-4775-2280
*Federal University of Minas Gerais, Brazil*

## ABSTRACT

*Vaccinology is the study of vaccines, their development, and their use. Classical vaccinology, practiced for centuries, relies on identifying and isolating a pathogen's protective antigens. Classical vaccinology has successfully developed vaccines against many diseases, including smallpox, measles, mumps, and diphtheria. However, it can be a time-consuming and labor-intensive process. Immunoinformatics offers a new approach to vaccine development that can address some of the limitations of classical vaccinology. This interdisciplinary field combines biology and computer science to explore the immune system's interactions. In the last decade, technological advances, such as big data, machine learning, and artificial intelligence, are enabling new advances in Immunoinformatics, potentially revolutionizing vaccine development. This chapter aims to provide a comprehensive overview of Immunoinformatics, its techniques, and its applications, exploring its foundations, advances, and future perspectives.*

## INTRODUCTION

Immunization is fundamental to public health, with vaccines playing a vital role in maintaining and preventing various diseases. Since the proper establishment of national immunization programs in several countries, many endemic diseases that were previously a public health problem have been eradicated (Excler et al., 2021). The World Health Organization says these programs save between 2 and 3 million lives annually. Vaccines have been one of the most successful public health achievements (*World Health Organization (WHO)*, n.d.). They are economical and practical tools that have evolved over the decades into many approaches to the production of immunogens, each with its advantages and disadvantages, with the aim of continuous improvement (Hajj Hussein et al., 2015). Vaccines can be divided into three generations according to the composition and strategies used to prepare the active antigen: 1st, 2nd, and 3rd generation vaccines, as shown in Figure 1 (Bagnoli et al., 2011; Zarei et al., 2016).

Throughout history, vaccines have been studied for significant advances to monitor their side effects and benefits relevant to their preventive use in each disease and guarantee their safe use. Vaccination began in at least the 15th century when people from different regions intentionally exposed themselves to the pathogen to prevent smallpox, which prompted studies in the area in later years. The doctor Edward Jenner, who in 1796 saw smallpox as the greatest threat to humanity, expands on this discovery. He carried out an experiment in which he observed that the inoculation of an 8-year-old boy, James Phipps, with biological material from cows infected with cowpox could prevent exposure to a more virulent pathogen; it was observed that he did not fall ill after infection with human smallpox, like in the first time he had been exposed to the pathogen, giving birth to vaccinology (Bagnoli et al., 2011).

At the end of the 19th century, Louis Pauster was one of the pioneers in the development of vaccines. His studies led to the beginning of the "Age of Microbiology," where there was a quest to characterize and understand the relationship between infections and hygiene and, thus, to visualize the presence of microorganisms. These organisms are "invisible to the naked eye." Louis Pasteur developed studies centered on the analysis of microorganisms and prophylactic measures against them, which led to the development of the rabies vaccine. This research involved attenuating the infectious agent, using the attenuated virus found in the brains of rabid dogs in the brains of laboratory rabbits one hundred times (Hilleman, 2000; Saliou & Girard, 2005).

These discoveries made it possible to study further the principles of vaccination, especially the attenuation process stemming from Pauster's projects, which were duly established by suggesting that it would be necessary to "isolate, inactivate and inject" the microorganism to provoke a long-term immune response, allowing the first vaccines to be developed (Hajj Hussein et al., 2015).

Based on these vaccination principles, a decade after Jenner's discovery, another work was carried out, thus marking the beginning of the first generation of vaccines, represented by those that use whole pathogens subjected to treatments that inactivate or weaken the microorganisms. An excellent example of this approach is the tuberculosis vaccine, which uses a bacterium derived from cattle, *Mycobacterium bovis* (BCG), and vaccines to prevent rubella (Soto et al., 2022). In addition, vaccines were developed against the causative agents of diseases such as typhoid fever, cholera, and plague, led by the Salmon, Kolle, and Haffkine vaccines, using heat-killed bacteria. In 1931, Goodpasture developed vaccines against yellow fever and influenza, with the innovation of growing viruses in embryonated chicken eggs (D'Amelio et al., 2016). In the middle of the century, cell culture techniques were developed to develop several viral vaccines, including Salk's polio vaccine and the measles and mumps vaccine (Ulschmid et al., 2023).

Second-generation vaccines include acellular vaccines that use purified toxoids, proteins, and polysaccharides, such as vaccines against tetanus, diphtheria, hepatitis B, meningococcus, and influenza (Hajj Hussein et al., 2015). Research in the 20th century focused on developing vaccines using bacterial toxins responsible for infections such as diphtheria and tetanus, following the discoveries of Roux, Yersin, Glenny, and Hopkins. Subunit vaccines using proteins and polysaccharides from pathogens were also introduced, and advances in genetic engineering contributed to the development of recombinant vaccines and the study of vaccine adjuvants (Cho et al., 2021).

In the early 1990s, research began using genetic material for vaccine research, using Nucleic Acid, such as DNA and RNA, and viral vector vaccines. These are very safe and have the advantage of being quicker to produce because they do not require the steps involved in traditional vaccines, such as the isolation and cultivation of large quantities of isolates, which often involves complex systems, avoiding purification and formulation steps after the production of antigens, since these steps can be simplified due to the more direct nature of the production process. Once inside the body, they act as a mold for producing specific proteins that generate a targeted immune response against the infection. These are third-generation vaccines like against SARS-CoV-2 and Ebola (Khoshnood et al., 2022; Y. Liu & Ye, 2022; Melo et al., 2022).

At the end of the 20th century, new technologies emerged to combat persistent pathogens. As a result of next-generation sequencing (WGS) methodologies, more genomes were being sequenced and deposited in public databases, which made it possible to analyze factors at a molecular and genomic level that could be used in the development and understanding of prophylactic measures such as vaccines and immunological factors associated with infections. In 1995, Craig Venter's pioneering work in genome sequencing enabled access to microbial genomes, leading to the emergence of 'reverse vaccinology, led by Rino Rappuoli. This method bypassed the need for microbial cultivation and revolutionized vaccine development by using computers to access microbial genomes directly (Saliou & Girard, 2005; Sette & Rappuoli, 2010).

*Figure 1. Timeline of first, second, and third generation vaccines with their respective classification according to the antigen used, as well as examples of vaccines for each generation*

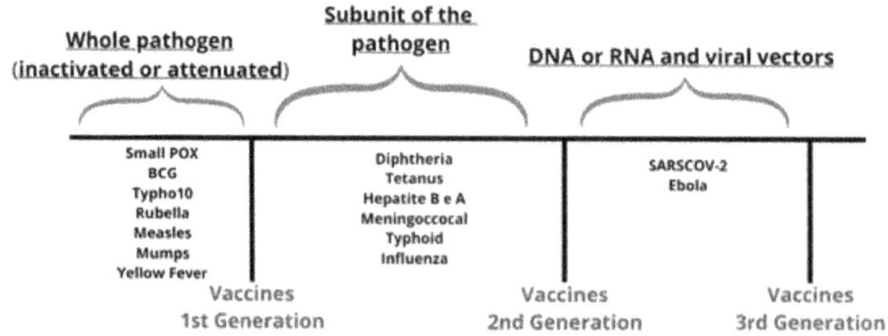

From an immunological point of view, a vaccine is a formulation that contains an inactive, fragmented, attenuated, or molecular part of a specific pathogen. This careful selection of the part is intended to ensure that it is not pathogenic and does not cause disease but stimulates an immune response in the host. When administered, the vaccine presents the immune system with a representation of the target antigen, allowing it to learn to recognize and produce specific antibodies against it (Canouï & Launay, 2019).

The process of antigen presentation by lymphocytes leads to the differentiation of cytotoxic cells or T-helper cells. In specific cases, the antigen is recognized directly by B cells, which activate the production of immunoglobulins (Baljon & Wilson, 2022). These immunoglobulins can neutralize or promote phagocytosis of the antigen. Through this controlled exposure to the antigen, the immune system is induced to develop an immunological memory. In this way, the body is prepared to mount a rapid and effective immune response when it encounters the real antigen in the future. The antibodies generated in response to vaccination are available to neutralize the invading antigen, which can protect the body from the infection or disease associated with the antigen (Markina et al., 2020).

The immune response is initiated when phagocytic cells, such as macrophages or dendritic cells, recognize the pathogen or its components after vaccine inoculation. These cells process and present the antigen to the class I major histocompatibility complex (MHC). The presentation of antigens to CD8 (cytotoxic) T lymphocytes occurs via the class I MHC, while CD4+ T lymphocytes recognize the antigens presented via the class II MHC. B lymphocytes can recognize antigens directly or in a T-dependent manner. The activation of cellular and humoral adaptive immunity leads to the production of cytokines and the activation of innate immune cells such as neutrophils and macrophages, as well as the production of immunoglobulins (Zhang et al., 2023).

Immunology represents one of the most frequently addressed fields of research in studying the body's responses to different substances, allowing the evaluation of various effective methods, and weighing up advantages and disadvantages for different groups of individuals. However, a significant challenge lies in the time required to conduct the research and reach the testing phase, where effectiveness and reactions, both positive and negative, are assessed. A prominent example of this impasse is the production of vaccines, where the necessary steps can take a long time before they are distributed to the public (Soubeyrand, 2018).

When the process is carried out on a bench, it tends to be more time-consuming and demanding since it depends on specific cultural conditions, precise temperature, and adequate collection, as well as meticulous handling according to standard operating procedures (Ishack & Lipner, 2021). Faced with these difficulties, bioinformatics has emerged as a tool to optimize bench testing time. Through computer programs, where experimental phases can be carried out more quickly, with results available in a matter of minutes or hours, compared to traditional methods that take days, this advance provides significant efficiency and faster access to the necessary information (Michalik et al., 2016).

Immunoinformatics is a constantly developing field encompassing several areas, such as biology, immunology, and bioinformatics. One of its objectives is understanding and comprehending the immune system, helping create and develop treatments, diagnose, and even prevent diseases. This large area uses technology to its advantage, enjoying robust analysis of biological data such as proteins, RNA, and DNA, immune response profiles, and even three-dimensional structures (Bahrami et al., 2019).

Using Immunoinformatics to optimize bench tests, results can be obtained more quickly, accurately, and with minimal error probability. This allows for faster correction of any problems that may arise during the testing phase, ultimately leading to faster and safer public access to vaccines and insights about the mechanisms of immune function and their application in the understanding of disease pathogenesis (Lancaster et al., 2020).

Regarding research in Immunoinformatics, several contributions can be cited and detailed, such as discovering biomarkers that help develop vaccines and diagnostic tests against various infections that humans may have. Bioinformatics has provided tools that have enabled *in-silico* mapping in some antigens (epitopes) from existing antigens (van der Poll et al., 2021). In this way, the antibodies will bind to the linear epitopes of the B cells and will interact with the segments of some amino acid residues located on the surface of the antigen. By doing this, it is possible to find specific antigenic regions that will connect with the individual's defense cells, increasing the chances of obtaining a better result because the specificity and sensitivity of the tests will be better. Vaccines will have fewer side effects.

This chapter covers Immunoinformatics and its applications, its basic principles, and concepts, applied methodologies, high-throughput technologies in antigen identification, precision medicine, personalized antigen design, and computational strategies for vaccine design. This will provide a better understanding of this interdisciplinary and evolving field, which is extremely important for public health. In doing so, we will discuss established methodologies in the field and the application of future methods, such as artificial intelligence, named Immunoinformatics 2.0.

## Foundations of Immunoinformatics

Immunoinformatics is a growing area that has boosted research in immunology. DNA sequencing techniques have been generating a large amount of data, which can provide valuable insights into the mechanisms of immune function and disease pathogenesis (Tomar & De, 2014). The scope of Immunoinformatics is enormous and comprises databases, molecular-level and organism-level models, genomics, and proteomics of the immune system, as well as genome-to-genome studies as shown in Figure 2 (Lim et al., 2023). Although evolving, combining immunology and computational sciences has provided relevant information that could not be achieved using traditional approaches alone. Once a considerable combinatorial nature exists of immunological data, immunoinformatic tools also need to become more efficient and accurate daily.

*Figure 2. Different applications of immunoinformatics*

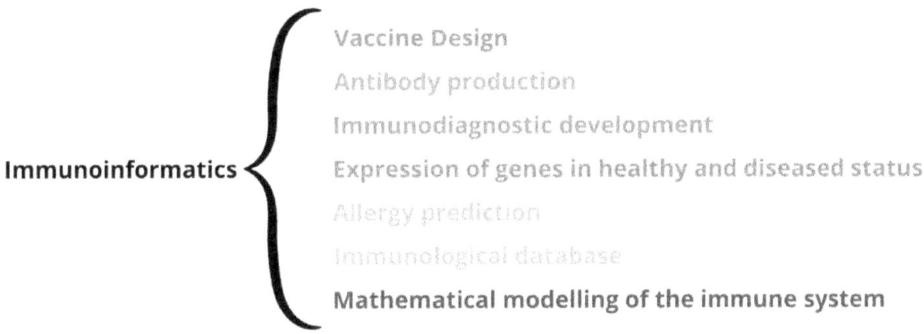


## Vaccine Design

Vaccines are essential for public health, mainly because they prevent the spreading of contagious and, often, fatal diseases worldwide. During the last years, most vaccines were designed using traditional methods, like those that contain attenuated strains of the pathogenic microorganism (smallpox in 1798) and those that contain only components of a pathogen or killed whole organisms (typhoid, 1896) (Bouazzaoui et al., 2021). Over the past years, several other methods have been developed, including virus-like particles (hepatitis B, 1986), viral vectors (Ebola, 2019), and nucleic acid-based RNA and DNA vaccines (SARS-CoV-2, 2020) (Pollard & Bijker, 2020). Immunoinformatic has been employed to design algorithms for mapping B- and T-cell epitopes, exploring potential binding sites, and consequently, leading to the development of new vaccines (Nielsen et al., 2020a). This methodology is termed 'reverse vaccinology,' several studies have been carried out using this approach to identify antigenic proteins. One of these studies gave rise to the 4CMenB vaccine against invasive serogroup B meningococcal (Masignani et al., 2019), which was approved in Europe in 2013 and introduced to the national immunization program of the UK in 2015.

## Immunological Database

With the exponential growth of biological and immunological data, it is essential to store and manage these data in a findable, accessible, interoperable, and reusable manner. Here (Table 1), there are some immunological databases which can be used:

*Table 1. Immunological databases*

| Database | Application | Reference |
|---|---|---|
| BCIpep | Determined liner B-cell epitopes | https://webs.iiitd.edu.in/raghava/bcipep/info.html |
| Epitome | Identify structure-inferred antigenic epitopes | https://ngdc.cncb.ac.cn/databasecommons/database/id/1612 |
| ExactAntigen | Monoclonal antibody products | https://www.hsls.pitt.edu/obrc/index.php?page= URL1163004330 |
| IEDB | Resource to predict, analyze, and characterize immune epitope | https://www.iedb.org |
| IPD | Study of polymorphic genes in the immune system | https://www.ebi.ac.uk/ipd/ |
| InnateDB | Facilitate analyses of innate immunity interactions and pathways | https://innatedb.com |
| MHCBN | MHC binding and non-binding peptides | https://webs.iiitd.edu.in/raghava/mhcbn/ |
| VIOLIN | vaccine investigation and online information network | https://ngdc.cncb.ac.cn/databasecommons/database/id/300 |

## Mathematical Modeling of The Immune System

The immune system is a complex network that comprises molecules, cells, tissues, and organs that stimulate the body to fight against dangerous foreign agents, infections, and other diseases, helping maintain the integrity of the organism (Hao Shi et al., 2020). This vital system has some interesting properties: each individual has an immune system (uniqueness); the immune cells are distributed throughout the body and are not subject to centralized control (distributed detection); perfect recognition is not necessary for the immune system to react against an element (imperfect detection); the immune system can learn and remember the structures of pathogens, so future responses to the pathogens can be targeted and much faster (adaptability) (Motta & Brusic, 2004).

Our knowledge of the immune system is still incomplete, and mathematical modeling has provided a better understanding of the principles and organization of this system, helping to find new treatments and therapies for several diseases. As mathematical models aimed at the immune system can model interactions of a significant number of elements ($10^6$ or higher), they represent an essential method that can approach the complexity of the human immune system (Handel et al., 2019). They can be used differently for exploration, prediction, and fitting, as shown in Figure 3.

*Figure 3. A schematic illustrating different ways to use the mathematical models*

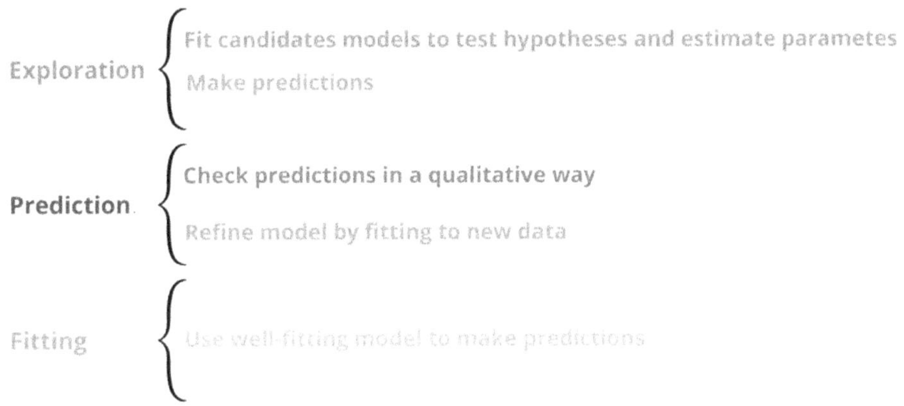


## Next-Generation Antigen Discovery

As explained in the previous section, Immunoinformatics or computational immunology is an area that links computer science and immunology together by applying computational resources and concepts to understand and carefully analyze the immunology data (Chatanaka et al., 2022). The main objective of Immunoinformatics is to evaluate the antigenicity and properties of identified protein vaccine candidates as antigen molecules, such as identification of MHCI-allele binding CTL epitopes, MHCII-allele binding HTL epitopes, and prediction of B-cell epitopes. Afterward, those epitopes may be carefully analyzed and filtered to find the best epitopes for further joining them and constructing a vaccine candidate. In doing so, several computational approaches are available that help Immunoinformatics researchers and vaccine scientists evaluate the antigenicity of the desired vaccine targets. In recent years, artificial intelligence (AI) and machine learning (ML) have directed the computational approaches to improve the existing algorithms for better immunogen prediction (Bravi, 2024; Kaushik et al., 2023). The continuous improvement in AI and ML algorithms is helping to generate large immune datasets for their collection and annotations, and the generated datasets provide more information about the immune responses (Bravi, 2024). The IEDB (Immune Epitope Database and Tools, https://www.iedb.org/) is a widely used database and tool that helps in mapping immunogenicity and identification of MHC I & II epitopes of antigen molecules with their continuous improvements in antigenic properties identification algorithms (recently added feature Next-generation Tool site at IEDB database) (Vita et al., 2019).

Modern advancements in high-throughput experimental techniques and next-generation sequencing platforms generate vast data in biomedical and vaccine development research (Kaushik et al., 2023). The generated datasets encourage AI and ML to predict more precise immunogenicity in vaccine development. In the current scenario, several vaccine research laboratories and companies are applying AI and ML to identify the immunogenic properties of vaccine candidates in their vaccine design and development (Blanco-González et al., 2023; Kaushik et al., 2023). In the future, genomics and proteomics datasets generated by high-throughput techniques of several infectious pathogens can help to build more accurate AI and ML-based models, and the models will help in better understanding of the immunogenicity of the pathogens based on cellular, molecular, and genetic levels in computational vaccine studies (Kaushik et al., 2023). They will allow researchers to screen vast libraries of potential antigens or test multiple candidates simultaneously.

The steps for successful vaccine candidate selection computationally start with the comprehensive analysis of the pathogen`s genomes and proteomes. The screening process of complete genomes and proteomes with various web and stand-alone programs to identify the small number of proteomes (proteins as vaccine targets) as vaccine-like candidates is called reverse vaccinology (RV) (L. Liu et al., 2023; Rappuoli, 2000). Identifying proteins as vaccine targets can start with identifying the common proteins (core proteome/genome). If using multiple strains, it should be screened against the host, and the proteins do not present in the host should be identified as non-host homologous. It should be carried out with subcellular localization and adhesion probability of the protein vaccine candidates (Figure 4) (Jaiswal et al., 2017a; Prado et al., 2022; Zarei et al., 2016). Further, the identified vaccine candidates should be evaluated with Immunoinformatics approaches, as explained in the above section, for immunogenicity properties (Jaiswal et al., 2017b).

*Figure 4. Schematic workflow of reverse vaccinology to identify vaccine candidates*

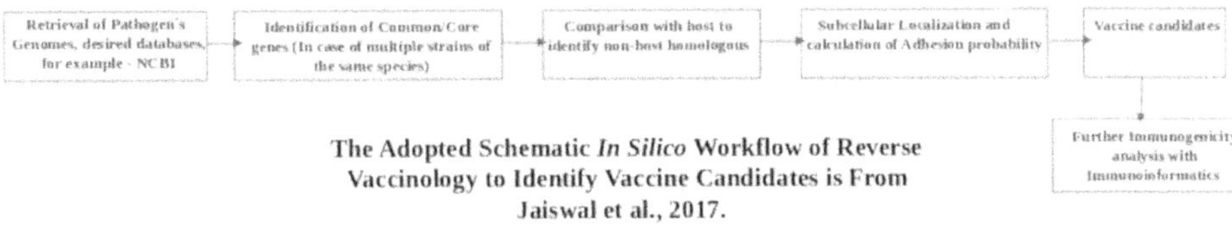


The computational methodologies in vaccine development, like RV and Immunoinformatics, are widely used approaches in the current scenario and can help precisely to the genetic and immunological characteristics of individual patients or specific patient subgroups with the help of genomic or proteomic data surveillance (Bravi, 2024; Guarra & Colombo, 2023). The RV and Immunoinformatics approaches can help in the selection or engineering of antigens that are most likely to elicit a strong and targeted immune response in a particular patient or population after the brief, genomic and immunogenicity profiling analysis of particular individual patients and identify antigens that match the patient's specific immune system—for example, recently developed novel peptide vaccine based on an Immunoinformatics approach against metallo-beta-lactamase (VIM and IMP) variants (Motamedi et al., 2023), identification of candidate epitopes and vaccine development for Early secreted antigens G and H (EsxG and EsxH) against pulmonary tuberculosis (Martinez-Olivares et al., 2023) and designing personalized vaccines targeting neoantigens against melanoma (Malaina et al., 2023).

## Immunoinformatics in Vaccine Design

### Need for Advancements in Vaccine Design

Vaccine development is a field with remarkable significance to the promotion of global health, with vaccines being responsible for eradicating diseases like smallpox and decreasing mortality and infection rates for various other conditions (Moyle & Toth, 2013b). Historically, it has been a slow and costly field of research, requiring extensive in vitro and in vivo work to test various vaccine preparations. These have ranged from attenuated live microorganisms to the isolation and application of different surface proteins as inactivated toxoids (Martinelli, 2022; Moyle & Toth, 2013a). Due to these limitations, many diseases still do not have available and accessible vaccines, as is the case for many Neglected Tropical Diseases (NTDs), as well as Emerging and Reemerging Infectious Diseases (ERID) (Oli et al., 2020b).

The previous segment of this chapter explored the use of Immunoinformatics to expedite the search for viable vaccine targets, thus accelerating vaccine development. Therefore, in this section, we will focus on specific challenges to the design of a vaccine preparation through Immunoinformatics.

An essential factor to consider in this scenario is that, for several significant infections, more traditional methods of vaccine development have failed to produce an effective immunogen, as is the case for HIV (Das et al., 2010; Moyle & Toth, 2013b). This factor points to developing vaccines designed more precisely to ensure and direct their immunogenic effects (Moyle & Toth, 2013b; Oli et al., 2020b). Another factor of significant relevance is the focus on developing tools to protect specific populations, such as designing vaccines to elicit responses from a majority of people or designing them to be safely applied to at-risk groups, such as the immunocompromised (Oli et al., 2020a).

Immunoinformatics techniques have turned to developing subunit or multi-epitope vaccines to address both requirements. These immunogens are designed *in silico*, delivering only the components required to elicit the desired immune response (Amos et al., 2023). This not only means that these vaccinal formulations provoke targeted responses from the immune system but also that unnecessary antigenic load, which would be present in an inactivated microorganism or even toxoid vaccine, is not present in the vaccine (Gomes et al., 2023a).

## Multi-epitope Vaccine Construction

Advances in multi-epitope vaccine developments have only been possible due to developments in computing and extensive databases accumulating data from previous immunological studies and sequencing efforts (Rawat et al., 2023a). These databases have allowed for the development of tools for predicting immunogenic epitopes in a protein sequence capable of binding to immune cell receptors and provoking adequate responses. Among them, tools such as IEDB-AR and NetMHC can provide methods for predicting epitopes for both Cytotoxic T Lymphocytes and Helper T Lymphocytes (Greenbaum et al., 2011; Jensen et al., 2018; Larsen et al., 2007; H. H. Lin et al., 2008; Martini et al., 2020). Other tools, such as ABCPred (Galanis et al., 2021) Ellipro (Ponomarenko et al., 2008), can predict linear and conformational B-Cell epitopes based on protein sequence and structure.

The peptides predicted by these resources can then be used to construct a chimeric protein composed of the immunogenic epitopes connected by linker peptides (Ayyagari et al., 2022; Chauhan et al., 2019; Chen et al., 2013; Martinelli, 2022; Moyle & Toth, 2013b). This construct may include an adjuvant protein sequence to improve its ability to provoke an immune response (Burton, 2017; Moyle & Toth, 2013b; Oli et al., 2020b) (Figure 5).

Immunoinformatics-based vaccine development has been directed to research efforts in various fields. Subtractive genomics and reverse vaccinology have been applied to the discovery of putative vaccine candidates in the genomes of bacteria such as *Treponema pallidum* and *Mycoplasma pneumoniae* (Jaiswal et al., 2017a; Rodrigues et al., 2019). Similarly, vaccine constructs designed via the immunoinformatics-based identification of immunogenic epitopes has been applied towards designing immunogens for varied infections. Among them, vaccines for dengue virus (Ali et al., 2017), yellow fever virus (Ali et al., 2017), *T. pallidum* (Gomes et al., 2022), and *M. pneumoniae* (Rodrigues et al., 2019) are only a few examples.

*Figure 5. Multi-epitope vaccine construction workflow contains a schematic representation of the epitope prediction and multi-epitope vaccine construction processes. This vaccine construct may then be subjected to further in silico testing before it is tested in vitro and in vivo*

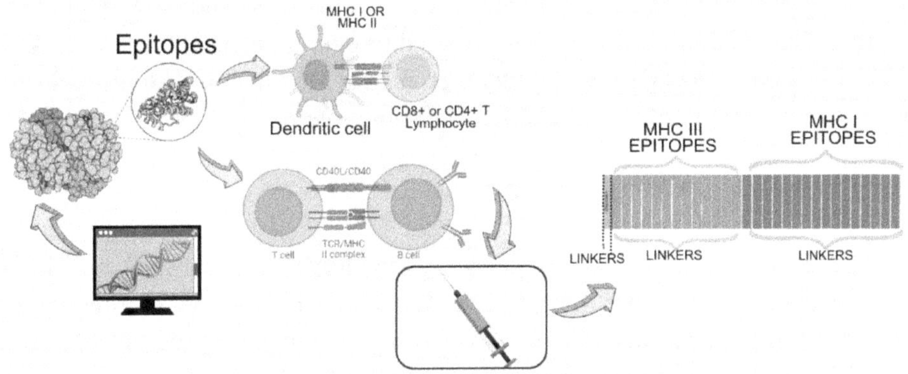


## Evaluating Vaccine Effectiveness

Given the importance of vaccination in efforts to combat disease, the development of new vaccines is a highly relevant field. In addition to its uses in identifying immunogenic sequences and designing immunogens that may serve as vaccines for a range of pathogens, Immunoinformatics may also be used to evaluate their effectiveness. Thus, the robust computational toolbox provided by bioinformatics allows us to effectively develop efficient and safe vaccines. Through the use of toxin databases such as Toxinpred (Gupta et al., 2013), and allergen databases such as AllerTOP (Dimitrov et al., 2014), it is possible to determine the safety of the vaccine. The IEDB-AR Population Coverage tool can predict whether the epitopes present in the construct are able to elicit responses from a significant portion of the population (Martini et al., 2020). Antigen databases such as VaxiJen (Doytchinova & Flower, 2007, 2008) may also be used to predict whether the vaccine construct shares common features with known antigens. Further, tools like the C-IMMSIM server allow us to mathematically model the immune system, making it possible to simulate the response the vaccine would elicit once applied to a living organism (Rapin et al., 2010).

Protein structure prediction, another developing field in bioinformatics, may also be used to determine whether the multi-epitope vaccine would be able to elicit an adequate immune response. Through the usage of powerful tools such as Alphafold (Jumper et al., 2021) and Phyre2 (Kelley et al., 2015), it is possible to predict the tertiary structure of a protein based on its sequence. Structural data can be used to determine the presence of conformational B-cell epitopes (Ponomarenko et al., 2008) and for molecular docking and dynamics. Molecular docking, using tools such as AutoDock Vina (Eberhardt et al., 2021) and SwarmDock (Torchala et al., 2013), allows to determine whether or not the vaccine is capable of binding with immune receptors. This binding can be further explored through molecular dynamics using programs such as GROMACS (Abraham et al., 2015), which uses force fields to determine the stability of the interaction between proteins.

## Interdisciplinary Collaboration

As explained before, Immunoinformatics is by itself an interdisciplinary field. It can collaborate with other disciplines as an approach to studying immunology and developing applications. One discipline is population genomics, which aims to identify variations in adaptive immune responses. The adaptive immune receptor repertoire sequencing (AIRR-seq) increased the knowledge of T cell receptors (TCRs) repertoire in humans. Still, most studies were made using samples from populations of European descent. Human genome databases are being built for non-European populations, such as the GenomeAsia 100K Project and the Human Heredity and Health in Africa (H3Africa) initiative, and the data can be used to investigate the genetic factors in vaccine-mediated immune responses (Peng et al., 2021). Once the association of genetic variation and immunity is established, the data can be used to design more efficient vaccines.

Another discipline is epidemiology. The association of epidemiology, genomics, and Immunoinformatics can be used to develop vaccines for pathogens and their variants quickly. A recent case study was the development of the Sars-Cov-2 vaccines. COVID-19 originated in December 2019 in Wuhan, China, and was declared a pandemic on 11 March 2020 (Chakraborty et al., 2023). In January 2020, Chinese researchers sequenced and deposited the first Sars-Cov-2 genome in GenBank's public database (Wu et al., 2020). After that, many researchers and pharmaceutical companies applied immunoinformatics to develop a vaccine for the virus. Between December 2020 and January 2021, USFDA (United States Food and Drug Administration) and EMA (European Medicines Agency) approved the two first vaccines, based on mRNA, from Pifzer/BioNTech (BNT162b) and Moderna (mRNA-1273). The former was developed and approved within eight months and the latter within a few days, aided by the knowledge from the previous research on SARS-CoV and MERS-CoV in which the spike glycoprotein was known as a vaccine target (Chakraborty et al., 2023). The reduced effectiveness against emerging variants, such as Delta and Omicron, showed it was necessary to create next-generation vaccines, such as multi-variant, mutation-proof, and pan-coronavirus vaccines. Immunoinformatics can provide multi-epitope and novel candidate vaccine constructs (Chakraborty et al., 2023).

Another discipline is cancer immunotherapy; Immunoinformatics can help develop vaccines for cancer. In this approach, DNA and RNA sequencing are applied to identify antigens exclusively expressed in cells of specific tumors that can be used to develop vaccines targeting shared or personalized antigens. Shared antigens with substantial immunologic or clinical achievements include ones from melanoma and breast, cervical, colorectal, and cancers. In contrast, customized antigens include ones from lymphoma, melanoma, and gastrointestinal, glandular, and epithelial cancers (M. J. Lin et al., 2022).

## Emergence of Immunoinformatics 2.0.

Immunoinformatics is a concept that emerged from the interdisciplinary between computer science and genetic immunology; as explained before, with the scientific-technical advancement is large the quantity of information, not only in the field of immunologics (Chatanaka et al., 2022); the data generated, organized information, about all of this knowledge, is indispensable to the progress, the cost reduction and the grow of scientific searches dynamism, promoting the creation of new treatments and prophylactic methods as of biotechnology therapeutic tools.

Given this, Immunoinformatics is a Bioinformatics area that transcends the immunogenetic through computational technologies (Nielsen et al., 2020b); in other words, the cellular genetic is what defers or not defers any actions that the cell does, by innumerable cellular structures that receive internal and external stimuli which transfer the information to the cellular nucleus. This structure compartmentalizes the DNA, which is responsible for delegating tasks to other cell organelles, enabling a cellular reaction in response to this stimulus, operating like a machine or a computer processor.

The immunological system handles a multitude of processes necessary for the defense of living organisms, with varying degrees of specialization and changing and adaptive biological processes, generating an enormous amount of complex data that pose a challenge to immunological research; as a countermeasure, various codes are programmed to automate activities that would be highly tedious manually (Mitra et al., 2022), act as filters and mediators of specific comparative and qualitative analyses of the stored information—facilitating work that has not yet been automated by artificial intelligence (AI) technology (Chatanaka et al., 2022).

Based on the current world, to determine the future of this field of study, we realize how crucial the development of analytical codes, especially those of artificial intelligence, will be in the field of Immunoinformatics, as we can see in many of the recent works related to vaccine research linked to Sars-Cov-2, through the Vaxign-ML (Bali & Bali, 2022), and which reinforce the need to direct efforts towards improving the conditions associated with the development of these technologies.

Database structures, one of the computational foundations of this field of science and fundamental to its continued growth, provide many of the tools needed to facilitate human work and increase the accuracy and speed of experimental results obtained at the bench (Mitra et al., 2022). AI, for example, is one of those tools that depends entirely on increasing the data available for improvement, in this case, for the analysis of B and T cells and the pathogen (De Groot et al., 2020; Khalid & Poh, 2023).

In particular, scientific research, especially in vaccinology, has significantly benefited from the development of bioinformatics through the use of computational analysis tools in proteomics, structural prediction and protein modeling, genomics, response prediction and gene evolution, and other omics to study targets in pathogens (Khalid & Poh, 2023); an example is the Sars-CoV-02 virus that triggered the 2020 pandemic and which, through these technologies, has made it possible to develop vaccines so quickly and even trace the host of this disease (Bali & Bali, 2022).

The new vaccines developed using Immunoinformatics have a significant differential that has boosted the development and production of new prophylactics. In contrast to traditional vaccines, those developed using reverse vaccinology have further intensified the application of these technologies, resulting in the identification of specific regions of antigens, called epitopes, used as multiple targets that generate a more intensified cellular response (Khalid & Poh, 2023). However, despite the evident importance of advances in the biological and health sciences, the great challenge of data storage that haunts the computational sciences still remains and is now expanding (Akash et al., 2024).

The dedication relative to expansion, standardizing, and organizing databases, as well as increasing server storage capacity and data processing speed, are some of the focus points to be highlighted (Gomes et al., 2023b; Holmes et al., 2010). Big data can be defined as databases of large size, variety, and high speed of data flow, either in the processing of the database itself or in the ingestion of new information (Al-Mekhlal & Ali Khwaja, 2019). In this context, the importance of this concept in Immunoinformatics is evident, given the exponential growth of data development in this area, which, as mentioned above, is one of the foundations of this scientific field.

Following the below approach, the evolution of hardware technologies is essential for the future of computational algorithms, data storage, processing speed and other requests that are in the present and will surge in future. We can quote technologies like Dream Chip, capable of allocating all memory and calculate in a single chip, but currently is being done through a chip system that each one has on-chip memory and a high speed switching they state(on/ off) (Radway et al., 2021). Heterogeneous Systems looks a high efficiency and a currently reality, don't existing a single CPU for administering a giant system, It is now taking on a more managerial role in the system, transferring activities to GPU units, only carrying out those activities that these units are not efficient at. Another relevant highlight are architectures based on the human brain, such as the TrueNorth Chip (Welser et al., 2018).

In addition, it is necessary to highlight the common languages used in developing these tools cited before used in Immunoinformatics researchers, and which, if known, can offer a range of possibilities both in the development of new tools and in the use of language libraries to support research. Current languages include Java, Python, Perl, R, Ruby, and C, which are widely used in Bioinformatics (Bonnal et al., 2019).

AI has rapidly ascended the fields of Bioinformatics and has been indispensable for the advancement of these studies; this is no different within Immunoinformatics, as mentioned above; in addition to the reverse vaccinology already mentioned, is the reuse or repositioning of drugs, characterized by new testing of drugs, approved or not, initially intended for another applicability, a method of lower financial cost and less strenuous in the development of traditional medicines (Bali & Bali, 2022; Mitra et al., 2022).

AIs have made technological evolution a possibility for us to experience scenarios previously only seen in science fiction films, as well as transcend the fiction we know. The complexity of defining limits for technology is so great that there are now models that simulate the learning of the human brain. Deep learning neural network models are receiving the most attention worldwide for their immense contribution to various areas of science, particularly multidisciplinary areas such as bioinformatics (Welser et al., 2018).

Neural networks are countless algorithms that make up the brain of the machine. Neurons are represented by modeled algorithms that are dispersed in layers and sublayers linked together to form an interconnected network. These network models vary in architecture according to the function they perform, presenting exceptional computational and algorithmic specificities, such as: recurrent neural networks

(RNN), capable of storing in their neurons a memory of the input data analyzed; and convolutional neural networks, whose neurons are capable of performing an analysis per point, since all their neurons in all layers and sub-layers are connected to each other, forming an extremely dense network (Dhillon & Verma, 2020; Staudemeyer & Morris, 2019).

In addition, there are architectures developed by deep learning that focus on parts of the information, which is a key factor in minimizing errors, especially in the analysis of long sequences; within the field of bioinformatics, there are tools that integrate all these technological architectures to obtain better performance, such as new software for the subcellular localization of proteins, like DeepLoc, for example. All this shows how variable and unpredictable the process of technological development is (De et al., 2022) (Figure 6).

*Figure 6. Diagram with some important hardware and software technologies that are important for the present and prospects of in silico bioinformatics and Immunoinformatics research*

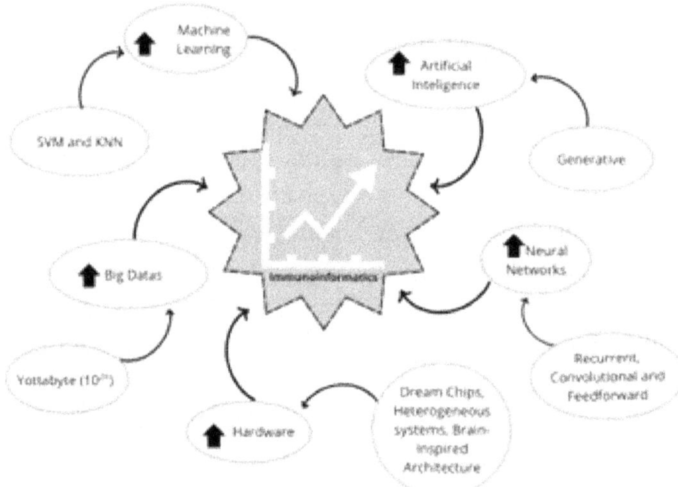


This approach contributes to indications that consider the variability of patients, i.e., it is individualized by mechanisms that can assess the heterogeneity and complexity of these different organisms to reduce possible toxicity and treatment efficacy. This occurs through integrating AI software and databases, such as virtual screening, which promotes an analysis of similarity, through statistical, mathematical coefficients, between chemical molecules of pharmacological use—allowing the determination and diagnosis of the proximity of closely related molecular structures, generating new more specific databases (Ghosh et al., 2023).

Thus, proposing a future perspective for Immunoinformatics is strictly linked to the perspective of technological evolution, which is an almost unimaginable prediction, given that in the 1990s, AI technologies spread quickly; this is evident if we compare the 2000s. CRT radios and televisions were still used, something that some young people have only seen in photographs, with today, where AI has been used to scan a person's potential infection in a physical space, as happened with Sars-CoV-02 (Kaushik et al., 2023).

## Future Horizons and Emerging Trends

Especially after the pandemic events, the attention to world-spreading diseases has increased, and vaccine development must be efficient and prompt to tackle emergent health crises (Shawan et al., 2023). With Bioinformatics and computational modeling advancements, conventional vaccinology is becoming obsolete, giving space to next-generation vaccines, such as mRNA and recombinant protein or peptide-based vaccines (Yahaya et al., 2023), that combined with the Bioinformatics and Immunoinformatics approach, opened new doors towards improved vaccine design.

The advancements made during the pandemic are starting to be explored with other organism infections with high death rates, such as different viruses (HIV, Ebola, Zika, Dengue), bacteria (*Staphylococcus aureus*) (Krishna et al., 2023) and fungi (*Candida albicans, P. brasiliensis)* (Inácio et al., 2023).Therefore, Immunoinformatics research is being explored in animal treatment with diseases such as leishmaniasis (both visceral and tegumentary), which, to date, no vaccine has yet been licensed for clinical use (EM & DN, 2010; Lari et al., 2022). Although this is a relatively recent methodology, the studies show the potential in developing vaccines that were not yet possible in the traditional way.

Relating to viruses, this approach in vaccine development has been expanded to organisms such as *Haemophilus influenza, which is known for the difficulty in vaccine production and treatment* since the strains have a lot of genetic and antigenic diversity, which can result in antibiotic resistance and ineffective vaccines. (AlChalabi et al., 2023). Therefore, advancements in Immunoinformatics can lead to the development of a vaccine that covers all strains. This new method of vaccinology is desirable since it is low-cost in terms of equipment and development, and it is being explored with Arboviruses (viruses transmitted by hematophagous arthropods). Dengue, Zika, and Chikungunya, which Aedes Aegypti commonly transmits, are endemic in tropical and subtropical countries (Soni et al., 2023), widely associated with markers of lower socioeconomic position, including lower education, income poverty and low healthcare coverage (Power et al., 2022). Therefore, the possibility of using cheaper next-generation vaccine models would be a viable option.

One of the biggest problems with actual Immunoinformatics is the lack of excellent and accurate programs to predict the effect and efficacy of the epitope vaccines in terms of adaptive and innate immunity scenarios after vaccination. This can be accomplished with advancement in the development of algorithms for Immunoinformatics and bioinformatics mixed with richer databases, which will help to perform a more accurate and faster calculation without any computational errors (Chakraborty et al., 2021). Therefore, one major challenge associated with vaccine design is the need for in vivo validations, which involves not just budget but discussions like animal suffering and the actual need to perform those tests. (Rawat et al., 2023b). With a better understanding combined with better programs, algorithms, and hardware, it can be possible to see, shortly, the development of software that not only builds the vaccine molecules but also simulates in organisms accurately and reliably.

## Ethical Considerations and Regulatory Landscape

Bioethics emerged from a necessity for society, i.e., parliaments, governments, and churches, to follow the burgeoning advancements in biotechnology and medical sciences. It is an interdisciplinary field of ethics that studies the values and morals of medical, animal, and environmental ethics. Bioethics establishes the boundaries of what is deemed acceptable or not, based on societal norms (Düwell, 2012; Pellegrino, 1992).

Bioethics is based on four principles (Figure 7): Autonomy, ensuring that patients possess the right and freedom to make their own decisions; Beneficence, which involves making decisions aimed at enhancing the patient's well-being and improvement; Non-maleficence, which entails avoiding harm to the patient, thus mitigating pain and suffering; and Justice, ensuring equal treatment and equitable access to healthcare resources (Varkey, 2021).

*Figure 7. The four principles of bioethics*

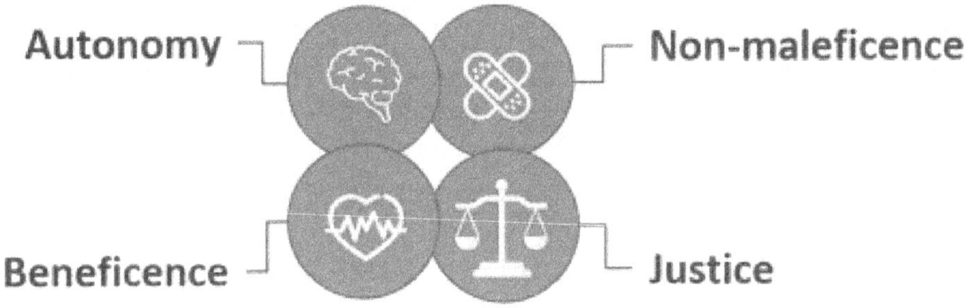


These principles also need to be fulfilled and respected in bioinformatics, as exemplified by transparency in the use of code, software reproducibility, and error correction. Furthermore, bioinformaticians are responsible for ensuring that the data used does not perpetuate biases, such as ethnic, racial, gender, or other biases. These biases, even if unintentional, can lead to distorted interpretations in the future. This is evident in the data used for machine learning algorithms. The quality of input data can significantly influence subsequent analysis, potentially resulting in conclusions that do not accurately represent the complete picture (Goodman, 2020).

An effective data management plan (DMP) is required to develop accurate research. DMP should address ethical and legal considerations, data nature and sources, storage procedures, backup strategies, and quality control measures. Additionally, DMPs must carefully assess data access, especially for research involving sensitive information such as human subjects, ensuring compliance with legal and ethical standards. The DMP can have bioinformaticians as a central development for this process (Kumuthini et al., 2020; Michener, 2015).

In some cases, the bioinformatician has not directly handled primary data, such as in Immunoinformatics. This aims to identify vaccinal targets by harnessing computational power with genetic, mathematical, and immunological knowledge (Oli et al., 2020c). Given the nature of the research and the advancement of molecular biology techniques, Immunoinformatics primarily deals with secondary data from genomic and protein sequence databases of microorganisms and their hosts (Tomar & De, 2010). Despite not directly handling primary data, Immunoinformatics is responsible for ensuring the accuracy of generated information, as the outcomes of these analyses may significantly impact public health. It is noteworthy that Immunoinformatics relies on experimental validation, highlighting the critical importance of data quality and the utilization of appropriate tools throughout all stages of future studies (Oli et al., 2020c).

In Immunoinformatics, one applied subject is personalized medicine. This plural concept considers the patient and their interactions with their genes and environment as well as cultural, educational, and other factors (Erdmann et al., 2021). This approach aims to facilitate more targeted treatment and diagnosis and may also help prevent certain diseases by analyzing each patient's genetic profile. In this context, personalized vaccinology emerges, aiming to assess how these factors can influence vaccine immunological response, efficacy, and safety. The goal of personalized vaccinology is not to create a unique vaccine for each individual but to identify patterns among groups that enable differentiation in dosage, route of administration, and vaccine formulation, for example (Poland et al., 2018).

To identify patterns in treating a population, these data must be representative, ensuring adherence to the bioethical principle of justice. However, the current representativeness of populations' genomic data is still not equitable since genomics studies are carried out on White/European ancestry populations (Green et al., 2023). There is a lack of data on the diversity of ethnic, racial, and minority groups, such as people with disabilities (hearing, visual) and also the LGBTQIA+ community (Lee, 2021). In this context, in addition to the issues already discussed involving data storage and data sharing, we have two pillars of bioethics as a challenge for implementing personalized medicine: justice and benevolence. The lack of representation of minorities raises issues of justice, and consequently, benevolence becomes a challenge: who will benefit from personalized medicine if it is not accessible to everyone? These challenges must be discussed in constant dialogues between researchers, including bioinformaticians, government, health entities, and the population, to promote the optimal utilization of this innovative approach.

## CONCLUSION

Immunoinformatics is a paradigm shift in immunogen development, leveraging computational approaches to access microbial genomes directly through bioinformatics. This approach optimizes researchers' time utilization and minimizes the costs associated with *in vitro* experimentation. Despite establishing robust methodologies within this domain, ongoing algorithmic refinement remains imperative, particularly with the advent of artificial intelligence (AI), marking the advent of Immunoinformatics 2.0.

The containment of contagious diseases through vaccination has historically underscored the severity of microbial threats to human health. The expeditious, productive, and promising deployment of immunogens translates to swift disease control and, consequently, diminished population susceptibility. Well-defined Immunoinformatics pipelines facilitate this endeavor by engendering multi-epitope proteins capable of eliciting immune responses, guided by mathematical models that emulate the intricacies of the immune system's constituent interactions. In this vein, Immunoinformatics facilitates the production of pathogen-targeted vaccine formulations. It mitigates the extra inclusion of proteins and peptides inherent in conventional vaccine compositions, such as those derived from inactivated micro-organisms.

Advancements in this methodology have engendered Immunoinformatics 2.0, ushering in novel perspectives and components that enhance efficiency, safety, and precision. The development of various servers and software tools, underpinned by programming languages like Java, Python, Perl, R, Ruby, and C, supplemented by biological insights gleaned from databases and scholarly literature, underscores this evolution. However, as comprehension of the intricate dynamics governing pathogen-host interactions burgeons, the necessity for novel pipelines becomes apparent, necessitated by the imperative to conduct *in silico* analyses capable of robustly accommodating diverse data modalities intrinsic to research projects. These tools aim to complement investigations concerning the behavior of distinct multi-epitope immunogens and simulate immune system dynamics via computational algorithms.

Several factors underpin the evolutionary trajectory of Immunoinformatics, including database standardization and expansion, alongside augmented storage capacity and processing velocity. Recent expansions in database architecture have been propelled by AI and machine learning algorithms, affording enhanced predictive capabilities and a nuanced understanding of vaccine model design, along with their prospective immunological attributes and ramifications. However, sustained investment in algorithmic refinement is warranted to illuminate the genetic, immunological, biochemical, epidemiological, molecular, and cellular underpinnings of pathogen-host interactions that remain enigmatic. Bioethical data sourcing, storage, and utilization, in adherence to pertinent legal and ethical norms, are paramount to ensuring the responsible dissemination of project-derived insights.

Cultivating novel algorithms that address common queries across disparate research domains necessitates practical application, scholarly scrutiny, and financial backing from external research consortia. Furthermore, the imperative of educational outreach initiatives aimed at familiarizing individuals with Immunoinformatics is underscored, fostering the emergence of a new cohort of student researchers within the field. The trajectory of Immunoinformatics hinges upon concerted efforts directed at both the scientific community and the public sphere, aimed at perpetually expanding databases and pathways, thereby facilitating the elucidation of extant and nascent biological interactions. As scientists who work with (and depend on) databases, we hope that the advances in automation triggered by AI will always allow the sciences, and especially Immunoinformatics and its immunogens, to progress.

## REFERENCES


Abraham, M. J., Murtola, T., Schulz, R., Páll, S., Smith, J. C., Hess, B., & Lindah, E. (2015). GROMACS: High performance molecular simulations through multi-level parallelism from laptops to supercomputers. *SoftwareX*, 1–2, 19–25. DOI: 10.1016/j.softx.2015.06.001

Akash, A., Bencurova, E., & Dandekar, T. (2024). How to make DNA data storage more applicable. *Trends in Biotechnology*, 42(1), 17–30. DOI: 10.1016/j.tibtech.2023.07.006 PMID: 37591721

Al-Mekhlal, M., & Ali Khwaja, A. (2019). A synthesis of big data definition and characteristics. *Proceedings - 22nd IEEE International Conference on Computational Science and Engineering and 17th IEEE International Conference on Embedded and Ubiquitous Computing, CSE/EUC 2019*, 314–322. DOI: 10.1109/CSE/EUC.2019.00067

AlChalabi, R., Al-Rahim, A., Omer, D., & Suleiman, A. A. J. (2023). Immunoinformatics design of multi-epitope peptide-based vaccine against Haemophilus influenzae strain using cell division protein. *Network Modeling and Analysis in Health Informatics and Bioinformatics*, 12(1), 1. DOI: 10.1007/s13721-022-00395-x PMID: 36465492

Ali, M., Pandey, R. K., Khatoon, N., Narula, A., Mishra, A., & Prajapati, V. K. (2017). Exploring dengue genome to construct a multi-epitope based subunit vaccine by utilizing immunoinformatics approach to battle against dengue infection. *Scientific Reports, 7*(1), 1–13. DOI: 10.1038/s41598-017-09199-w

Amos, E. O., Araoyinbo, O. S., Akinleye, E. O., Alakanse, S. O., Bamikole, A. O., & Kolawole, O. M. (2023). Immunoinformatics design of multi-epitope vaccine using surface cell antigen OmpB and heat shock protein GroEL against rickettsioses. *Informatics in Medicine Unlocked*, 43, 101411. DOI: 10.1016/j.imu.2023.101411

Ayyagari, V. S., Venkateswarulu, T. C., Abraham Peele, K., & Srirama, K. (2022). Design of a multi-epitope-based vaccine targeting M-protein of SARS-CoV2: An immunoinformatics approach. *Journal of Biomolecular Structure & Dynamics*, 40(7), 2963–2977. DOI: 10.1080/07391102.2020.1850357 PMID: 33252008

Bagnoli, F., Baudner, B., Mishra, R. P. N., Bartolini, E., Fiaschi, L., Mariotti, P., Nardi-Dei, V., Boucher, P., & Rappuoli, R. (2011). Designing the Next Generation of Vaccines for Global Public Health. DOI: 10.1089/omi.2010.0127

Bahrami, A. A., Payandeh, Z., Khalili, S., Zakeri, A., & Bandehpour, M. (2019). Immunoinformatics: In Silico Approaches and Computational Design of a Multi-epitope, Immunogenic Protein. *International Reviews of Immunology*, 38(6), 307–322. DOI: 10.1080/08830185.2019.1657426 PMID: 31478759

Bali, A., & Bali, N. (2022). Role of artificial intelligence in fast-track drug discovery and vaccine development for COVID-19. *Novel AI and Data Science Advancements for Sustainability in the Era of COVID*, 19, 201–229. DOI: 10.1016/B978-0-323-90054-6.00006-4

Baljon, J. J., & Wilson, J. T. (2022). Bioinspired vaccines to enhance MHC class-I antigen cross-presentation. *Current Opinion in Immunology*, 77, 102215. DOI: 10.1016/j.coi.2022.102215 PMID: 35667222

Blanco-González, A., Cabezón, A., Seco-González, A., Conde-Torres, D., Antelo-Riveiro, P., Piñeiro, Á., & Garcia-Fandino, R. (2023). The Role of AI in Drug Discovery: Challenges, Opportunities, and Strategies. *Pharmaceuticals (Basel, Switzerland)*, 16(6), 891. DOI: 10.3390/ph16060891 PMID: 37375838

Bonnal, R. J. P., Yates, A., Goto, N., Gautier, L., Willis, S., Fields, C., Katayama, T., & Prins, P. (2019). Sharing programming resources between bio* projects. *Methods in Molecular Biology (Clifton, N.J.)*, 1910, 747–766. DOI: 10.1007/978-1-4939-9074-0_25 PMID: 31278684

Bouazzaoui, A., Abdellatif, A. A. H., Al-Allaf, F. A., Bogari, N. M., Al-Dehlawi, S., & Qari, S. H. (2021). Strategies for Vaccination: Conventional Vaccine Approaches Versus New-Generation Strategies in Combination with Adjuvants. *Pharmaceutics*, 13(2), 1–20. DOI: 10.3390/pharmaceutics13020140 PMID: 33499096

Bravi, B. (2024). Development and use of machine learning algorithms in vaccine target selection. *Npj Vaccines, 9*(1), 1–14. DOI: 10.1038/s41541-023-00795-8

Burton, D. R. (2017). What Are the Most Powerful Immunogen Design Vaccine Strategies? Reverse Vaccinology 2.0 Shows Great Promise. *Cold Spring Harbor Perspectives in Biology*, 9(11), a030262. Advance online publication. DOI: 10.1101/cshperspect.a030262 PMID: 28159875

Canouï, E., & Launay, O. (2019). Histoire et principes de la vaccination. *Revue des Maladies Respiratoires*, 36(1), 74–81. DOI: 10.1016/j.rmr.2018.02.015 PMID: 30579659

Chakraborty, C., Bhattacharya, M., & Dhama, K. (2023). SARS-CoV-2 Vaccines, Vaccine Development Technologies, and Significant Efforts in Vaccine Development during the Pandemic: The Lessons Learned Might Help to Fight against the Next Pandemic. *Vaccines, 11*(3), 682. DOI: 10.3390/vaccines11030682

Chakraborty, C., Sharma, A. R., Bhattacharya, M., & Lee, S. S. (2021). Lessons Learned from Cutting-Edge Immunoinformatics on Next-Generation COVID-19 Vaccine Research. *International Journal of Peptide Research and Therapeutics*, 27(4), 2303–2311. DOI: 10.1007/s10989-021-10254-4 PMID: 34276266

Chatanaka, M. K., Ulndreaj, A., Sohaei, D., & Prassas, I. (2022). Immunoinformatics: Pushing the boundaries of immunology research and medicine. *Immunoinformatics (Amsterdam, Netherlands)*, 5, 100007. DOI: 10.1016/j.immuno.2021.100007

Chauhan, V., Rungta, T., Goyal, K., & Singh, M. P. (2019). Designing a multi-epitope based vaccine to combat Kaposi Sarcoma utilizing immunoinformatics approach. *Scientific Reports, 9*(1), 1–15. DOI: 10.1038/s41598-019-39299-8

Chen, X., Zaro, J. L., & Shen, W. C. (2013). Fusion protein linkers: Property, design and functionality. *Advanced Drug Delivery Reviews*, 65(10), 1357–1369. DOI: 10.1016/j.addr.2012.09.039 PMID: 23026637

Cho, H. K., Park, S. E., Kim, Y. J., Jo, D. S., Kim, Y. K., Eun, B. W., Lee, T. J., Lee, J., Lee, H., Kim, K. H., Cho, E. Y., Ahn, J. G., & Choi, E. H. (2021). Recommendation for use of diphtheria and tetanus toxoids and acellular pertussis, inactivated poliovirus, Haemophilus influenzae type b conjugate, and hepatitis B vaccine in infants. *Clinical and Experimental Pediatrics*, 64(12), 602–607. DOI: 10.3345/cep.2021.00507 PMID: 34098627

D'Amelio, E., Salemi, S., & D'Amelio, R. (2016). Anti-Infectious Human Vaccination in Historical Perspective. *International Reviews of Immunology*, 35(3), 260–290. DOI: 10.3109/08830185.2015.1082177 PMID: 26606466

Das, A. T., Jeeninga, R. E., & Berkhout, B. (2010). Possible applications for replicating HIV-1 vectors. DOI: 10.2217/hiv.10.20

De, A., Correia, S., Colombini, E. L., De, A., & Colombini, E. L. (2022). Attention, please! A survey of neural attention models in deep learning. *Artificial Intelligence Review, 55*(8), 6037–6124. DOI: 10.1007/s10462-022-10148-x

De Groot, A. S., Moise, L., Terry, F., Gutierrez, A. H., Hindocha, P., Richard, G., Hoft, D. F., Ross, T. M., Noe, A. R., Takahashi, Y., Kotraiah, V., Silk, S. E., Nielsen, C. M., Minassian, A. M., Ashfield, R., Ardito, M., Draper, S. J., & Martin, W. D. (2020). Better epitope discovery, precision immune engineering, and accelerated vaccine design using Immunoinformatics tools. *Frontiers in Immunology*, 11, 527882. DOI: 10.3389/fimmu.2020.00442 PMID: 32318055

Dhillon, A., & Verma, G. K. (2020). Convolutional neural network: A review of models, methodologies and applications to object detection. *Progress in Artificial Intelligence*, 9(2), 85–112. DOI: 10.1007/s13748-019-00203-0

Dimitrov, I., Bangov, I., Flower, D. R., & Doytchinova, I. (2014). AllerTOP v.2—A server for in silico prediction of allergens. *Journal of Molecular Modeling*, 20(6), 2278. Advance online publication. DOI: 10.1007/s00894-014-2278-5 PMID: 24878803

Doytchinova, I. A., & Flower, D. R. (2007). VaxiJen: A server for prediction of protective antigens, tumour antigens and subunit vaccines. *BMC Bioinformatics*, 8(1), 1–7. DOI: 10.1186/1471-2105-8-4 PMID: 17207271

Doytchinova, I. A., & Flower, D. R. (2008). Bioinformatic Approach for Identifying Parasite and Fungal Candidate Subunit Vaccines. *The Open Vaccine Journal*, 1(1), 22–26. https://www.jenner.ac.uk/VaxiJen. DOI: 10.2174/1875035400801010022

Düwell, M. (2012). Bioethics: Methods, theories, domains. *Bioethics: Methods, Theories, Domains*, 1–308. DOI: 10.4324/9780203084205

Eberhardt, J., Santos-Martins, D., Tillack, A. F., & Forli, S. (2021). AutoDock Vina 1.2.0: New Docking Methods, Expanded Force Field, and Python Bindings. *Journal of Chemical Information and Modeling*, 61(8), 3891–3898. DOI: 10.1021/acs.jcim.1c00203 PMID: 34278794

EM, M., & DN, L. (2010). Treatment of visceral leishmaniasis. *Journal of Global Infectious Diseases, 2*(2), 280. DOI: 10.4103/0974-777X.62883

Erdmann, A., Rehmann-Sutter, C., & Bozzaro, C. (2021). Patients' and professionals' views related to ethical issues in precision medicine: A mixed research synthesis. *BMC Medical Ethics*, 22(1), 1–18. DOI: 10.1186/s12910-021-00682-8 PMID: 34465328

Excler, J. L., Saville, M., Berkley, S., & Kim, J. H. (2021). Vaccine development for emerging infectious diseases. *Nature Medicine, 27*(4), 591–600. DOI: 10.1038/s41591-021-01301-0

Galanis, K. A., Nastou, K. C., Papandreou, N. C., Petichakis, G. N., Pigis, D. G., & Iconomidou, V. A. (2021). Linear B-Cell Epitope Prediction for In Silico Vaccine Design: A Performance Review of Methods Available via Command-Line Interface. *International Journal of Molecular Sciences*, 22(6), 1–19. DOI: 10.3390/ijms22063210 PMID: 33809918

Ghosh, A., Larrondo-Petrie, M. M., & Pavlovic, M. (2023). Revolutionizing Vaccine Development for COVID-19: A Review of AI-Based Approaches. *Information, 14*(12), 665. DOI: 10.3390/info14120665

Gomes, L. G. R., Dutra, J. C. F., Profeta, R., Dias, M. V., García, G. J. Y., Rodrigues, D. L. N., Neto, A. G., Aburjaile, F. F., Tiwari, S., Soares, S. C., Azevedo, V., & Jaiswal, A. K. (2023a). Systematic review of reverse vaccinology and immunoinformatics data for non-viral sexually transmitted infections. *Anais da Academia Brasileira de Ciências*, 95(suppl 2), e20230617. DOI: 10.1590/0001-3765202320230617 PMID: 38055447

Gomes, L. G. R., Rodrigues, T. C. V., Jaiswal, A. K., Santos, R. G., Kato, R. B., Barh, D., Alzahrani, K. J., Banjer, H. J., Soares, S. de C., Azevedo, V., & Tiwari, S. (2022). In Silico Designed Multi-Epitope Immunogen "Tpme-VAC/LGCM-2022" May Induce Both Cellular and Humoral Immunity against Treponema pallidum Infection. *Vaccines*, 10(7), 1019. Advance online publication. DOI: 10.3390/vaccines10071019 PMID: 35891183

Goodman, K. W. (2020). Ethics in Health Informatics. *Yearbook of Medical Informatics*, 29(1), 26–31. DOI: 10.1055/s-0040-1701966 PMID: 32303095

Green, S., Prainsack, B., & Sabatello, M. (2023). Precision medicine and the problem of structural injustice. *Medicine, Health Care, and Philosophy*, 26(3), 433–450. DOI: 10.1007/s11019-023-10158-8 PMID: 37231234

Greenbaum, J., Sidney, J., Chung, J., Brander, C., Peters, B., & Sette, A. (2011). Functional classification of class II human leukocyte antigen (HLA) molecules reveals seven different supertypes and a surprising degree of repertoire sharing across supertypes. *Immunogenetics*, 63(6), 325–335. DOI: 10.1007/s00251-011-0513-0 PMID: 21305276

Guarra, F., & Colombo, G. (2023). Computational Methods in Immunology and Vaccinology: Design and Development of Antibodies and Immunogens. *Journal of Chemical Theory and Computation*, 19(16), 5315–5333. DOI: 10.1021/acs.jctc.3c00513 PMID: 37527403

Gupta, S., Kapoor, P., Chaudhary, K., Gautam, A., Kumar, R., & Raghava, G. P. S. (2013). In Silico Approach for Predicting Toxicity of Peptides and Proteins. *PLoS One*, 8(9), e73957. DOI: 10.1371/journal.pone.0073957 PMID: 24058508

Hajj Hussein, I., Chams, N., Chams, S., El Sayegh, S., Badran, R., Raad, M., Gerges-Geagea, A., Leone, A., & Jurjus, A. (2015). Vaccines Through Centuries: Major Cornerstones of Global Health. *Frontiers in Public Health*, 3, 167945. DOI: 10.3389/fpubh.2015.00269 PMID: 26636066

Handel, A., La Gruta, N. L., & Thomas, P. G. (2019). Simulation modelling for immunologists. *Nature Reviews Immunology, 20*(3), 186–195. DOI: 10.1038/s41577-019-0235-3

Hilleman, M. R. (2000). Vaccines in historic evolution and perspective: A narrative of vaccine discoveries. *Vaccine*, 18(15), 1436–1447. DOI: 10.1016/S0264-410X(99)00434-X PMID: 10618541

Holmes, C., McDonald, F., Jones, M., Ozdemir, V., & Graham, J. E. (2010). Standardization and Omics Science: Technical and Social Dimensions Are Inseparable and Demand Symmetrical Study. DOI: 10.1089/omi.2010.0022

Inácio, M. M., Moreira, A. L. E., Cruz-Leite, V. R. M., Mattos, K., Silva, L. O. S., Venturini, J., Ruiz, O. H., Ribeiro-Dias, F., Weber, S. S., Soares, C. M. de A., & Borges, C. L. (2023). Fungal Vaccine Development: State of the Art and Perspectives Using Immunoinformatics. *Journal of Fungi, 9*(6), 633. DOI: 10.3390/jof9060633

Ishack, S., & Lipner, S. R. (2021). Bioinformatics and immunoinformatics to support COVID-19 vaccine development. *Journal of Medical Virology*, 93(9), 5209–5211. DOI: 10.1002/jmv.27017 PMID: 33851735

Jaiswal, A. K., Tiwari, S., Jamal, S. B., Barh, D., Azevedo, V., & Soares, S. C. (2017a). An In Silico Identification of Common Putative Vaccine Candidates against Treponema pallidum: A Reverse Vaccinology and Subtractive Genomics Based Approach. *International Journal of Molecular Sciences, 18*(2), 402. DOI: 10.3390/ijms18020402

Jensen, K. K., Andreatta, M., Marcatili, P., Buus, S., Greenbaum, J. A., Yan, Z., Sette, A., Peters, B., & Nielsen, M. (2018). Improved methods for predicting peptide binding affinity to MHC class II molecules. *Immunology*, 154(3), 394–406. DOI: 10.1111/imm.12889 PMID: 29315598

Jumper, J., Evans, R., Pritzel, A., Green, T., Figurnov, M., Ronneberger, O., Tunyasuvunakool, K., Bates, R., Žídek, A., Potapenko, A., Bridgland, A., Meyer, C., Kohl, S. A. A., Ballard, A. J., Cowie, A., Romera-Paredes, B., Nikolov, S., Jain, R., Adler, J., … Hassabis, D. (2021). Highly accurate protein structure prediction with AlphaFold. *Nature, 596*(7873), 583–589. DOI: 10.1038/s41586-021-03819-2

Kaushik, R., Kant, R., & Christodoulides, M. (2023). Artificial intelligence in accelerating vaccine development - current and future perspectives. *Frontiers in Bacteriology*, 2, 1258159. DOI: 10.3389/fbrio.2023.1258159

Kelley, L. A., Mezulis, S., Yates, C. M., Wass, M. N., & Sternberg, M. J. E. (2015). The Phyre2 web portal for protein modeling, prediction and analysis. *Nature Protocols, 10*(6), 845–858. DOI: 10.1038/nprot.2015.053

Khalid, K., & Poh, C. L. (2023). The Promising Potential of Reverse Vaccinology-Based Next-Generation Vaccine Development over Conventional Vaccines against Antibiotic-Resistant Bacteria. *Vaccines, 11*(7), 1264. DOI: 10.3390/vaccines11071264

Khoshnood, S., Ghanavati, R., Shirani, M., Ghahramanpour, H., Sholeh, M., Shariati, A., Sadeghifard, N., & Heidary, M. (2022). Viral vector and nucleic acid vaccines against COVID-19: A narrative review. *Frontiers in Microbiology*, 13, 984536. Advance online publication. DOI: 10.3389/fmicb.2022.984536 PMID: 36118203

Krishna, C., Kumar, D., & Kushwaha, D. S. (2023). A Comprehensive Survey on Pandemic Patient Monitoring System: Enabling Technologies, Opportunities, and Research Challenges. *Wireless Personal Communications*, 131(3), 2125–2172. DOI: 10.1007/s11277-023-10535-9 PMID: 37360140

Kumuthini, J., Chimenti, M., Nahnsen, S., Peltzer, A., Meraba, R., McFadyen, R., Wells, G., Taylor, D., Maienschein-Cline, M., Li, J. L., Thimmapuram, J., Murthy-Karuturi, R., & Zass, L. (2020). Ten simple rules for providing effective bioinformatics research support. *PLoS Computational Biology*, 16(3), e1007531. DOI: 10.1371/journal.pcbi.1007531 PMID: 32214318

Lancaster, E. M., Jablons, D., & Kratz, J. R. (2020). Applications of Next-Generation Sequencing in Neoantigen Prediction and Cancer Vaccine Development. *Genetic Testing and Molecular Biomarkers*, 24(2), 59–66. DOI: 10.1089/gtmb.2018.0211 PMID: 30907630

Lari, A., Lari, N., & Biabangard, A. (2022). Immunoinformatics Approach to Design a Novel Subunit Vaccine Against Visceral Leishmaniasis. *International Journal of Peptide Research and Therapeutics*, 28(1), 34. DOI: 10.1007/s10989-021-10344-3 PMID: 34931120

Larsen, M. V., Lundegaard, C., Lamberth, K., Buus, S., Lund, O., & Nielsen, M. (2007). Large-scale validation of methods for cytotoxic T-lymphocyte epitope prediction. *BMC Bioinformatics*, 8(1), 1–12. DOI: 10.1186/1471-2105-8-424 PMID: 17973982

Lee, S. S. J. (2021). Obligations of the "Gift": Reciprocity and Responsibility in Precision Medicine. *The American Journal of Bioethics*, 21(4), 57–66. DOI: 10.1080/15265161.2020.1851813 PMID: 33325811

Lim, C. P., Kok, B. H., Lim, H. T., Chuah, C., Abdul Rahman, B., Abdul Majeed, A. B., Wykes, M., Leow, C. H., & Leow, C. Y. (2023). Recent trends in next generation immunoinformatics harnessed for universal coronavirus vaccine design. *Pathogens and Global Health*, 117(2), 134–151. DOI: 10.1080/20477724.2022.2072456 PMID: 35550001

Lin, H. H., Zhang, G. L., Tongchusak, S., Reinherz, E. L., & Brusic, V. (2008). Evaluation of MHC-II peptide binding prediction servers: applications for vaccine research. *BMC Bioinformatics, 9*(12), 1–10. DOI: 10.1186/1471-2105-9-S12-S22

Lin, M. J., Svensson-Arvelund, J., Lubitz, G. S., Marabelle, A., Melero, I., Brown, B. D., & Brody, J. D. (2022). Cancer vaccines: the next immunotherapy frontier. *Nature Cancer, 3*(8), 911–926. DOI: 10.1038/s43018-022-00418-6

Liu, L., Yu, W., Cai, K., Ma, S., Wang, Y., Ma, Y., & Zhao, H. (2023). Identification of vaccine candidates against rhodococcus equi by combining pangenome analysis with a reverse vaccinology approach. *Heliyon*, 9(8), e18623. Advance online publication. DOI: 10.1016/j.heliyon.2023.e18623 PMID: 37576287

Liu, Y., & Ye, Q. (2022). Nucleic Acid Vaccines against SARS-CoV-2. *Vaccines, 10*(11), 1849. DOI: 10.3390/vaccines10111849

Malaina, I., Martínez, L., Montoya, J. M., Alonso, S., Boyano, M. D., Asumendi, A., Izu, R., Sanchez-Diez, A., Cancho-Galan, G., & de la Fuente, , I. (2023). A Universal Antigen-Ranking Method to Design Personalized Vaccines Targeting Neoantigens against Melanoma. *Life (Chicago, Ill.)*, 13(1), 155. DOI: 10.3390/life13010155 PMID: 36676104

Markina, Y. V., Gerasimova, E. V., Markin, A. M., Glanz, V. Y., Wu, W. K., Sobenin, I. A., & Orekhov, A. N. (2020). Sialylated Immunoglobulins for the Treatment of Immuno-Inflammatory Diseases. *International Journal of Molecular Sciences*, 21(15), 1–17. DOI: 10.3390/ijms21155472 PMID: 32751832

Martinelli, D. D. (2022). In silico vaccine design: A tutorial in immunoinformatics. *Healthcare Analytics*, 2, 100044. DOI: 10.1016/j.health.2022.100044

Martinez-Olivares, C. E., Hernández-Pando, R., & Mixcoha, E. (2023). In silico EsxG EsxH rational epitope selection: Candidate epitopes for vaccine design against pulmonary tuberculosis. *PLoS One*, 18(4), e0284264. DOI: 10.1371/journal.pone.0284264 PMID: 37079575

Martini, S., Nielsen, M., Peters, B., & Sette, A. (2020). The Immune Epitope Database and Analysis Resource Program 2003–2018: Reflections and outlook. *Immunogenetics*, 72(1–2), 57–76. DOI: 10.1007/s00251-019-01137-6 PMID: 31761977

Masignani, V., Pizza, M., & Moxon, E. R. (2019). The Development of a Vaccine Against Meningococcus B Using Reverse Vaccinology. *Frontiers in Immunology*, 10(MAR), 751. Advance online publication. DOI: 10.3389/fimmu.2019.00751 PMID: 31040844

Melo, A. R. da S., de Macêdo, L. S., Invenção, M. da C. V., de Moura, I. A., da Gama, M. A. T. M., de Melo, C. M. L., Silva, A. J. D., Batista, M. V. de A., & Freitas, A. C. de. (2022). Third-Generation Vaccines: Features of Nucleic Acid Vaccines and Strategies to Improve Their Efficiency. *Genes, 13*(12), 2287. DOI: 10.3390/genes13122287

Michalik, M., Djahanshiri, B., Leo, J. C., & Linke, D. (2016). Reverse Vaccinology: The Pathway from Genomes and Epitope Predictions to Tailored Recombinant Vaccines. *Methods in Molecular Biology (Clifton, N.J.)*, 1403, 87–106. DOI: 10.1007/978-1-4939-3387-7_4 PMID: 27076126

Michener, W. K. (2015). Ten Simple Rules for Creating a Good Data Management Plan. *PLoS Computational Biology*, 11(10), e1004525. DOI: 10.1371/journal.pcbi.1004525 PMID: 26492633

Mitra, D., Mitra, D., Sabri Bensaad, M., Sinha, S., Pant, K., Pant, M., Priyadarshini, A., Singh, P., Dassamiour, S., Hambaba, L., Panneerselvam, P., & Das Mohapatra, P. K. (2022). Evolution of bioinformatics and its impact on modern bio-science in the twenty-first century: Special attention to pharmacology, plant science and drug discovery. *Computational Toxicology (Amsterdam, Netherlands)*, 24, 100248. DOI: 10.1016/j.comtox.2022.100248

Motamedi, H., Alvandi, A., Fathollahi, M., Ari, M. M., Moradi, S., Moradi, J., & Abiri, R. (2023). In silico designing and immunoinformatics analysis of a novel peptide vaccine against metallo-beta-lactamase (VIM and IMP) variants. *PLoS One*, 18(7), e0275237. DOI: 10.1371/journal.pone.0275237 PMID: 37471423

Motta, S., & Brusic, V. (2004). *Mathematical Modelling of the Immune System*. DOI: 10.1007/978-3-642-18734-6_10

Moyle, P. M., & Toth, I. (2013a). Modern subunit vaccines: Development, components, and research opportunities. *ChemMedChem*, 8(3), 360–376. DOI: 10.1002/cmdc.201200487 PMID: 23316023

Nielsen, M., Andreatta, M., Peters, B., & Buus, S. (2020a). Immunoinformatics: Predicting Peptide-MHC Binding. *Annual Review of Biomedical Data Science*, 3(1), 191–215. DOI: 10.1146/annurev-biodatasci-021920-100259 PMID: 37427310

Nielsen, M., Andreatta, M., Peters, B., & Buus, S. (2020b). Immunoinformatics: Predicting Peptide-MHC Binding. *Annual Review of Biomedical Data Science, 3*, 191–215. https://doi.org/DOI: 10.1146/ANNUREV-BIODATASCI-021920-100259/CITE/REFWORKS

Oli, A. N., Obialor, W. O., & Ositadimma, M. (2020a). Immunoinformatics and vaccine development: An overview. *ImmunoTargets and Therapy*, 9, 13–30. DOI: 10.2147/ITT.S241064 PMID: 32161726

Oli, A. N., Obialor, W. O., Ositadimma, M., Ifeanyichukwu, Odimegwu, D. C., Okoyeh, J. N., Emechebe, G. O., Adejumo, S. A., & Ibeanu, G. C. (2020c). Immunoinformatics and Vaccine Development: An Overview. *ImmunoTargets and Therapy, 9*, 13–30. DOI: 10.2147/ITT.S241064

Pellegrino, E. D. (1992). Intersections of Western Biomedical Ethics and World Culture: Problematic and Possibility. *Cambridge Quarterly of Healthcare Ethics*, 1(3), 191–196. DOI: 10.1017/S0963180100000360 PMID: 11651289

Peng, K., Safonova, Y., Shugay, M., Popejoy, A. B., Rodriguez, O. L., Breden, F., Brodin, P., Burkhardt, A. M., Bustamante, C., Cao-Lormeau, V. M., Corcoran, M. M., Duffy, D., Fuentes-Guajardo, M., Fujita, R., Greiff, V., Jönsson, V. D., Liu, X., Quintana-Murci, L., Rossetti, M., … Mangul, S. (2021). Diversity in immunogenomics: the value and the challenge. *Nature Methods, 18*(6), 588–591. DOI: 10.1038/s41592-021-01169-5

Poland, G. A., Ovsyannikova, I. G., & Kennedy, R. B. (2018). Personalized vaccinology: A review. *Vaccine*, 36(36), 5350–5357. DOI: 10.1016/j.vaccine.2017.07.062 PMID: 28774561

Pollard, A. J., & Bijker, E. M. (2020). A guide to vaccinology: from basic principles to new developments. *Nature Reviews Immunology, 21*(2), 83–100. DOI: 10.1038/s41577-020-00479-7

Ponomarenko, J., Bui, H. H., Li, W., Fusseder, N., Bourne, P. E., Sette, A., & Peters, B. (2008). ElliPro: A new structure-based tool for the prediction of antibody epitopes. *BMC Bioinformatics*, 9(1), 1–8. DOI: 10.1186/1471-2105-9-514 PMID: 19055730

Power, G. M., Vaughan, A. M., Qiao, L., Clemente, N. S., Pescarini, J. M., Paixão, E. S., Lobkowicz, L., Raja, A. I., Souza, A. P., Barreto, M. L., & Brickley, E. B. (2022). Socioeconomic risk markers of arthropod-borne virus (arbovirus) infections: A systematic literature review and meta-analysis. *BMJ Global Health*, 7(4), 7735. DOI: 10.1136/bmjgh-2021-007735 PMID: 35428678

Prado, L. C. da S., Giacchetto Felice, A., Rodrigues, T. C. V., Tiwari, S., Andrade, B. S., Kato, R. B., Oliveira, C. J. F., Silva, M. V., Barh, D., Azevedo, V. A. C., Jaiswal, A. K., & Soares, S. C. (2022). New putative therapeutic targets against Serratia marcescens using reverse vaccinology and subtractive genomics. *Journal of Biomolecular Structure & Dynamics*, 40(20), 10106–10121. DOI: 10.1080/07391102.2021.1942211 PMID: 34192477

Radway, R. M., Sethi, K., Chen, W. C., Kwon, J., Liu, S., Wu, T. F., Beigne, E., Shulaker, M. M., Wong, H. S. P., & Mitra, S. (2021). The Future of Hardware Technologies for Computing: N3XT 3D MOSAIC, Illusion Scaleup, Co-Design. *Technical Digest - International Electron Devices Meeting, IEDM,* 25.4.1-25.4.4. DOI: 10.1109/IEDM19574.2021.9720647

Rapin, N., Lund, O., Bernaschi, M., & Castiglione, F. (2010). Computational Immunology Meets Bioinformatics: The Use of Prediction Tools for Molecular Binding in the Simulation of the Immune System. *PLoS One*, 5(4), 9862. DOI: 10.1371/journal.pone.0009862 PMID: 20419125

Rappuoli, R. (2000). Reverse vaccinology. *Current Opinion in Microbiology*, 3(5), 445–450. DOI: 10.1016/S1369-5274(00)00119-3 PMID: 11050440

Rawat, S. S., Keshri, A. K., Kaur, R., & Prasad, A. (2023a). Immunoinformatics Approaches for Vaccine Design: A Fast and Secure Strategy for Successful Vaccine Development. *Vaccines, 11*(2), 221. DOI: 10.3390/vaccines11020221

Rawat, S. S., Keshri, A. K., Kaur, R., & Prasad, A. (2023b). Immunoinformatics Approaches for Vaccine Design: A Fast and Secure Strategy for Successful Vaccine Development. *Vaccines*, 11(2), 221. DOI: 10.3390/vaccines11020221 PMID: 36851099

Rodrigues, T. C. V., Jaiswal, A. K., De Sarom, A., Oliveira, L. D. C., Oliveira, C. J. F., Ghosh, P., Tiwari, S., Miranda, F. M., Benevides, L. D. J., Azevedo, V. A. D. C., & Soares, S. D. C. (2019). Reverse vaccinology and subtractive genomics reveal new therapeutic targets against Mycoplasma pneumoniae: A causative agent of pneumonia. *Royal Society Open Science*, 6(7), 190907. Advance online publication. DOI: 10.1098/rsos.190907 PMID: 31417766

Saliou, P., & Girard, M. (2005). De Jenner et Pasteur à la vaccinologie. *Therapie*, 60(3), 201–204. DOI: 10.2515/therapie:2005024 PMID: 16128259

Sette, A., & Rappuoli, R. (2010). Reverse vaccinology: Developing vaccines in the era of genomics. *Immunity*, 33(4), 530–541. DOI: 10.1016/j.immuni.2010.09.017 PMID: 21029963

Shawan, M. M. A. K., Sharma, A. R., Halder, S. K., Al Arian, T., Shuvo, M. N., Sarker, S. R., & Hasan, M. A. (2023). Advances in Computational and Bioinformatics Tools and Databases for Designing and Developing a Multi-Epitope-Based Peptide Vaccine. *International Journal of Peptide Research and Therapeutics*, 29(4), 60. DOI: 10.1007/s10989-023-10535-0 PMID: 37251529

Shi, H., Yan, K. K., Ding, L., Qian, C., Chi, H., & Yu, J. (2020). Network Approaches for Dissecting the Immune System. *iScience*, 23(8), 101354. DOI: 10.1016/j.isci.2020.101354 PMID: 32717640

Soni, S., Gill, V. J. S., Anusheel, , Singh, J., Chhabra, J., Gill, G. J. S., & Bakshi, R. (2023). Dengue, Chikungunya, and Zika: The Causes and Threats of Emerging and Re-emerging Arboviral Diseases. *Cureus*, 15(7). Advance online publication. DOI: 10.7759/cureus.41717 PMID: 37575782

Soto, J. A., Gálvez, N. M. S., Andrade, C. A., Ramírez, M. A., Riedel, C. A., Kalergis, A. M., & Bueno, S. M. (2022). BCG vaccination induces cross-protective immunity against pathogenic microorganisms. *Trends in Immunology*, 43(4), 322–335. DOI: 10.1016/j.it.2021.12.006 PMID: 35074254

Soubeyrand, B. (2018). De la fabrication d'un vaccin à sa mise à disposition en pharmacie. *Revue des Maladies Respiratoires*, 35(10), 1005–1019. DOI: 10.1016/j.rmr.2018.07.003 PMID: 30266457

Staudemeyer, R. C., & Morris, E. R. (2019). *Understanding LSTM — a tutorial into Long Short-Term Memory Recurrent Neural Networks*. https://arxiv.org/abs/1909.09586v1

Tomar, N., & De, R. K. (2010). Immunoinformatics: An integrated scenario. *Immunology*, 131(2), 153–168. DOI: 10.1111/j.1365-2567.2010.03330.x PMID: 20722763

Tomar, N., & De, R. K. (2014). Immunoinformatics: A brief review. *Methods in Molecular Biology (Clifton, N.J.)*, 1184, 23–55. DOI: 10.1007/978-1-4939-1115-8_3 PMID: 25048118

Torchala, M., Moal, I. H., Chaleil, R. A. G., Fernandez-Recio, J., & Bates, P. A. (2013). SwarmDock: A server for flexible protein–protein docking. *Bioinformatics (Oxford, England)*, 29(6), 807–809. DOI: 10.1093/bioinformatics/btt038 PMID: 23343604

Ulschmid, C. M., Patel, J., Pan, A. Y., Liegl, M., & Holland, K. E. (2023). Intralesional measles, mumps, and rubella vaccine after failure of intralesional Candida antigen for the treatment of recalcitrant pediatric warts. *Pediatric Dermatology*, 40(6), 1057–1059. DOI: 10.1111/pde.15415 PMID: 37596908

van der Poll, T., Shankar-Hari, M., & Wiersinga, W. J. (2021). The immunology of sepsis. *Immunity*, 54(11), 2450–2464. DOI: 10.1016/j.immuni.2021.10.012 PMID: 34758337

Varkey, B. (2021). Principles of Clinical Ethics and Their Application to Practice. *Medical Principles and Practice : International Journal of the Kuwait University. Medical Principles and Practice*, 30(1), 17–28. DOI: 10.1159/000509119 PMID: 32498071

Vita, R., Mahajan, S., Overton, J. A., Dhanda, S. K., Martini, S., Cantrell, J. R., Wheeler, D. K., Sette, A., & Peters, B. (2019). The Immune Epitope Database (IEDB): 2018 update. *Nucleic Acids Research*, 47(D1), D339–D343. DOI: 10.1093/nar/gky1006 PMID: 30357391

Welser, J., Pitera, J. W., & Goldberg, C. (2018). Future Computing Hardware for AI. *Technical Digest - International Electron Devices Meeting, IEDM,* 1.3.1-1.3.6. DOI: 10.1109/IEDM.2018.8614482

*World Health Organization (WHO).* (n.d.). Retrieved March 28, 2024, from https://www.who.int/home

Wu, F., Zhao, S., Yu, B., Chen, Y. M., Wang, W., Song, Z. G., Hu, Y., Tao, Z. W., Tian, J. H., Pei, Y. Y., Yuan, M. L., Zhang, Y. L., Dai, F. H., Liu, Y., Wang, Q. M., Zheng, J. J., Xu, L., Holmes, E. C., & Zhang, Y. Z. (2020). A new coronavirus associated with human respiratory disease in China. *Nature, 579*(7798), 265–269. DOI: 10.1038/s41586-020-2008-3

Yahaya, A.-F., Khalid, A. A., Lim, K., Poh, H. X., Aziz, A., Khalid, K., Lim, H. X., & Poh, C. L. (2023). Development of Next Generation Vaccines against SARS-CoV-2 and Variants of Concern. *Viruses, 15*(3), 624. DOI: 10.3390/v15030624

Zarei, A. E., Almehdar, H. A., & Redwan, E. M. (2016). Hib Vaccines: Past, Present, and Future Perspectives. *Journal of Immunology Research*, 2016(1), 7203587. DOI: 10.1155/2016/7203587 PMID: 26904695

Zhang, T., Aipire, A., Li, Y., Guo, C., & Li, J. (2023). Antigen cross-presentation in dendric cells: From bench to bedside. *Biomedicine and Pharmacotherapy*, 168, 115758. DOI: 10.1016/j.biopha.2023.115758 PMID: 37866002

## KEY TERMS AND DEFINITIONS

**Antigen:** Molecule or substance that is recognized by the immune system and capable of generating an immune response.

**Artificial Intelligence:** Is a branch of computer science that focuses on the development of intelligent machines. They include machine learning, deep learning, and natural language processing.

**Bioinformatic:** Is an interdisciplinary field that combines biology and computer science to analyze and interpret biological data.

**Immunoinformatic:** Integrates immunology with bioinformatics and computational biology for the study of immune factors in the development of vaccines and targeted therapies, for example.

**Multi-Epitope:** A chimeric protein with several subunits, several epitopes.

**Prophilaxis:** Measures to prevent certain diseases and infections from occurring to guarantee protection.

**Reverse Vaccinology:** Is a bioinformatics approach used in vaccine development. It involves the identification of potential vaccine candidates by analyzing the entire genome of a pathogen to predict antigenic proteins.

**Vaccine:** Biological prophylactic measure to prevent and mitigate infections against pathogens and ensure health for the host.

# Chapter 4
# Computational Modeling and Machine Learning in Bioinformatics

**Diego Mariano**
https://orcid.org/0000-0002-5899-2052
*Universidade Federal de Minas Gerais, Brazil*

**Lucas Moraes dos Santos**
*Universidade Federal de Minas Gerais, Brazil*

**Raquel Cardoso de Melo-Minardi**
*Universidade Federal de Minas Gerais, Brazil*

## ABSTRACT

*In recent years, machine learning has revolutionized the world. Indeed, bioinformatics has benefited from machine learning techniques to make new scientific discoveries, such as producing new biotechnological products, discovering new drugs, and understanding the mechanism of action of diseases, among others. In this chapter, you will learn about machine learning methodologies, the types of machine learning, and the traditional algorithms used to build models. You will also learn about some strategies used for the computational modeling of biological problems with a special focus on the representation of structures in the data pre-processing stage.*

DOI: 10.4018/979-8-3693-3192-7.ch004

# 1. INTRODUCTION

Artificial intelligence has brought profound changes to society, especially in life sciences. New high-performance technologies like Next-Generation Sequencing (NGS) and Cryo-electron microscopy (cryo-EM) equipment have increased biological data collection. In this context, machine learning (ML) techniques have allowed knowledge extraction, leading to new scientific discoveries (Dos Santos; Mariano; De Melo-Minardi, 2024). A popular example is the Alphafold algorithm, which uses machine learning to solve the protein folding problem (Jumper et al., 2021).

We like to think that machine learning is the art of predicting the future based on observations made in the past or even deeper observations made in the data itself. Machine learning focuses on extracting knowledge by applying statistical methods to extensive data sets collected from previous observations (Mitchell, 1997).

In this case, it is common for machine learning pipelines to have a feature engineering step (*e.g.*, dimensionality reduction, kernel methods, etc.) or feature selection before applying the learning algorithm. This approach makes it possible to build models capable of making high-performance future predictions (Dos Santos; Mariano; De Melo-Minardi, 2024).

Additionally, ML-based techniques have great applicability in biological sciences, from discovering the mechanism of action of diseases to the production of new drugs (Patel et al., 2020). However, before we talk about the applicability of these algorithms, you need to know the types of machine learning.

## 1.1 Types of Machine Learning

Traditionally, machine learning methods are divided into supervised, unsupervised, semi-supervised, and reinforcement learning (Duda; Hart; Stork, 2012).

In supervised learning, labels are used in the training and testing stages (Cunningham; Cord; Delany, 2008). Thus, models are built to detect patterns based on the answers they already have. Examples of supervised learning techniques are classification and regression. In classification, the model tries to predict groups, while in regression, the model tries to predict discrete values.

Meanwhile, in unsupervised learning, only features are used in pattern detection. Examples of unsupervised machine learning techniques are clustering and dimensionality reduction.

Semi-supervised learning methodologies bring together characteristics of supervised and unsupervised learning. This technique is generally used when little labeled data is available (Zhu, 2005).

Reinforcement learning is a type of machine learning where software (called an agent) learns to perform actions based on a cumulative reward system. Reinforcement learning is most commonly seen in deep learning systems (which will be presented in another chapter).

## 1.2 Supervised Learning

Supervised ML approaches can be divided into four steps:

I. data pre-processing;
II. separation of data sets for training and testing;
III. training the predictive model;
IV. evaluation of the output.

In the following sections, we will discuss the importance of these steps. Let's first discuss data preprocessing and how this step plays a fundamental role in modeling biological problems.

# 2. DATA PRE-PROCESSING

Computational modeling of biological problems involves representing the structures of molecules in a way that can be used as input for artificial intelligence methodologies. Sequence representation is the most commonly used way to represent the structures of nucleic acids and proteins. In this form of representation, each nucleotide or amino acid is represented as a one-letter code.

## 2.1 Sequence Representations

Bioinformatics generally involves the analysis of molecules related to the central dogma of molecular biology. The central dogma of molecular biology is related to the flow of hereditary information responsible for maintaining life. This process has three main molecules: DNA (deoxyribonucleic acid), RNA (ribonucleic acid), and protein. DNA is the base molecule responsible for storing the genetic code in genes that encode the main molecules of living beings. In the process of life, DNA is transcribed into RNA, which is translated into protein (Crick, 1970). Proteins are molecules responsible for various functions in living beings, such as cellular respiration, defense against invaders, catalyzing chemical reactions, regulating metabolism, and muscle movement, among other functions (Nelson et al., 2014). Furthermore, in recent times, the literature has highlighted the importance of non-coding RNAs.

From a practical point of view, these molecules are tiny physical structures composed of a limited set of molecules, which in turn are composed of atoms. From a computational point of view, these molecules can be represented by letter patterns. For example, DNA, despite having a double helix structure, is composed of a combination of four types of nucleotide bases: adenine, cytosine, guanine, and thymine (Nelson; Lehninger; Cox, 2008). Thus, we can represent them by a four-letter pattern:

RNA molecules differ only by the replacement of thymine by uracil, represented by the letters:

Finally, proteins are composed of a combination of amino acid residues linked together through peptide bonds. Dayhoff (1972) proposed that there are 20 most common types of amino acids, which could be represented by either a one-letter or three-letter code (Table 1).

*Table 1. Table of one-letter (1L) and three-letter (3L) amino acid codes*

| Amino acid | 1L | 3L |
|---|---|---|
| Alanine | A | Ala |
| Cysteine | C | Cys |
| Aspartate | D | Asp |
| Glutamate | E | Glu |
| Phenylalanine | F | Phe |
| Glycine | G | Gly |
| Histidine | H | His |
| Isoleucine | I | Ile |
| Lysine | K | Lys |
| Leucine | L | Leu |
| Methionine | M | Met |
| Asparagine | N | Asn |
| Proline | P | Pro |
| Glutamine | Q | Gln |
| Arginine | R | Arg |
| Serine | S | Ser |
| Threonine | T | Thr |
| Valine | V | Val |
| Tryptophan | W | Trp |
| Tyrosine | Y | Tyr |

Source: adapted from (Iupac; Org; Comm, 1975).

The traditional way of representing fragments of DNA, RNA, or proteins is through sequences. Observe an example:

```
>DNA sequence

GTCGATCGAACG
```

This sequence represents a DNA fragment stored in FASTA format. The first line, starting with the ">" symbol, indicates the sequence title in this format. The following lines indicate the nucleotide sequence of this DNA fragment.

This same storage pattern can be used for RNA sequences:

```
>RNA sequence

ACGAUCGUACG
```

Or even for protein sequences:

```
>Protein sequence

MSEKRVQPLDK
```

## 2.2 1-Hot Encoding

Sequences represent the primary structure of the molecules involved in the central dogma of molecular biology. From a computational point of view, they are fundamental structures for representing these molecules.

In bioinformatics, sequence comparison is done using alignment algorithms. For example, they can be performed using dynamic programming algorithms such as the Needleman-Wunsch algorithm (used for global alignment) and the Smith-Waterman algorithm (used for local alignment). BLAST is one of the best-known tools that implement alignment algorithms (Altschul et al., 1990). The following example demonstrates how two sequences (ATCGATCAT and TCGTCTTA) can be aligned:

```
ATCGATCAT-
```

```
-TCG-TCTTA
```

The "-" symbol indicates gaps in the alignment, while different letters indicate mismatches and are highlighted in bold.

Sequences are potentially useful in machine learning tasks as they are a simplified representation of the molecule's primary structure. However, their format must be standardized for artificial intelligence tasks so that algorithms can use them as input.

Consider, for example, that algorithms need binary inputs (0 or 1). DNA sequences could be represented using four bits:

```
A = 0001

T = 0010

C = 0100

G = 1000
```

Therefore, the binary representation of the sequence GTCGATCGAACG could be done as follows:

```
1000 0010 0100 1000 0001 0010 0100 1000 0001 0001 0100 1000
```

A two-bit representation would also be possible:

```
A = 00

T = 01

C = 10

G = 11
```

To represent RNA, simply replace T with U.

To represent proteins, each representative code would have 20 bits:

*Table 2. Binary matrix*

| | A | C | D | E | F | G | H | I | K | L | M | N | P | Q | R | S | T | V | W | Y |
|---|---|---|---|---|---|---|---|---|---|---|---|---|---|---|---|---|---|---|---|---|
| **A** | 0 | 0 | 0 | 0 | 0 | 0 | 0 | 0 | 0 | 0 | 0 | 0 | 0 | 0 | 0 | 0 | 0 | 0 | 0 | 1 |
| **C** | 0 | 0 | 0 | 0 | 0 | 0 | 0 | 0 | 0 | 0 | 0 | 0 | 0 | 0 | 0 | 0 | 0 | 0 | 1 | 0 |
| **D** | 0 | 0 | 0 | 0 | 0 | 0 | 0 | 0 | 0 | 0 | 0 | 0 | 0 | 0 | 0 | 0 | 0 | 1 | 0 | 0 |
| **E** | 0 | 0 | 0 | 0 | 0 | 0 | 0 | 0 | 0 | 0 | 0 | 0 | 0 | 0 | 0 | 0 | 1 | 0 | 0 | 0 |
| **F** | 0 | 0 | 0 | 0 | 0 | 0 | 0 | 0 | 0 | 0 | 0 | 0 | 0 | 0 | 0 | 1 | 0 | 0 | 0 | 0 |
| **G** | 0 | 0 | 0 | 0 | 0 | 0 | 0 | 0 | 0 | 0 | 0 | 0 | 0 | 0 | 1 | 0 | 0 | 0 | 0 | 0 |
| **H** | 0 | 0 | 0 | 0 | 0 | 0 | 0 | 0 | 0 | 0 | 0 | 0 | 0 | 1 | 0 | 0 | 0 | 0 | 0 | 0 |
| **I** | 0 | 0 | 0 | 0 | 0 | 0 | 0 | 0 | 0 | 0 | 0 | 0 | 1 | 0 | 0 | 0 | 0 | 0 | 0 | 0 |
| **K** | 0 | 0 | 0 | 0 | 0 | 0 | 0 | 0 | 0 | 0 | 0 | 1 | 0 | 0 | 0 | 0 | 0 | 0 | 0 | 0 |
| **L** | 0 | 0 | 0 | 0 | 0 | 0 | 0 | 0 | 0 | 0 | 1 | 0 | 0 | 0 | 0 | 0 | 0 | 0 | 0 | 0 |
| **M** | 0 | 0 | 0 | 0 | 0 | 0 | 0 | 0 | 0 | 1 | 0 | 0 | 0 | 0 | 0 | 0 | 0 | 0 | 0 | 0 |
| **N** | 0 | 0 | 0 | 0 | 0 | 0 | 0 | 0 | 1 | 0 | 0 | 0 | 0 | 0 | 0 | 0 | 0 | 0 | 0 | 0 |
| **P** | 0 | 0 | 0 | 0 | 0 | 0 | 0 | 1 | 0 | 0 | 0 | 0 | 0 | 0 | 0 | 0 | 0 | 0 | 0 | 0 |
| **Q** | 0 | 0 | 0 | 0 | 0 | 0 | 1 | 0 | 0 | 0 | 0 | 0 | 0 | 0 | 0 | 0 | 0 | 0 | 0 | 0 |
| **R** | 0 | 0 | 0 | 0 | 0 | 1 | 0 | 0 | 0 | 0 | 0 | 0 | 0 | 0 | 0 | 0 | 0 | 0 | 0 | 0 |
| **S** | 0 | 0 | 0 | 0 | 1 | 0 | 0 | 0 | 0 | 0 | 0 | 0 | 0 | 0 | 0 | 0 | 0 | 0 | 0 | 0 |
| **T** | 0 | 0 | 0 | 1 | 0 | 0 | 0 | 0 | 0 | 0 | 0 | 0 | 0 | 0 | 0 | 0 | 0 | 0 | 0 | 0 |
| **V** | 0 | 0 | 1 | 0 | 0 | 0 | 0 | 0 | 0 | 0 | 0 | 0 | 0 | 0 | 0 | 0 | 0 | 0 | 0 | 0 |
| **W** | 0 | 1 | 0 | 0 | 0 | 0 | 0 | 0 | 0 | 0 | 0 | 0 | 0 | 0 | 0 | 0 | 0 | 0 | 0 | 0 |
| **Y** | 1 | 0 | 0 | 0 | 0 | 0 | 0 | 0 | 0 | 0 | 0 | 0 | 0 | 0 | 0 | 0 | 0 | 0 | 0 | 0 |

We call this type of representation **1-hot encoding** (Defresne; Barbe; Schiex, 2021).

The problem with this type of representation is that as the sequence grows, the greater the amount of space needed to represent it. For example, the following code could represent the sequence MSEKRVQPLDK:

```
    00000000010000000000 00001000000000000000
00000000000000001000 00000000000100000000 00000100000000000000
00100000000000000000 00000010000000000000 00000001000000000000
00000000001000000000 00000000000000000100 00000000000100000000
```

Another problem with this approach is that sequences of different sizes will have different-sized binary codes, which can pose a problem when compared. Many machine learning algorithms require their inputs to be the same size. This can be solved through another representation technique: the sliding window.

## 2.3 Learned Embedding: Sliding Window (k-mer)

Many machine learning algorithms require their inputs to be fixed in size. This can be a problem when dealing with sequences of different sizes. Thus, to standardize them, we can break the sequences into fragments of size k. Then, we can analyze all possible combinations of these fragments and see which are present. First, let's learn how to break down sequences using the sliding window technique.

Notice how we can break the sequence ATCGATCAT into fragments of size k=3:

```
ATCGATCAT

ATC------

-TCG-----

--CGA----

---GAT---

----ATC--

-----TCA-

------CAT
```

We call each fragment a k-mer. This example produced seven k-mers: ATC, TCG, CGA, GAT, ATC, TCA, and CAT. A sequence with n letters will generate n-k+1 subsequences of size k (k-mers). In other words, a sequence with nine letters will generate seven subsequences (9-3+1) of size k=3.

Note that each fragment overlaps the others by k-1 letters, *i.e.*, for k=3, the overlap will be size 2.

Now, see how we can represent this sequence in binary. Consider that, for k=3 and four possible nucleotides, they generate 64 combinations ($4^3$).

First, a vector of possible combinations is filled with zeros. In the following, the table will be filled with one if the k-mer is present in the sequence ATCGATCAT (Table 3).

*Table 3. K-mer representation*

| **AAA** | **0** | **TAA** | **0** | **CAA** | **0** | **GAA** | **0** |
|---|---|---|---|---|---|---|---|
| AAT | 0 | TAT | 0 | CAT | 1 | GAT | 1 |
| AAC | 0 | TAC | 0 | CAC | 0 | GAC | 0 |
| AAG | 0 | TAG | 0 | CAG | 0 | GAG | 0 |
| ATA | 0 | TTA | 0 | CTA | 0 | GTA | 0 |
| ATT | 0 | TTT | 0 | CTT | 0 | GTT | 0 |
| ATC | 1 | TTC | 0 | CTC | 0 | GTC | 0 |
| ATG | 0 | TTG | 0 | CTG | 0 | GTG | 0 |
| ACA | 0 | TCA | 1 | CCA | 0 | GCA | 0 |
| ACT | 0 | TCT | 0 | CCT | 0 | GCT | 0 |
| ACC | 0 | TCC | 0 | CCC | 0 | GCC | 0 |
| ACG | 0 | TCG | 1 | CCG | 0 | GCG | 0 |
| AGA | 0 | TGA | 0 | CGA | 1 | GGA | 0 |
| AGT | 0 | TGT | 0 | CGT | 0 | GGT | 0 |
| AGC | 0 | TGC | 0 | CGC | 0 | GGC | 0 |
| AGG | 0 | TGG | 0 | CGG | 0 | GGG | 0 |

Note that the "ATC" k-mer appears twice in the sequence. In the binary matrix, it will be represented by a single value. However, a variation of this matrix can be made using the count of times the value appears.

Note that we can represent the sequence ATCGATCAT as a single numeric vector:

```
000000100000000000000000100100000100000000001000010000000000
0000
```

The same strategy can be applied to proteins. However, note that the number of combinations grows exponentially as k increases.

```
Vector size = 20k
```

For example, for k=2, the identifier vector will have 400 values:

```
AA = 0

AC = 0

AD = 0

[...] +395 combinations

YW = 0

YY = 0
```

For k=3, there will be 8000 values:

```
AAA = 0

AAC = 0

AAD = 0
```

[...] +7995 combinations

```
YYW = 0

YYY = 0
```

For example, the sequence MSEKRVQPLDK would have the following k-mers (k=3):

```
MSEKRVQPLDK

MSE--------

-SEK-------

--EKR------

---KRV-----

----RVQ----

-----VQP---

------QPL--

-------PLD-

--------LDK
```

Therefore, its representative vector would only have 1s in the fields corresponding to the nine k-mers: MSE, SEK, EKR, KRV, VQP, QPL, PLD, and LDK. All other 7991 k-mers would be filled with 0s.

```
000 [...] 1 [...] 1 [...] 1 [...] 1 [...] 1 [...] 1 [...] 1 [...] 1
[...] 1 [...] 1 [...] 000
```

Sequences are good ways of representing biological data, but they do not represent a good level of abstraction, especially when dealing with three-dimensional structures, such as proteins. Hence, we need better forms of representation, which will be discussed in the next section.

## 2.4 Structural Representations

Sequence representation generates a loss of information, as in the case of three-dimensional structures. In this context, finding differences and similarities between 3D structures of proteins and other molecules has great applicability in life sciences. For example, proteins with similar shapes probably have similar functions. Therefore, other types of representation have been proposed, such as secondary structure alignments, distance maps, voxel representation, and graph-based signatures (Defresne; Barbe; Schiex, 2021).

## 2.5 Graphs

A graph G=(V, E) is a data structure composed of a set of vertices (V), also called nodes, connected by edges (E) (Tutte, 2001). Edges can also have orientations. In this case, the edge is called an arc, and the graph is called a digraph or a directed graph. Additionally, graphs can have weights associated with edges. In bioinformatics, graphs can be used in various tasks. For example, in genome assembly, they are used to model similarity relationships between reads. Many genome assembly algorithms use De Bruijn graph approaches to represent overlaps between sequences and reconstruct the original genome (Hunt et al., 2014; Mariano et al., 2015; Miller; Koren; Sutton, 2010). In structural bioinformatics, they can be used to model the structure of molecules. For example, atoms can be modeled as vertices, while chemical bonds can be modeled as edges.

*Figure 1. The peptide (A) is modeled as a graph (B) (Wikimedia Commons (CC-BY 4.0))*

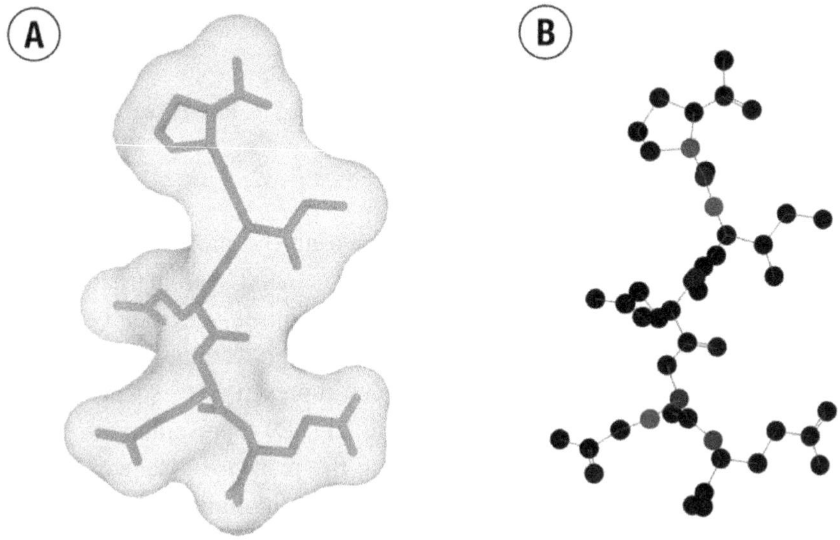


In genome assembly, graph-based approaches have been used to reconstruct the original sequence of a genome from DNA fragments (Mariano et al., 2016). One example is De Bruijn graph-based assembly algorithms. In this method, DNA fragments are divided into k-mers, subsequences of length k. Then, a node is created for each k-mer, and the overlaps between the k-mers are represented as directed edges between the nodes. The final objective of the algorithm is to build a path that passes through all the sequences and thus manages to reconstruct the original genome. For example, SPAdes is a genome assembly tool that uses the De Bruijn graph approach (Bankevich et al., 2012).

Another example of using graphs is in tools for building protein-protein interaction (PPI) networks. In this approach, proteins are represented by nodes, while interactions between different proteins are represented by edges. Examples of programs that use similar approaches are STRING (Szklarczyk et al., 2015) and Cytoscape (Shannon et al., 2003).

Other examples of use are graph-based structural signatures, such as CSM and aCSM (Pires et al., 2011, 2013). Graph-based signatures are essential for comparing 3D structures as they generate good representations of molecules. In practice, these techniques convert macromolecules into one-dimensional numerical vectors based on the positions of their atoms, pharmacochemical properties, and interactions between them. Hence, numeric vectors are input in machine learning algorithms for model building. Based on these models, new molecules can be converted into similar vectors and compared with previously constructed models. These models can be built using various algorithms, such as k-nearest neighbors (KNN), decision trees, random forests, neural networks, support vector machines (SVM), XGBoost, etc. It is important to highlight that good computational modeling is an essential basic step towards building a good model.

## 2.6 Distance Maps

Mathematical approaches applied to understanding the properties of proteins have provided insights relevant to structural bioinformatics. These approaches focus on a data representation restricted to structural information, specifically pairwise distances within a biomolecule. For example, the information contained in the structure of a protein is encoded in the internal distances, represented by square matrices known as distance matrices. These two-dimensional (2D) representations serve as the principal geometrical descriptors, containing the pairwise distances between residues in a protein, and are used to infer protein-protein interactions.

Given a set of $N$ points in a three-dimensional (3D) space, where each point $p_i$ is defined by its coordinates $(x_i, y_i, z_i)$, the distance matrix $D$ is an $N \quad N$ matrix defined as:

$$D = [d_{ij}] \text{ where } d_{ij} = \left\{ {}^{0}\sqrt{(x_j - x_i)^2 + (y_j - y_i)^2 + (z_j - z_i)^2} \right\} \text{ if } i$$

$$= j, \text{otherwise.}$$

In the context of structural biology, $d_{ij}$ denotes the Euclidean distance between the $i$th and the $j$th amino acid residue. The matrix is symmetric, with $d_{ij} = d_{ji}$ for all $i, j$, and its diagonal elements are zero, indicating that the distance from any residue to itself is zero. The complete distance matrix $D$ can be represented as:

$$D = \begin{bmatrix} 0 & 1_{\{1\neq2\}} \cdot \sqrt{(x_2 - x_1)^2 + (y_2 - y_1)^2 + (z_2 - z_1)^2} & \cdots & 1_{\{1\neq n\}} \cdot \sqrt{(x_2 - x_1)^2 + (y_n - y_1)^2 + (z_n - z_1)^2} \\ 1_{\{2\neq1\}} \cdot \sqrt{(x_1 - x_2)^2 + (y_1 - y_2)^2 + (z_1 - z_2)^2} & 0 & \cdots & 1_{\{2\neq n\}} \cdot \sqrt{(x_n - x_z)^2 + (y_n - y_2)^2 + (z_n - z_2)^2} \\ \vdots & \vdots & \ddots & \vdots \\ 1_{\{n\neq1\}} \cdot \sqrt{(x_1 - x_n)^2 + (y_1 - y_n)^2 + (z_1 - z_n)^2} & 1_{\{n\neq2\}} \cdot \sqrt{(x_2 - x_n)^2 + (y_2 - y_1)^2 + (z_2 - z_n)^2} & \cdots & 0 \end{bmatrix}$$

where $d_{ij}$ is the Euclidean distance between the $i^{th}$ and the $j^{th}$ residue. Generally, the coordinates of the atoms of $C_\alpha$ (carbon-$\alpha$) are input to the method (although all atoms can be used). A *graphical representation* of the distances between all pairs of amino acid residues or atoms within a molecule encapsulated in a distance matrix is known as a *distance map* (DM).

Figure 2 presents the structure of a protein (PDB ID: 2LZM) alongside the distance map constructed based on the distances between all versus all carbon-$\alpha$.

*Figure 2. Example of distance matrix for a lysozyme with 164 amino acids (PDB: 2LZM)*

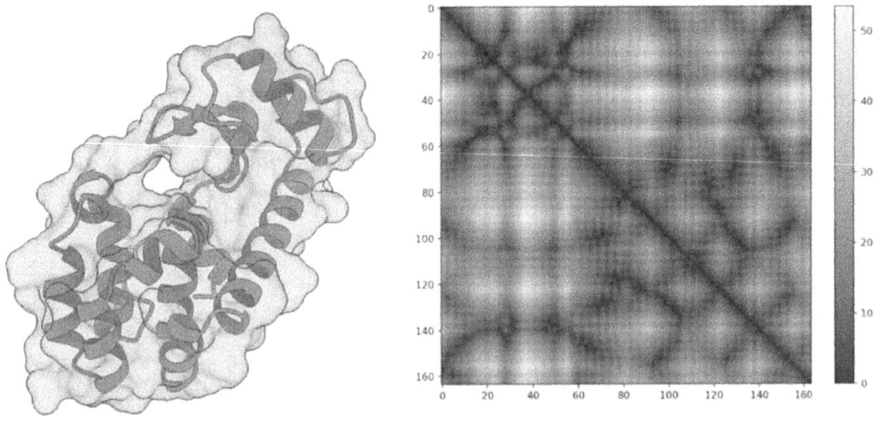


## 2.7 Voxels

Voxels ("volume" and "pixel") are a type of biomolecular representation in which the three-dimensional space is divided into a grid of polygons. When representing proteins and other macromolecules, values are assigned to each polygon, indicating the presence of atoms. Occasionally, the type and characteristics of each atom may be used in this representation.

# 3. SEPARATION OF DATA SETS FOR TRAINING AND TESTING

A fundamental aspect of supervised learning is the division of databases into separate training and testing sets. The training set is used to build the learning model in this case. On the other hand, the test set is used to test the constructed model. The main goal is to find out how good the built model is.

In this case, the model is used to predict the test data labels. As the labels are already known in advance, we can check the percentage of correct answers.

A model with good accuracy is a model that can be used successfully to predict data without knowing the labels.

In general, the training dataset is larger than the testing dataset. For example, in the work of XYZ, 70% of the database was used for training and 30% for testing.

Another strategy is k-fold cross-validation. In this process, the database is divided into $k$ groups. Then, training is done using k-1 groups, and testing is repeated with the remaining group. This process is repeated k times so that each group is used in the test. Thus, the final result will be the average of the k training and testing sessions.

For instance, in 5-fold cross-validation, the dataset is divided into five groups (k=5). The model is trained using the four last groups in the first step, and the test is made with the first one. Then, the model is trained using groups 1, 3, 4, and 5, and the test is made with group 2. Then, training is made using 1, 2, 4, and 5, and test with 3, and so on. Finally, the final result is the average for all training and test steps.

## 4. TRAINING THE PREDICTIVE MODEL

A model is a mathematical function that takes an input value and returns a class (classification model) or discrete value (regression model) corresponding to the input data. Models can be built based on different learning algorithms. There are distance-based algorithms, such as KNN. There are also models based on tree algorithms, such as decision trees, random forests, and gradient boosting. Another very popular strategy is the Support Vector Machine (SVM) approach.

### 4.1 KNN

K-Nearest Neighbors (KNN) is a machine learning method that uses the $K^{th}$ nearest neighbor to delimit a class based on the distances. The KNN approach suggests that, given an $n$-dimensional dataset with known classes, new instances tend to have coordinates closer to individuals of the same class in the feature space. For example, if we delimit K=3, the class of a new instance can be established based on the classes of its three closest neighbors (Mariano, 2020). In general, the value of K should be a prime number greater than two, as this will reduce the chances of a tie in the class choice voting stage. However, each problem can use different values of K. It is important to highlight that the parameter choice, such as K values, is a fundamental step for the success of a model built using KNN (Guo et al., 2003).

The first step of the KNN algorithm is the construction of a distance matrix between points. For this, the Euclidean distance between the input feature vectors is used. Thus, in an $n$-dimensional space, the Euclidean distance $d$ between two points $(i, j)$ is given by the equation:

$$d(i,j) = \sqrt{(i_1 - j_1)^2 + (i_2 - j_2)^2 + \cdots + (i_n - j_n)^2}$$

where $i = (i_1, i_2, \ldots, i_n)$ and $j = (j_1, j_2, \ldots, j_n)$ are vectors representing the coordinates of points in $n$-dimensional space.

Finally, for each test point, the k-nearest neighbors are defined. Hence, based on the class of these neighbors, the class of the test point will be delimited.

Figure 3 presents an example of implementing the KNN algorithm using K=5. In this case, several instances of two classes, A and B, were arranged in this two-dimensional environment (represented by blue and green dots). Next, two new instances are added (represented by gray and black dots).

*Figure 3. KNN example (with k=5) (Mariano, 2020)*

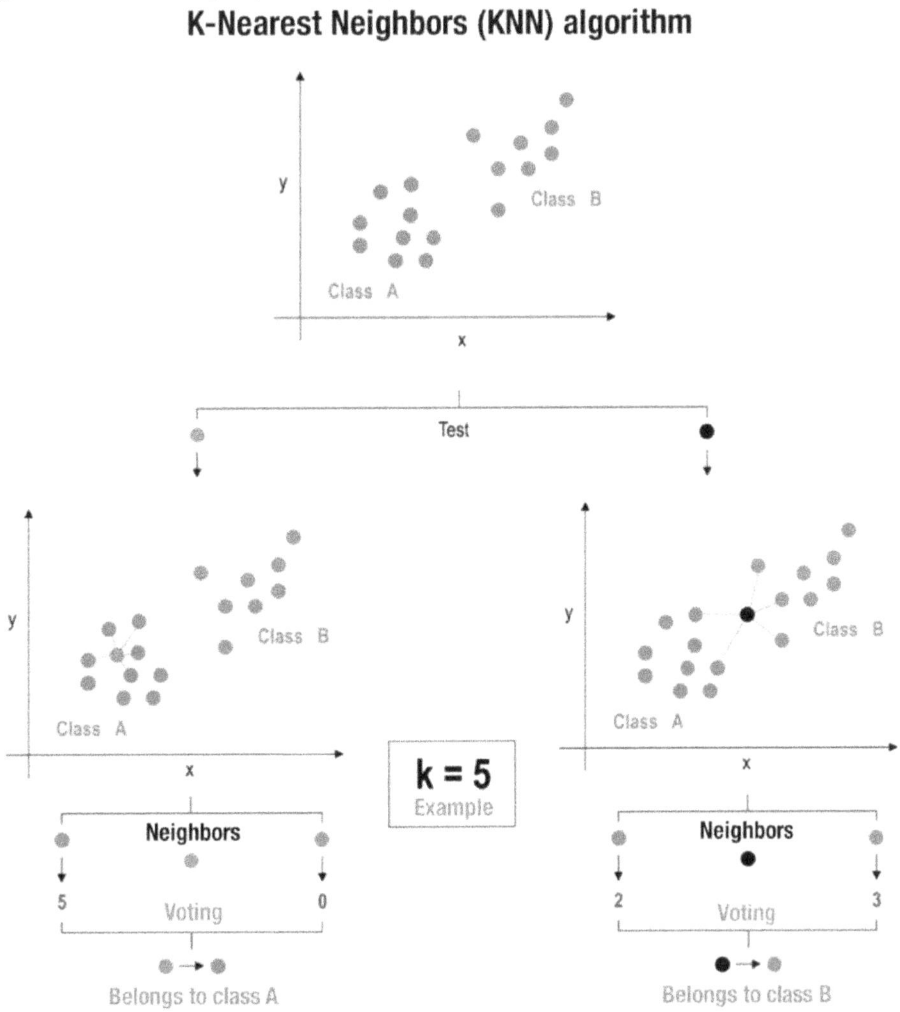

We want to know what classes the new points are. To do this, we can calculate the Euclidean distance between the new points and all other points. As we used k=5 as a parameter, we must evaluate the classes of the five closest points. The class that appears at these points will identify the class of the new points.

In this example, the five neighbors of the gray point belong to class A. Therefore, it also tends to belong to class A. On the other hand, of the five points closest to the black point, three belong to class B, while only two belong to class A. Therefore, he tends to belong to class B.

## 4.2 SVM

Support Vector Machines (SVMs) is a linear discriminant algorithm that transforms pattern vectors into a high-dimensional feature space. In this space, an ideal separating hyperplane, known as a maximum margin hyperplane, is designed to best distinguish between two classes by having the broadest possible margin between the closest points (support vectors) of any category (Webb; Copsey; Cawley, 2011).

Let $x$ be a set of features $\{x_i, i = 1 \ldots n\}$ assigned to one of two classes, $\omega_1$ and $\omega_2$, with labels $y = \pm 1$. Then, the linear discriminant function, $g(x) = w^T x + w_0$, represents a hyperplane, and the decision rule for classifying an instance of x into one of two classes is based on the sign of $g(x)$, that is:

$$\begin{cases} g(x) > 0 \text{ then } x \epsilon w_1 \text{ with label } y_i = +1 \\ g(x) < 0 \text{ then } x \in w_2 \text{ with label } y_i = -1 \end{cases}$$

Hence, $w$ is a weighted vector that determines the direction of the hyperplane and is related to the importance of each feature $x_i$, while $w_0$ is a bias term that shifts the decision limit from the origin. Thus, the objective of the algorithm is to find the optimal values of $w$ and $w_0$ that minimize the loss function, which quantifies the discrepancy between the predicted and true labels in the training set, *i.e.*, $\min_{w,w_0}(w, w_0)$ (Webb; Copsey; Cawley, 2011).

This can be achieved by determining the **hyperplane with maximum margin**, *i.e.*, the sum of the distances between $g(x)$ for the closest instance referring to class $\omega_1$ and for the instance nearest class $\omega_2$. A variation of the perceptron introduces the concept of margin where, for b>0, we have:

$$y_i(w^T x_i + w_0) \geq b$$

Based on this, all points $x_i$ are at a distance greater than $b/|w|$ of the separation hyperplane. Margin maximization is related to a solution that minimizes $|w|$, subject to the restrictions. Considering $b = 1$, the distance between each hyperplane resulting from the above relationship and the separating hyperplane, $g(x) = 0$, is $1/|w|$, and the margin is $2/|w|$. These hyperplanes are canonical, and their points are called support vectors (Webb; Copsey; Cawley, 2011).

$$H_1 : w^T x_i + w_0 \geq +1$$

$$H_2 : w^T x_i + w_0 \leq -1$$

Alternatively, a standard approach to optimization problems with equality and inequality constraints is the Lagrange Formalism, which represents the primary form of the objective function, represented by:

$$L_p = \frac{1}{2} w^T w - \sum_{i=1}^{n} \alpha_i \left( y_i (w^T x_i + w_0) - 1 \right),$$

where $\{\alpha_i, i = 1 \ldots n; \alpha_i \geq 0\}$. The objective function $L_p$ refers to the dual Lagrangian problem, where the objective is to maximize $L_p$ with respect to $\alpha_i$, while minimizing it with respect to $w$ and $w_0$. Consequently, only support vectors will have non-zero Lagrange multipliers ($\alpha_i > 0$) (Webb; Copsey; Cawley, 2011).

## 4.3 Trees

The decision tree algorithm uses a representation decision model similar to a tree to separate data into smaller groups (Ferrari; De Castro Silva, 2017). A tree can be represented as a graph in which the vertices represent decisions, while the edges represent the branches of a decision. The final vertices represent the decisions made and are called leaves.

We will use the iris flower database to illustrate the following example. This database is composed of 150 samples measuring petals and sepals from three classes of iris flowers: setosa, versicolor, and virginica (Figure 4).

*Figure 4. Three species of iris flowers (Wikipedia Commons (CC-BY 4.0))*

Figure 5 presents an example of a decision tree for classifying iris flowers dataset (Anderson, 1936) based on the width and length of the petals and sepals. The initial node, also called the "root", shows the first decision to be made. In this example, the root evaluates whether the petal width is less than or equal to 0.6 cm or greater than 0.6 cm. Thus, depending on this decision, new decisions may be made based on other attributes (branches), or a final classification may be presented (leaf node).

*Figure 5. Example of a decision tree built using the iris flower database (Mariano, 2020)*

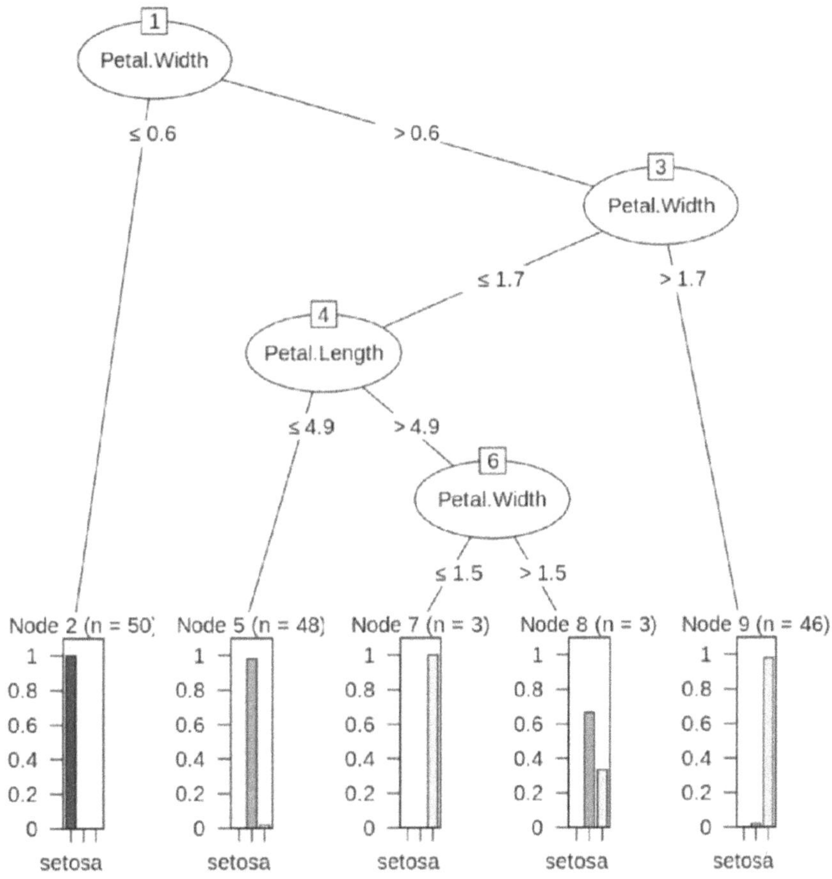


Some algorithms use decision trees as a basis, such as random forests and gradient boosting.

Random forest (RF) algorithm uses several decision trees selected from random samples in the training and prediction stages. RF is an ensemble algorithm, that is, an algorithm that combines several different models to form a more robust model. By combining several trees built using random samples from the input database, the algorithm tends to reduce overfitting, that is, when a model is better adjusted to make predictions for a specific data set. However, this requires a higher processing cost.

Gradient Boosting (GB) is an ensemble algorithm based on decision trees, which transforms weak learning models into strong models by improving learning through errors. GB uses multiple trees to incrementally improve the prediction and build a tree that has a minimum error, that is, using gradient descent to minimize residual errors (Wade; Glynn, 2020). Examples of gradient boosting are GBM (Gradient Boosting Machine), XGBoost (eXtreme Gradient Boosting), LightGBM, and CatBoost.

## 4.4 Tools and Libraries Used for Machine Learning Projects

Machine learning models can be built using various tools and programming languages. For example, the Python programming language has become popular as a starting point for implementing machine learning projects. This is because many machine-learning libraries, such as the scikit-learn library, have been integrated into Python development (Pedregosa et al., 2011). Furthermore, several user-friendly interface tools have allowed the construction of models through a graphical interface. An example is the Orange Data Mining tool (Demšar et al., 2004).

Table 4 lists tools and libraries commonly used in machine learning tasks from the pre-processing stage to model construction.

*Table 4. List of tools and libraries commonly used in machine learning tasks*

| Tool/ Library | Description | Link |
|---|---|---|
| Scikit-learn | Machine learning library in Python that offers a wide range of ML algorithms. | https://scikit-learn.org |
| TensorFlow | Open source library developed by Google for deep learning tasks. | https://www.tensorflow.org |
| Keras | A high-level API for neural networks, which runs on top of TensorFlow or other backends. | https://keras.io |
| PyTorch | Open-source machine learning library maintained by Facebook. | https://pytorch.org |
| StatsModels | Python library that provides classes and functions for estimating many different statistical models. | https://www.statsmodels.org |
| Dask-ML | Library for distributed machine learning in Python. | https://ml.dask.org |
| Orange | Visual and easy-to-use data mining platform with a graphical interface. | https://orangedatamining.com |
| LEMONADE | Open source platform for machine learning and data analysis. | https://www.lemonade.org.br |

## 5. EVALUATION OF THE OUTPUT

When evaluating the performance of a classification system, considering solely the probability of error as a metric often proves inadequate (Koutroumbas; Theodoridis, 2008). In this case, classifier evaluation can be done by comparing the performance of several classifiers trained on the same data set. For this, some complementary metrics are necessary that characterize the chosen classifier model, in terms of performance. These metrics can be extracted from the *confusion matrix* (WEBB & COPSEY, 2011).

A confusion matrix $A$ can be defined such that its element $a_{ij}$ is the number of samples whose true class label is $j$ and are classified as class $i$ (Koutroumbas; Theodoridis, 2008). The table presents a confusion matrix for the 2-class problem (positive - P and negative - N). It shows the number of true and false predictions for each class. True positives (TP) correspond to the number of instances of the positive class correctly predicted by the classifier to belong to the positive class. False positives (FP), in turn, represent the number of instances of the negative class incorrectly predicted by the classifier to belong to the positive class. True negatives (TN) are correctly predicted negative instances, and false negatives (FN) are positive instances predicted to be negative. Thus, the total number of positive and negative samples can be estimated by, respectively,

$$P = TP + FN$$

$$N = FP + TN$$

1. Accuracy (*Acc*): is the percentage of samples from the class $j$ that were correctly classified. Given a problem with $M$ classes, and $N$ is the number of samples from the test set. *Acc* can be estimated from the confusion matrix $A$ (Koutroumbas; Theodoridis, 2008).

$$Acc = \sum_{j=1}^{M} A_{jj}.$$

According to (Webb; Copsey; Cawley, 2011), for the case of a two-class problem, the accuracy of a classifier can be estimated using:

$$Acc = \frac{TP + TN}{P + N}.$$

2. Recall (*Rec*): the percentage of positive samples accurately classified within this category is estimated by

$$Rec = \frac{TP}{P}.$$

3. Specificity (*Spec*): the percentage of negative samples accurately classified in this category is estimated by

$$Spec = \frac{TN}{N}.$$

4. Precision (*Prec*): the percentage of positive samples correctly classified relative to the total number of samples identified as positive is estimated by

$$Prec = \frac{TP}{TP + FP}.$$

5. F1-measure (F1): the F1-Score is the harmonic mean of precision and recall, which takes into account false positives and false negatives:

$$F1 - measure = \frac{2}{\frac{1}{Prec} + \frac{1}{Rec}} = 2\frac{Prec \cdot Rec}{Prec + Rec} = \frac{2TP}{2TP + FP + FN}.$$

These metrics, such as precision, recall, specificity, and F-score, can be used depending on the research objective. The following table summarizes the formulas for obtaining these measurements (Table 5):

*Table 5. Metrics used to evaluate the models*

| Method | Equation |
|---|---|
| **Recall** | TP / (TP+FN) |
| **Specificity** | TN / (FP+TN) |
| **Accuracy** | (TP+TN) / P+N |
| **Precision** | TP / (TP+FP) |
| **F-score** | 2 x (Pr x Se) / (Pr + Se) |

TP: true positives; FN: false negatives; FP: false positives; TN: true negatives; Pr: precision; Se: sensitivity; N: total number of elements. Source: adapted from (Mariano, 2021).

As previously discussed, to evaluate the performance of classifiers, some technique is used to divide the feature space into training and testing data. One of the techniques applied is known as leave-one-out cross-validation (LOOCV). In this approach, an instance is excluded, the classifier is trained on the remaining $N - 1$ samples, and the excluded example is tested by the model. This operation is repeated $N$ times to test all $N$ samples. Then, the number of incorrect samples is counted to estimate the error. As each test sample is excluded from the training sample set, independence is maintained between the training and testing sets. In this case, all $N$ samples are tested, and $N-1$ samples are used for classifier design. Thus, available samples are more effectively used. Also, it is not necessary to worry about the dissimilarity between the design and test distributions (Fukunaga, 2013).

One disadvantage of the method is that it is computationally costly for large classes, as $N$ classifiers must be designed (one classifier to test each sample). However, it is approximately unbiased, although at the cost of an increase in the estimator's variance (Fukunaga, 2013; Webb; Copsey; Cawley, 2011).

Cross-validation (CV) is characterized by a heuristic approach to optimize a classifier's performance by accurately estimating its generalization error. This validation error provides an estimate of the accuracy of the final classifier, serving as a reliable proxy for the classifier's effectiveness on an unknown test set. The method employs a partitioning of the training data, often using a specific percentage $\gamma < 0.5$, to form this validation set. This approach not only aids in evaluating the model's current performance but also in fine-tuning the hyperparameters, thereby minimizing the generalization error. Such a technique ensures that the model is neither overfitting nor underfitting, enhancing its predictive accuracy on new, unseen data (Duda; Hart; Stork, 2012).

An alternative version of this approach is known as $k$-fold cross-validation ($k$-Fold CV). Basically, the technique consists of randomly partitioning the training set into $k$ mutually exclusive subsets of the same size ($n/k$), where $n$ is the total number of training samples. A subset is used for validation, and the remaining $k-1$ are used for parameter estimation. This process is carried out $k$-times, alternating the validation subset in a circular fashion. Performance, in the end, is estimated based on the average of the $k$ error rates corresponding to each of the partitions. In this case, $k = 5$ was used because, in this way, it is possible to guarantee that $\gamma \geq 1$, frequently recommended and effective in most applications (Duda; Hart; Stork, 2012).

## 6. CONCLUSION

In this chapter, you learned about the fundamentals of machine learning and its potential use in bioinformatics for analyzing biological data. You also learned about some strategies used for the computational modeling of biological problems focusing on DNA, RNA, and protein sequences, in addition to strategies for representing macromolecule structures.

We hope that the techniques presented in this chapter can guide future researchers on the methodologies that can be adopted to model and solve problems using ML approaches. In the next chapters, you will learn more about newer machine-learning approaches, such as deep learning algorithms.

## REFERENCES


Altschul, S. F., Gish, W., Miller, W., Myers, E. W., & Lipman, D. J. (1990). Basic local alignment search tool. *Journal of Molecular Biology*, 215(3), 403–410. DOI: 10.1016/S0022-2836(05)80360-2 PMID: 2231712

Anderson, E. (1936). The species problem in Iris. *Annals of the Missouri Botanical Garden*, 23(3), 457–509. DOI: 10.2307/2394164

Bankevich, A., Nurk, S., Antipov, D., Gurevich, A. A., Dvorkin, M., Kulikov, A. S., Lesin, V. M., Nikolenko, S. I., Pham, S., Prjibelski, A. D., Pyshkin, A. V., Sirotkin, A. V., Vyahhi, N., Tesler, G., Alekseyev, M. A., & Pevzner, P. A. (2012, May). SPAdes: A New Genome Assembly Algorithm and Its Applications to Single-Cell Sequencing. *Journal of Computational Biology*, 19(5), 455–477. DOI: 10.1089/cmb.2012.0021 PMID: 22506599

Crick, F. (1970). Central dogma of molecular biology. *Nature*, 227(5258), 561–563. DOI: 10.1038/227561a0 PMID: 4913914

Cunningham, P., Cord, M., & Delany, S. J. (2008). Supervised learning. In *Machine learning techniques for multimedia: case studies on organization and retrieval* (pp. 21–49). Springer. DOI: 10.1007/978-3-540-75171-7_2

Dayhoff, M. O. (1972). *Atlas of protein sequence and structure*. National Biomedical Research Foundation.

Defresne, M., Barbe, S., & Schiex, T. (2021, January). Protein Design with Deep Learning. *International Journal of Molecular Sciences*, 22(21), 11741. DOI: 10.3390/ijms222111741 PMID: 34769173

Demšar, J. (2004). Orange: From experimental machine learning to interactive data mining. In European Conference On Principles Of Data Mining And Knowledge Discovery. Springer.

Dos Santos, L. M., Mariano, D., & De Melo-Minardi, R. C. (2024). O impacto da Inteligência artificial nas ciências da vida através da Bioinformática. *Revista da Universidade Federal de Minas Gerais, 30.*

Duda, R. O., Hart, P. E., & Stork, D. G. (2012). *Pattern Classification*. Wiley.

Ferrari, D. G., & De Castro Silva, L. N. (2017). *Introdução a mineração de dados*. Saraiva Educação S.A.

Fukunaga, K. (2013). *Introduction to statistical pattern recognition*. Elsevier.

Guo, G. (2003). KNN model-based approach in classification. In *On the move to meaningful internet systems 2003: coopis, doa, and odbase: otm confederated international conferences, coopis, doa, and odbase 2003, catania, sicily, italy, november 3-7, 2003. Proceedings*. Springer. DOI: 10.1007/978-3-540-39964-3_62

Hunt, M., Newbold, C., Berriman, M., & Otto, T. D. (2014). A comprehensive evaluation of assembly scaffolding tools. *Genome Biology*, 15(3), R42. DOI: 10.1186/gb-2014-15-3-r42 PMID: 24581555

Iupac, C., Org, N., & Comm, I.-I. (1975). Nomenclature of α-amino acids. *Biochemistry*, 14(2), 449–462. DOI: 10.1021/bi00673a036

Jumper, J. (2021). Highly accurate protein structure prediction with AlphaFold. *Nature, 596*(7873), 583–589.

Koutroumbas, K., & Theodoridis, S. (2008). *Pattern recognition*. Academic Press.

Mariano, D. (2020). Classificação de dados. In *Data Mining. 1*. Sagah.

Mariano, D. (2021). Métricas de avaliação em machine learning: acurácia, sensibilidade, precisão, especificidade e F-score. In *Revista Bioinfo. 1* (Vol. 1). Alfahelix. DOI: 10.51780/978-6-599-275326-15

Mariano, D. C., Pereira, F. L., Ghosh, P., Barh, D., Figueiredo, H. C. P., Silva, A., Ramos, R. T. J., & Azevedo, V. A. C. (2015). MapRepeat: An approach for effective assembly of repetitive regions in prokaryotic genomes. *Bioinformation*, 11(6), 276–279. DOI: 10.6026/97320630011276 PMID: 26229287

Mariano, D. C. B. (2016). SIMBA: a web tool for managing bacterial genome assembly generated by Ion PGM sequencing technology. *BMC Bioinformatics, 17*(18), 456.

Miller, J. R., Koren, S., & Sutton, G. (2010, June). Assembly algorithms for next-generation sequencing data. *Genomics*, 95(6), 315–327. DOI: 10.1016/j.ygeno.2010.03.001 PMID: 20211242

Nelson, D. L. (2014). Princípios de bioquímica de Lehninger. *The Art of Medication*.

Nelson, D. L., Lehninger, A. L., & Cox, M. M. (2008). *Lehninger Principles of Biochemistry*. W. H. Freeman.

Patel, L., Shukla, T., Huang, X., Ussery, D. W., & Wang, S. (2020). Machine learning methods in drug discovery. *Molecules (Basel, Switzerland)*, 25(22), 5277. DOI: 10.3390/molecules25225277 PMID: 33198233

Pedregosa, F. (2011). Scikit-learn: Machine learning in Python. *The Journal of Machine Learning Research, 12*, 2825–2830.

Pires, D. E. (2011). Cutoff Scanning Matrix (CSM): structural classification and function prediction by protein inter-residue distance patterns. *BMC Genomics, 12*(4).

Pires, D. E. V., de Melo-Minardi, R. C., da Silveira, C. H., Campos, F. F., & Meira, W.Jr. (2013, April 1). aCSM: Noise-free graph-based signatures to large-scale receptor-based ligand prediction. *Bioinformatics (Oxford, England)*, 29(7), 855–861. DOI: 10.1093/bioinformatics/btt058 PMID: 23396119

Shannon, P., Markiel, A., Ozier, O., Baliga, N. S., Wang, J. T., Ramage, D., Amin, N., Schwikowski, B., & Ideker, T. (2003). Cytoscape: A software environment for integrated models of biomolecular interaction networks. *Genome Research*, 13(11), 2498–2504. DOI: 10.1101/gr.1239303 PMID: 14597658

Szklarczyk, D., Franceschini, A., Wyder, S., Forslund, K., Heller, D., Huerta-Cepas, J., Simonovic, M., Roth, A., Santos, A., Tsafou, K. P., Kuhn, M., Bork, P., Jensen, L. J., & von Mering, C. (2015). STRING v10: Protein–protein interaction networks, integrated over the tree of life. *Nucleic Acids Research*, 43(D1), D447–D452. DOI: 10.1093/nar/gku1003 PMID: 25352553

Tutte, W. T. (2001). *Graph theory* (Vol. 21). Cambridge University Press.

Wade, C., & Glynn, K. (2020). *Hands-On Gradient Boosting with XGBoost and scikit-learn: Perform accessible machine learning and extreme gradient boosting with Python*. Packt Publishing.

Webb, A. R., Copsey, K. D., & Cawley, G. (2011). *Statistical pattern recognition* (Vol. 2). Wiley Online Library. DOI: 10.1002/9781119952954

Zhu, X. J. (2005). Semi-supervised learning literature survey. Academic Press.

# Chapter 5
# Deep Learning in Bioinformatics:
## Principles and Applications

**Alessandra G. Cioletti**
https://orcid.org/0009-0008-3486-3548
*Federal University of Minas Gerais, Brazil*

**Frederico C. Carvalho**
*Federal University of Minas Gerais, Brazil*

**Lucas M. Dos Santos**
*Federal University of Minas Gerais, Brazil*

**Raquel C. M. Minardi**
*Federal University of Minas Gerais, Brazil*

## ABSTRACT

*The power of deep learning-based models to process and interpret complex biological data has opened new avenues for research and discovery, offering innovative solutions to longstanding and emerging challenges in bioinformatics. This chapter explores how deep learning is researchers solve some of the most pressing challenges in bioinformatics today. From predicting protein structures to speeding the development of new drugs, these models are at the forefront of bioinformatic research, redefining the boundaries of the. The authors explore applications where deep learning has made significant contributions and the synergy between deep learning techniques and bioinformatics, illustrating how this powerful collaboration is helping to advance medical research, drug discovery, and beyond. The current challenges and limitations of such methods are also covered in this chapter. The goal is to provide insights into the transformative impact of deep learning on bio-*

DOI: 10.4018/979-8-3693-3192-7.ch005

*informatics, underscoring its potential to revolutionize our understanding of life.*

## 1. INTRODUCTION

Bioinformatics employs a variety of tools to achieve its goal of understanding biological processes through computational methods. Among these tools, artificial intelligence (AI) models and techniques, such as machine learning and deep learning, have proven themselves particularly effective in addressing complex problems with poorly understood rules and mechanisms, as illustrated in Figure 1. These AI techniques have enabled significant breakthroughs in bioinformatics challenges such as protein structure prediction, genomic sequence analysis, and drug discovery. By leveraging the power of AI, bioinformatics continues to advance our understanding of biology, providing new insights and accelerating research in molecular biology and genetics.

With the advent of omics, the abundance of data brought answers but also raised new questions and challenges. A great example is high-throughput sequencing, a technology that emerged in the late 20th century that gave us genomics and enabled large-scale endeavors such as the human genome project (International Human Genome Sequencing Consortium, 2001). By reducing the costs and increasing the speed of obtaining genetic sequences of organisms, fields such as phylogenetics (Zhou et al., 2013) have advanced significantly and continue to do so, enabling us to perform more complex analyses and better understand evolution and its dynamics (John et al., 2021). However, this fast-paced increase in available data also brings challenges, such as: (i) How do you interpret all this data? (ii) How do you find patterns in it? Moreover, (iii) how to correlate multiple data layers in multi-omics studies?

*Figure 1. Deep learning in artificial intelligence*

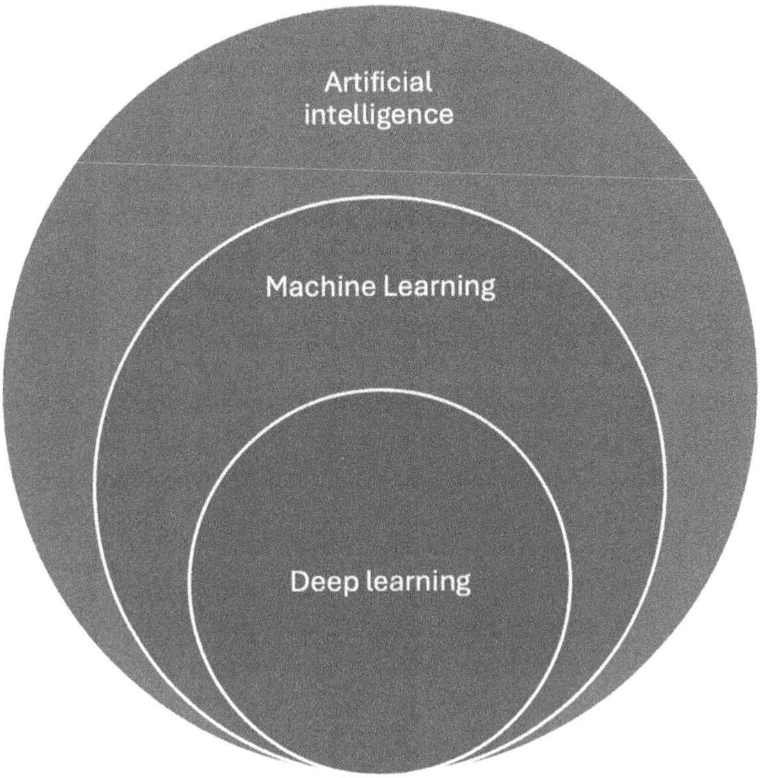


To solve these questions, many researchers have turned to deep learning models. Their varied architectures have successfully extracted meaningful patterns from data, as they can handle high-dimensional data representations while learning high-level abstractions (Shen et al., 2022). Furthermore, the advent of new models and frameworks and advancements in hardware availability and capability have contributed to their widespread adoption in bioinformatics while mitigating significant technological challenges, such as scalability and training on large datasets.

Adopting deep learning models has allowed bioinformatics to achieve significant advancements in challenges such as protein structural modeling and a better understanding of how different molecules interact in systems. Understanding how a protein sequence relates to its shape and function has been a long-standing question in biology. It motivated the scientific community to hold the CASP (Critical

Assessment of Structure Prediction) biannual challenge. Up until 2020, the best-performing models in the field were the Template-based models, which compared protein sequences to those with known structures and used this information to model every part of the sequence provided. At CASP13, which happened that year, however, AlphaFold achieved unprecedented accuracy in predicting a protein's 3D structure from its sequence using a deep learning-based approach (Senior et al., 2020)

Since then, these models have evolved rapidly, and their capability far surpasses that of previous methods. Applications like AlphaFold 2 (Jumper et al., 2021) and, more recently, ESM-2 (Lin et al., 2023) have become the state of the art in modeling protein structures directly from the primary amino acid sequences. A notable example is Evoformer, a key AlphaFold 2 deep learning component that uses an initial hypothetical structure to enhance co-evolutionary analysis. This analysis, in turn, helps to refine the hypothesis about the structure. In this iterative process, both representations – sequence and structure – exchange information, and this cycle continues until the neural network reaches a reliable inference. Similarly, ESM-2, a new family of protein language models based on generative pre-trained transformers (GPT), leverages residue coevolution information in sequences to produce atomic-level predictions about proteins.

The advances in the field of generative AI led to the emergence of a specific class of models known as generative models, which have been overlapping consolidated DL architectures. These models aim to model a distribution from training data and produce similar results. Researchers have discovered that when trained on native protein sequences and structures, these models demonstrate extraordinary capabilities beyond basic protein modeling; they effectively design entirely new molecular structures. This approach has exhibited potential in creating custom molecules by exploring sequence space and their structural and functional properties (Callaway, 2023b), which could aid in future drug design endeavors.

Finally, deep learning has also impacted drug discovery, accelerating various stages and processes in identifying and developing new compounds, such as applying geometric DL models in property prediction, molecular design, and intermolecular interactions (Liu et al., 2024).

In this chapter, we discuss how deep learning is reshaping bioinformatics research and solving important challenges. We focus primarily on two problems: protein structure predictions and drug design. In the following sections, we briefly introduce deep learning models and their common applications in bioinformatics. Section 3 explores some state-of-the-art models for protein structure prediction and drug discovery. Section 4 explains some of the challenges deep learning models and applications in bioinformatics face and the solutions being developed to tackle them.

## 2. INTRODUCTION TO DEEP LEARNING IN BIOINFORMATICS

In recent years, the term artificial intelligence (AI) has surged in popularity, and it is used by the scientific community and professionals outside academia to make news headlines. This rise is attributable primarily to developing and deploying sophisticated generative models. Among these, GPT-4 (Bubeck et al., 2023), the model behind ChatGPT, and state-of-the-art diffusion models designed for image synthesis (Yang et al., 2022) are pivotal contributions. These innovations are not sudden but the fruits of extensive, cumulative research within machine learning and deep learning.

At the heart of these advancements lies machine learning, a domain within AI that develops algorithms and statistical models to enable computers to identify patterns and make decisions from data without needing explicit programming. The learning process, called training, allows these models to perform various tasks, the most common being classification, clustering, and regression (Zaki & Meira, 2020). Models can also be classified based on how they learn. In supervised learning, the model receives a dataset containing labeled data, meaning that for each input, there is a known output the model must predict. After training, the model can predict outputs for new inputs not present in the original training data. On the other hand, unsupervised learning provides the model only with inputs, from which it must learn patterns to define its parameters (Zaki & Meira, 2020).

Traditional machine learning models, such as decision trees, support vector machines, and k-nearest neighbors, have been widely applied to bioinformatics problems including, but not limited to, gene expression analysis, protein-protein interaction prediction, and sequence alignment (Auslander et al., 2021). For example, decision trees can classify different types of cancer based on gene expression profiles, while support vector machines can predict protein functions by analyzing sequence data (Srivastava et al., 2017). Despite their success, these traditional models often struggle with biological data's high dimensionality and complexity, requiring extensive feature engineering and domain knowledge to perform effectively.

Deep learning models address many of these limitations by automatically learning hierarchical data representations, making them particularly suited for complex bioinformatics tasks. One of the first such models was the multi-layer perceptrons (MLP), which combine many smaller units called perceptrons inspired by human neurons' functioning. Such configurations were called neural networks for their similarity with the connections of neurons in the human brain. This enabled their use to describe nonlinear patterns and laid the groundwork for what would evolve into deep learning architectures. This early deep-learning architecture has shown remarkable potential to learn patterns in data and classify complex and non-linear scenarios while enabling complex problem-solving to advanced input-output mapping (Singh & Banerjee, 2019)

In bioinformatics, neural networks such as the MLP have been extensively employed to help advance the field. From gene expression analysis (Bhukya & Ashok, 2020) to predicting protein-protein interactions (Sun et al., 2017), these computational models have proven invaluable in helping scientists make sense of the complex biological data available. Their adaptability and learning capacity make them particularly suited for analyzing high-dimensional and heterogeneous biological datasets, thereby driving forward research and discovery in the landscape of biological sciences.

*Figure 2. Example of application of an MLP to gene expression data classification*

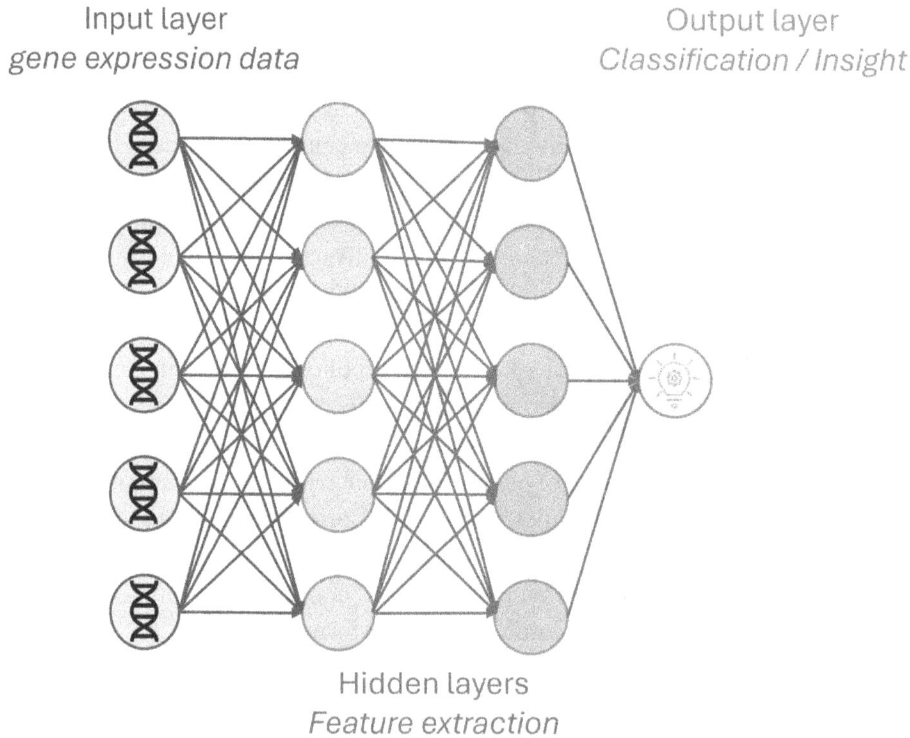


However, only some of the data available can be handled by a single neural network architecture. Thus, new ones have been proposed, each tailored to different tasks and challenges (Santos et al., 2023). For instance, architectures such as convolutional neural networks (CNNs) specialize in processing data with a grid or matrix topology, such as images (Goodfellow et al., 2016). These architectures are inspired by the hierarchy of the human visual cortex, where initial neurons respond to elementary patterns in specific sub-regions of visual stimuli. In contrast, higher-level neurons synthesize information from these simpler cells, identifying more complex patterns (Seonwoo et al., 2017). The core operation in a CNN is convolution, which involves a small matrix of weights, known as a kernel or filter, sliding over the input data to produce a feature map. Each kernel position on the input data computes a weighted sum of the input elements, capturing local patterns such as edges or textures in images. Unlike traditional neural networks, where each individual neuron is connected to every input, CNNs leverage sparse interactions and parameter sharing. Sparse interactions mean that each neuron in a convolutional layer is connected only to a small, localized input region, reducing the number of parameters and computational cost.

Because of how this architecture works, CNNs have been applied to problems such as predicting the 3D structure of proteins and their interaction with other biomolecules (Torrisi & Pollastri, 2020) as well as the prediction of protein function from structure (Gao et al., 2020). Such applications were possible thanks to novel representations of the proteins, such as atom-atom distance maps (Defresne et al., 2021), which capture important spatial features of the protein's 3D structure while generating an image-like output where each pixel at position (A, B) represents the distance between atom A and atom B. This results in a symmetrical matrix or image since the distance between A and B is the same as between B and A.

*Figure 3. Example of how convolutional neural network work*

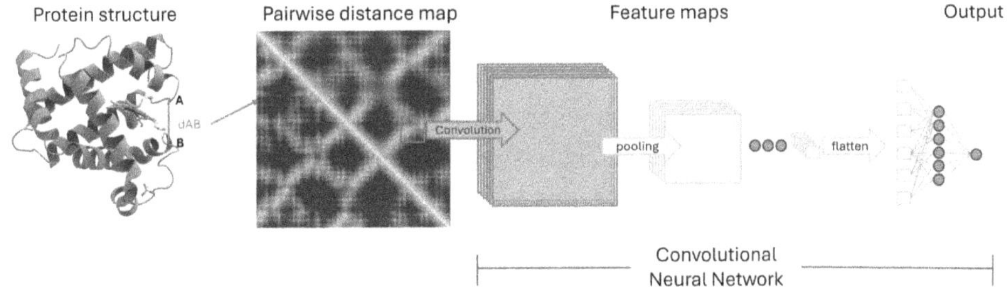


The sequential nature of many types of biological data, such as DNA, RNA, and amino acid sequences, makes them particularly amenable to analysis using techniques from natural language processing (NLP). By treating these sequences as sentences or phrases, we can apply NLP methods to transform how we analyze genomic and proteomic data (Zeng et al., 2015). This approach leverages models like recurrent neural networks (RNNs), a class of neural networks particularly suited for sequential data, as they possess an internal state that allows them to capture temporal dependencies.

Unlike traditional feedforward neural networks, RNNs have connections that form directed cycles, enabling information to persist. This architecture allows RNNs to process sequences of inputs, updating their internal state at each time step based on the current input and the previous state. The core operation in an RNN involves taking an input vector x at a time step t and combining it with the previous hidden state ht-1 to produce the current hidden state ht, typically through an activation function such as the hyperbolic tangent (tanh) or rectified linear unit (ReLU). This process can be mathematically expressed as equation 1, where W1 and W2 are weight matrices, b is a bias term, and is the activation function.

$$h_t=\sigma(W_h h_{t-1} + W_x x_t+b) \text{ (eq1)}$$

By iteratively applying this process, RNNs can maintain a memory of previous inputs, making them powerful for tasks like language modeling, time series prediction, and sequence labeling. In bioinformatics, these models excel at tasks such as gene annotation, mutation effect prediction, and understanding the regulatory mechanisms of gene expression. More modern variants of the RNN architecture, such as the Long-short-term memory (LSTM) network, the Gated Recurring Unit (GRU) network, and the bi-directional LSTM (Bi-LSTM), have been extensively used in the literature to solve important biological. For example, a CNN-LSTM-based deep learning method has been developed for Autism Spectrum Disorders risk RNA identification, effectively combining convolutional neural networks (CNNs) and LSTM to predict relevant RNA sequences (Fan et al., 2021). Additionally, bidirectional LSTM-RNN and GRU models have been employed to identify splice sites and predict exons in genomic sequences, demonstrating significant improvements in genome annotation accuracy (Canatalay & Osman, 2021).

Transformer architectures have revolutionized the way we handle sequential data. Unlike RNNs, transformers do not sequentially process data but instead rely on a mechanism called self-attention, which allows them to consider all parts of the input data simultaneously (Vaswani et al.,2017). This enables transformers to capture long-range dependencies more effectively and handle larger datasets more efficiently. Transformers have been successfully applied to various bioinformatics tasks, such as protein structure prediction and genomic sequence analysis. Notably, models like AlphaFold, which utilize transformer-based architectures, have achieved unprecedented accuracy in predicting protein folding, marking a significant breakthrough in computational biology (Jumper et al., 2021).

Another class of advanced deep-learning models is the generative models, such as diffusion models (Ho et al., 2020). They differ from the previously explained models, for they can generate data instead of just returning a classification or prediction. They work by gradually adding noise to the training data and then learning to reverse this process, effectively denoising the data and generating new samples that closely resemble the original data. This method has been particularly useful in generating molecular structures and simulating biological processes at a high level of detail (Watson et al., 2023).

## 3. NOTABLE APPLICATIONS OF DEEP LEARNING IN BIOINFORMATICS

The versatility and diversity of deep learning models and techniques have catalyzed significant breakthroughs in various bioinformatics problems. In this section, we will delve into state-of-the-art applications of deep learning in two critical areas: protein structure prediction and drug discovery. These applications demonstrate the practical utility of deep learning models and highlight their potential to transform and accelerate research in molecular biology and pharmacology. These advancements exemplify how deep learning is reshaping the landscape of bioinformatics and paving the way for future innovations.

### 3.1. Protein Structure Prediction

Predicting protein structures from their sequences is considered a fundamental problem in bioinformatics, and it has remained one of the main challenges in the field for decades. The functionality of a protein is intrinsically linked to its folding structure; therefore, predicting its final form from a given sequence is vital for developing potential inhibitors, understanding intermolecular interactions, and more. In the late 1960s, Cyrus Levinthal proposed the Levinthal's paradox (Levinthal, 1968), highlighting the complexity of protein folding by estimating that a typical protein could assume around 10300different conformations (Levinthal, 1969). Although recent advances in experimental techniques for structure determination have improved accuracy, these methods remain complex and time-consuming. However, deep learning has brought new, powerful tools to this challenge, offering promising solutions to predict protein structures from sequences with remarkable accuracy.

One of the key drivers of the recent advances in protein structure prediction efforts is the Critical Assessment of Techniques for Protein Structure Prediction (CASP). Established in 1994 and happening every two years since its first edition, CASP aims to objectively assess the accuracy of computational methods by challenging researchers to predict protein structures that have been experimentally determined but not yet publicly released (Callaway, 2023a). This competition has become a pivotal platform for the bioinformatics and deep learning communities, fostering collaboration and innovation. The continuous feedback loop provided by CASP, where predictions are rigorously evaluated against experimental data, has driven methodological improvements and the adoption of cutting-edge techniques.

In recent CASP competitions, deep learning-based methods, such as AlphaFold, developed by DeepMind, have demonstrated unprecedented accuracy in predicting protein folding (Jumper et al., 2021). These breakthroughs are attributed to the ability of deep learning models to capture complex patterns and interactions within amino acid sequences, leading to highly accurate structural predictions. This demonstrates that CASP has served as both a benchmark for progress and a catalyst for adopting and integrating advanced computational techniques, accelerating molecular biology and pharmacology discoveries.

### 3.1.1. AlphaFold

Developed by the DeepMind research team, the AI system known as AlphaFold outperformed all other competitors in terms of accuracy during the 13th Critical Assessment of Protein Structure Prediction (CASP). The core component of AlphaFold is a convolutional residual network, which is trained using 3D structures from the Protein Data Bank (PDB) to predict protein distance maps (distograms) from amino acid sequences. Using so-called dilated convolutions, which use non-contiguous receptive fields that cover larger spatial extents compared to traditional convolutions, these networks can effectively capture long-range interactions across the protein sequence, enabling a more accurate and comprehensive prediction of folding (Alquraishi, 2019; Senior, Evans, et al., 2020).

In 2020, a new version of the program, named AlphaFold 2 (AF2) (Jumper et al., 2021), achieved a remarkable accuracy score of 92.4 (on a scale from 0 to 100) in the Global Distance Test (GDT) during CASP14 (mean TM-Score of 0.85). AlphaFold 2 benefited from novel large-scale protein language models (PLMs) pre-trained from self-supervised learning on billions of proteins. These have been the core of computational methods for predicting 3D protein structures with high precision, representing the state-of-the-art in this field. A notable example is Evoformer, a key AF2 deep learning component that uses an initial hypothetical structure to enhance co-evolutionary analysis. This analysis, in turn, helps to refine the hypothesis about the structure. In this iterative process, both representations – sequence and structure – exchange information, and this cycle continues until the neural network reaches a reliable inference (Jumper et al., 2021).

In AF2, the hypothetical structure is modeled as a spatial graph, where amino acid residues are represented by nodes (or vertices) and their connections by edges. This graph representation approach is fundamental for analyzing the physical interactions between molecules as well as their evolutionary dynamics. In the most recent version of AF2, presented at CASP14, a transformer was used to analyze and interpret the structure of this graph. This process involves updating the graph, employing evolutionarily related sequences, multiple sequence alignments (MSA), and a detailed representation of amino acid residue pairs to improve the accuracy of the modeled structure.

Recently, leveraging the capabilities of AF2, researchers at the University of Missouri developed MULTICOM (Liu, Guo, Wu, et al. 2023), which aims to enhance three-dimensional protein structure prediction. In CASP 15, which happened in 2022, MULTICOM demonstrated significant advancements over its previous performance in CASP 14 and the standard AlphaFold 2 predictor. The system increases input diversity through MSAs and structural models, implements a novel alignment-based refinement framework developed by Foldseek, and integrates tertiary and quaternary predictions for protein assemblies. These improvements have allowed MULTICOM to achieve greater accuracy, with its refined version, MULTICOM refine, ranking 3rd out of 47 server predictors and achieving improved GDT-TS and TM-scores, outperforming AF 2 by 9.6% and 8.2%, respectively (Ozden, Kryshtafovych & Karaca, 2023). These advances highlight MULTICOM's goal to refine AF 2-based predictions and address challenges in predicting structures for complex targets involving protein-protein interactions (Liu, Guo, Wu, et al., 2023).

### 3.1.2. ESMFold

In 2022, researchers from Meta AI unveiled ESMFold, highlighting its remarkable capability to model proteins significantly faster than AlphaFold2, which typically finalizes predictions in a matter of minutes. A key factor contributing to this speed advantage is ESMFold's reliance solely on the protein sequence for input, as opposed to AlphaFold2's multiple sequence analysis (MSA). Although MSA enhances the accuracy of predictions, it also increases computational demands.

ESM-2 (Lin et al., 2023), represents an innovative family of protein language models that utilize generative pre-trained transformers (GPT) to infer protein structures directly from their primary sequences. This model harnesses residue co-evolution information within sequences to generate atomic-level predictions, significantly enhancing the accuracy of structure inference (Lin et al., 2023; Zhang et al., 2023). The training of ESM-2 employs a masked language modeling approach, where random amino acids in the sequence are masked, and the model predicts these masked amino acids based on the surrounding context. This method improves the model's understanding of sequence relationships and facilitates more accurate and detailed structural predictions. By leveraging these advanced techniques, ESM-2 offers a powerful tool for advancing our understanding of protein folding and function, potentially accelerating discoveries in molecular biology and bioinformatics.

The dataset used to train this model contains millions of sequences derived from the UniRef database. During the training, it develops specific attention patterns that mimic the interactions between amino acids within the sequence. The researchers extracted attention maps and discovered that these representations align with the tertiary structures of proteins. They then applied an equivariant transformer to convert these maps into spatial representations, which detail the positions of the atoms within the proteins. This process leverages the model's internal representation to accurately reconstruct the structure of a protein at atomic resolution (Lin et al., 2023; Zhang et al., 2023).

ESMFold demonstrates competitive accuracy compared to AF2 and RoseTTA-Fold on the CAMEO and CASP14 datasets, achieving a mean TM-score of 0.83 on CAMEO (versus AF2's 0.88 and RoseTTAFold's 0.82) and 0.68 on CASP14 (versus AF2's 0.85). ESMFold's distinct advantage lies in its ability to generate accurate predictions directly from single sequences, thereby eliminating the necessity for MSAS, which is integral to the functionality of both AF2 and RoseTTAFold. This MSA-independent approach allows ESMFold to produce predictions up to 60 times faster than AF2 for shorter sequences (Lin et al., 2023).

Furthermore, the model exhibits strong correlations between its predicted Local Distance Difference Test (pLDDT) scores and actual LDDT scores, with Pearson correlations of 0.79 on a random subset of approximately 4,000 metagenomic sequences and demonstrating high calibration accuracy on CAMEO (mean pLDDT > 0.9). Additionally, ESMFold's performance strongly correlates with its language model's perplexity, with correlations of -0.55 on the CAMEO test set and -0.67 on CASP14, indicating that enhanced sequence understanding significantly improves prediction accuracy. Consequently, ESMFold is especially valuable for large-scale structural predictions, offering a practical balance between speed and accuracy, and is particularly advantageous for characterizing the vast number of uncharacterized protein structures in metagenomic datasets (Lin et al., 2023).

### 3.1.3. RFDiffusion

A study developed by researchers at the University of Washington introduced a novel generative model for protein design from scratch named RoseTTAFold diffusion (RFdiffusion) (Watson et al., 2023). Recently, the software tool DALL-E has garnered attention for its ability to generate entirely novel high-quality images, leveraging a technique known as diffusion. Diffusion models (DIFF) (Sohl-Dickstein et al., 2015) are specialized in adding and removing noise and have been successful in performing tasks such as generating images from text, producing high-quality samples (Anstine and Isayev, 2023). These networks aim to model a conditional probability distribution of data based on sequential steps.

In this case, the input structure (e.g., image) is gradually transformed through an iterative procedure known as diffusion. This process involves the introduction of Gaussian noise, which shifts the data towards a target distribution where the resulting structure has no similarity to the initial structure. During training, the network learns to remove noise from a maximum likelihood estimate, in which the model aims to maximize the probability of generating input from noisy data (Sohl-Dickstein et al., 2015). The RF diffusion technique modifies the RoseTTAFold structure prediction network to accommodate specific design requirements through a technique known as conditioning. This involves refining the model by removing noise according to these predefined constraints, leading to outstanding capabilities in designing the protein backbone.

RFdiffusion demonstrates superior performance over traditional protein design methods across a range of complex challenges. These challenges encompass the design of protein monomers with specific topological constraints, the creation of protein binders, the design of symmetric oligomers, scaffolding for enzyme active sites, and the construction of symmetric motifs for therapeutic and metal-binding proteins. Notable achievements of RFdiffusion include the creation of a picomolar-level binder through purely computational means and the development of innovative symmetric assemblies, which have been experimentally validated using electron microscopy (Watson et al., 2023; Krishna et al., 2024).

RFdiffusion's strengths are particularly evident in its ability to generate diverse protein structures and its robust performance in designing novel proteins with specific functional sites. For instance, RFdiffusion achieved a significant *in silico* success rate with a mean TM-score versus PDB structures exceeding 0.5 for designed TIM barrels and NTF2 folds, further validated using AF2 and ESMFold predictions. These comparisons highlight AF2's superior accuracy in traditional benchmarks, while RFdiffusion excels in designing novel, diverse proteins and maintaining high structural fidelity, particularly when employing fold-conditioning and incorporating coarse-grained fold information (Watson et al., 2023).

## 3.2. Drug Discovery

The drug development process is a long, expensive, and uncertain process. Developing a new drug costs an average estimate of US$0.8 billion to US$2.3 billion per approved drug (CBO, 2021). This value includes expenses with the newly approved drug but with all others being developed and not reaching the market. Data from

2019 points to pharmaceutical industries spending around 83 billion on research and development, with a percentage of 15.7% in the pre-clinical phase, a stage where bioinformatics plays a crucial role (CBO, 2021; PhRMA, 2020).

Bioinformatics operates in several areas related to drug discovery, seeking to accelerate high-throughput screening (HTS) and identification of new drug targets, facilitate the characterization of side effects, and predict toxicity, drug interactions, and microbiological resistance to drugs. The increase in genomic, proteomic, and epigenetic data, as well as databases of protein and RNA structures, small molecules, and metabolites, combined with the availability of high performance, have opened up space for the use of deep learning in drug development (Figure 4), allowing the carrying out more realistic and informative protein-ligand docking and virtual screening experiments (Kim et al., 2021; Chen et al., 2018; Xia, 2017).

*Figure 4. Deep learning application*

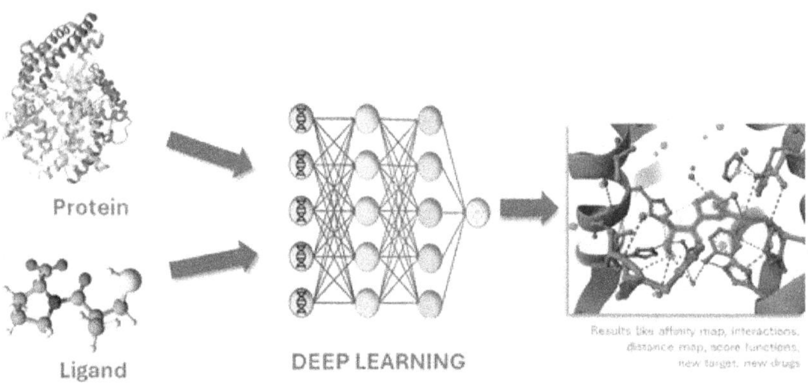


Neural network architectures are generally built in the Python programming language, and the leading deep learning packages available are open source, the most used being TensorFlow, PyTorch, Keras, and Caffe. The main deep learning models used in drug discovery are convolutional neural network (CNN), recurrent neural network (RNN), graph neural network (GNN), graph attention network (GAT), graph convolution network (GCN), generative adversarial network (GAN) and autoencoder (AE), variational autoencoder (VAE) (Kim et al., 2021).

### 3.2.1. Drug Representations

The input data for drug discovery are small molecules or macromolecules that can be represented by descriptors. Molecule representation can be done through strings (SMILES), pharmacophoric fingerprints (FPs), graphs, or voxels (Figures 5, 6).

*Figure 5. Molecular representations*

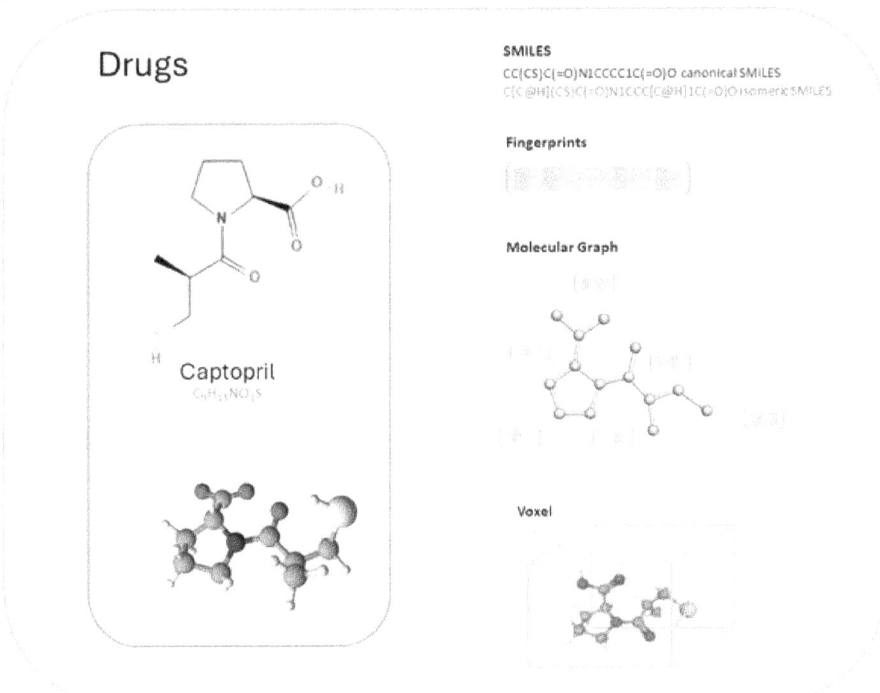


#### *SMILES*

SMILES (Simplified Molecular Input Line System) is a chemical notation representing the molecular structure by a linear string of symbols, like natural language (WEINNGER, 1988). Because it is based on a sequence, it can be used as a "sentence" in various deep learning models, using an approach like that used in natural language processing (NLP). It is used to learn representations in autoencoder models.

### *Pharmacophore Fingerprints*

The fingerprint (FP) is a representation of the chemical molecule in the form of a vector of bits, where each bit counts as a presence (1) or absence (0). These impressions have the pharmacophoric characteristics of the structure. FP can be used to evaluate the similarity between chemical compounds based on common substructures and the similarity between binding sites. There are two main types of FP, structural or hashed, and two molecular (MPF) or interaction (IFP) subclasses according to the information it encodes (Fassio et al., 2022). Two commonly used fingerprints are extended connectivity fingerprints (ECFP) and functional connectivity fingerprints (FCFP), also known as circular or Morgan fingerprints. Hashed FPs are created by a hash function, requiring no predefined substructures. FCFP is widely used in deep learning models for predicting drug-target interactions (DTI) (Kim et al., 2021).

### *Molecular Graphs*

One of the most used forms of molecular representation is the molecular graph. It is based on mapping atoms and bonds as a set of nodes and edges, where connectivity between atoms is represented using an adjacency matrix. (David et al., 2020). Graph chemistry theory considers that chemical structures represented as graphs contain the information necessary to model and provide information about various biological phenomena. This information can be in the form of bond length and other atomic properties (weighted chemical graphs), as well as bond valence (multi-edge reduced graphs) (Lo et al., 2018). Graphs are used in deep learning models GNN, GCN, and GAT.

*Figure 6. Proteins representations*

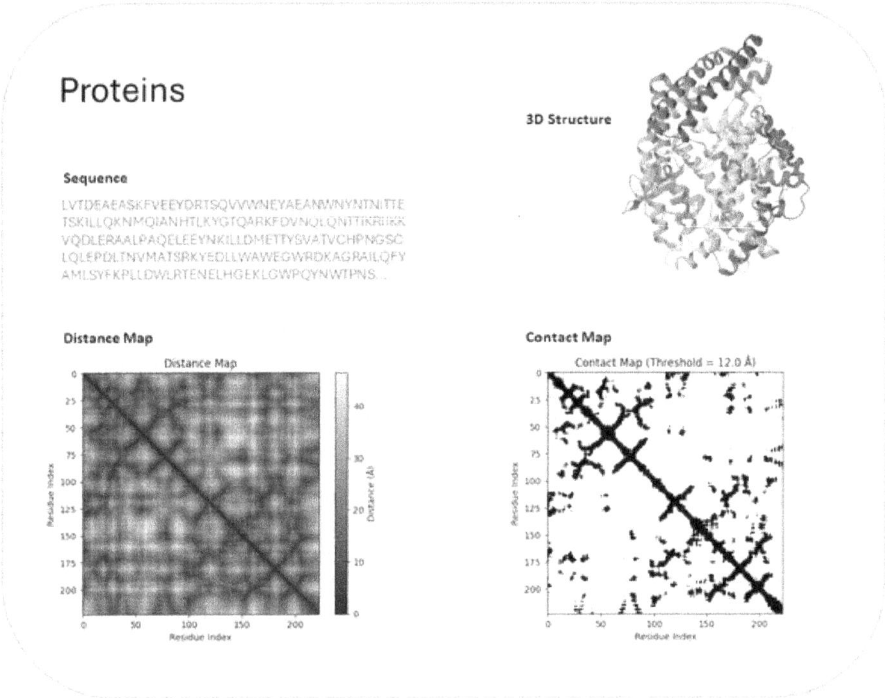


### *Distance Maps*

Mathematical methodologies in structural bioinformatics focus on representing protein structures through pairwise distances within biomolecules, encoded as square matrices referred to as *distance matrices*. These two-dimensional matrices, which capture the Euclidean distances between residues, serve as crucial geometrical descriptors for deducing protein-protein interactions. Given a set of N points in three-dimensional space, where each point is defined by coordinates (xi, yi, zi), the distance matrix is an N x N symmetric matrix, with diagonal elements of zero, indicating the distance from any residue to itself is zero. Typically, the coordinates of carbon atoms are used, but all atoms can be included (Kloczkowski et al. 2009). This graphical representation of distances between residues within a molecule can be referred to as distance map (DM), or contact map (CM).

### Voxel

The voxel is a combination of "volume" and "pixel", that is, it is a representation that extends a 2D image to a 3D dimension, where a value is assigned to the geographic location of the atom and the rest is filled with zeros. Resolution is an essential factor that must be observed when using the voxel; accuracy is lost as the resolution is reduced. Conversely, when the resolution is increased, the data size and training speed will increase, resulting in a greater computational demand for memory and processing. It is a representation widely used in CNN neural networks for connection prediction in DTI (Kim et al., 2021).

Targets, in turn, can be represented by the linear sequence of amino acid residues, the 3D structure, or the relationship between the ligand and the target.

## 3.2.2. Predicting Drug-Target Interactions (DTI)

In their review of methodologies, Kim et al. (2021) divides the works on using deep learning for DTI into three groups according to the characteristics of the data input: ligand-based approach, target structure-based approach, ligand-interaction-based approach, and target. Table 1 presents the main deep learning models and their inputs.

### Ligand-Based Approach

For years, the interaction between ligands and receptors has been explained using the "key-lock" paradigm, where receptors are considered "locks" and ligands are the "keys" that fit into them. This model suggests that the recognition of a ligand by a receptor is highly dependent on the chemical structure of the ligand. In medicinal chemistry, it is well-established that molecules with similar structures often exhibit similar properties. Leveraging this principle, the ligand-based approach in drug discovery utilizes virtual screening methodologies to identify compounds from large chemical libraries that are structurally similar in 2D or 3D to known ligands of target proteins. By doing so, it builds on the foundational knowledge that structural similarity often translates to functional similarity, facilitating the identification of potential drug candidates.

*Table 1. Deep learning methods for DTI prediction*

| Aproach | DL model | Input | Task |
|---|---|---|---|
| Ligand-based | CNN | smiles, image 2D | classification |
| | GCN | smiles, molecular graph | regression |
| Structure-based | CNN | smiles, voxel, fingerprint | regression/ classification |
| | GCN | fingerprint, molecular graph | regression/ classification |
| | GNN | 3D molecular graph | regression |
| Relationship-based | CNN/ AE | similarity | classification |

QSAR is a ligand-based virtual screening method that relies on the principle that compounds with similar physicochemical properties exhibit similar activities. CNNs an MLP are among the main neural network architectures used alongside this representation. Recent studies have proposed a deepQSAR, where the chemical descriptors used in traditional QSAR are replaced by molecular structure data in the form of numerical vectors in high-dimensional spaces, which are used together with neural network models. This application allows the transfer of knowledge, which results in greater effectiveness in solving multi-objective tasks. However, this result relies on the curation of models and input data. This evolution of QSAR to deepQSAR has made it possible to expand QSAR's performance to the scope of molecular docking (Tropsha et al, 2024).

### *Target Structure-Based Approach*

This approach uses data on protein and ligand structures, such as ligand scoring functions and binding pose generation, to estimate the binding affinity of protein-ligand interaction. Historically, it has been carried out using molecular docking, but in recent years, works relying on deep learning models achieved results comparable to those of Autodock Vina. These works use a convolutional neural network set as a scoring function (Ragoza et al., 2017). An example is IGN (Interaction Graph Net) which uses two GNN, one GNN to predict the intramolecular interactions of the protein-ligand complex and the other to learn the intermolecular interactions and uses the results to predict the binding affinity (Jiang et al., 2021). Other examples that use deep learning for scoring functions are GNINA 1.0 (McNutt et al., 2021) which uses CNNs and RMTScore and GenScore, new scoring functions that use several graph transformer layers to learn ligand and protein representations (Shen et a., 2023, Zhang et al., 2024).

### *Relationship-Based Approach*

The relationship-based approach uses neural networks to learn and associate other available data about the ligand, target, and interactions. This data can be genomic, toxicological, pharmacological, drug interactions, and other data from available databases such as PharmGKB, DrugBank, CTD, and BindingDB. This data contains precious information that, if well explored, allows us to improve the understanding of the affinity of the protein-ligand bond and greater precision and optimization of the pharmacophore before it passes to clinical stages (Born & Manica, 2021). One example is DLDTI, which uses convolutional neural networks to predict how a proposed drug will interact with its intended target and identify new target proteins of known drugs, based on a selection of structural characteristics of the drug and protein properties (Zhao et al., 2020). One application that encompasses protein-ligand interaction is the use of deep learning to predict distance matrix and generate PL binding poses. For examples, there is PoseNetDiMa (Mahmoud et al., 2020), TankBind (Lu et al, 2022), and EquiBind (Stark et al., 2022) which uses an equivariant graph neural network (EGNN) to predict movement of atoms and more recently DiffDock (Corso et al., 2023) which uses generative algorithms to generate different conformations with acceptable lengths and angles.

## 4. CHALLENGES AND LIMITATIONS

As discussed before, deep learning has been a transformative force in bioinformatics, providing powerful tools to analyze and interpret the ever-growing amount of biological data generated by modern research. However, integrating deep learning within bioinformatics has its challenges. Some of the most prominent challenges are the availability and quality of data, the interpretability and explainability of the models themselves, and the scalability and computational resources required to support these data-intensive analyses. Each of these challenges needs to be addressed to harness the potential of deep learning in bioinformatics fully.

Data availability and quality are critical, as deep learning models require large, well-annotated datasets to achieve high accuracy. However, such datasets are often scarce due to the high costs and privacy concerns associated with generating and sharing biological data. Furthermore, the interpretability and explainability of deep learning models are essential for gaining the trust of biologists and clinicians, as these models often operate as "black boxes," making it difficult to understand how they arrive at their predictions. Finally, the scalability of deep learning models and the availability of computational resources are significant hurdles, as training these models is computationally intensive and requires substantial memory and processing power.

Addressing these challenges involves concerted efforts in data curation, developing interpretability techniques, and adopting scalable computational solutions such as cloud computing and efficient model architectures.

## 4.1. Data Availability and Quality Issues in Bioinformatics

Deep learning models require large quantities of high-quality data to be effectively trained. In bioinformatics, the scarcity of well-annotated datasets poses a significant challenge. The prohibitive costs of generating high-quality biological data, privacy concerns, and intellectual property issues often limit data availability. Furthermore, available datasets may suffer from inconsistencies in annotation, missing values, or batch effects that introduce noise and bias, ultimately impacting model performance. Addressing these issues requires concerted efforts in data curation, developing data collection and annotation standards, and implementing robust data preprocessing techniques to ensure the highest possible data quality (Goodfellow et al., 2016; Lecun et al., 2015).

## 4.2. Interpretability and Explainability of Deep Learning Models

One of the significant advantages of deep learning models is their ability to learn patterns from data without explicit programming. This advantage, however, leads to one of its major drawbacks: the reduced interpretability and explainability of their predictions, especially for models with complex architectures. Such models often operate as "black boxes," making understanding how input features relate to outputs is difficult. This lack of transparency can hinder the trust and adoption of these models by biologists and clinicians who require clear explanations of model decisions (Lipton, 2016).

Various efforts have been made to improve the explainability of deep learning models. In bioinformatics, where understanding the biological underpinnings of model predictions is crucial, efforts to improve model interpretability include techniques such as layer-wise relevance propagation and attention mechanisms. These approaches aim to shed light on the inner workings of models and provide insights into the biological relevance of the features they identify (Montavon et al., 2018). Karim et al. (2024) have also explored the use of explainable AI methods in bioinformatics, emphasizing the need for transparency and interpretability in model predictions to foster trust among researchers and clinicians. Some of the currently used techniques to mitigate this problem are the probing techniques, such as gradient-based methods and the attention mechanism, which aim to analyze and interpret the internal workings of black-box models, and perturbation techniques, which are methods used to interpret and understand the behavior of deep learning models by systematically modifying the input data and observing the resulting changes in the model's output (Karim et al. 2023).

Recent advances have highlighted the importance of moving from merely explainable AI to interactive AI, which allows users to engage with and understand model predictions more effectively. For example, Lin et al. (2024) discuss various interactive AI techniques that enable users to interact with models to gain a deeper understanding of the decision-making process. This shift towards interactive AI represents a significant step forward in making deep learning models more transparent and user-friendly, thereby increasing their acceptance and utility in bioinformatics.

### 4.3. Scalability and Computational Resources

Another critical challenge in developing and deploying deep learning models is their scalability and the availability of computational resources. Training deep learning models is computationally intensive, requiring substantial memory, processing power, and often specialized hardware like GPUs or TPUs. This demand significantly burdens research institutions, particularly those with limited access to high-performance computing facilities. Additionally, processing large-scale biological datasets, such as whole-genome sequences or high-resolution medical images, exacerbates these demands, leading to prolonged training times and increased costs. Solutions to address these challenges include adopting cloud computing services, which offer scalable computational resources, and exploring more efficient model architectures that maintain performance while reducing computational load (Dean et al., 2012; Barroso et al., 2019).

## 5. CONCLUSION

This chapter explored some of the main techniques based on deep learning, and their applications in the bioinformatics challenges of protein structure prediction and drug discovery. The adoption of computational models based on artificial intelligence has the potential to transform the landscape of bioinformatics research, signifying a substantial departure from conventional approaches. The growing volume and complexity of data from biological systems drive the development of innovative models in bioinformatics. Consequently, the integration of AI methods in biology is already catalyzing significant progress in various research and computational areas, underscoring the importance of interdisciplinary collaboration.

Looking ahead, the future of AI in bioinformatics holds tremendous potential. Future work will likely focus on enhancing the interpretability and transparency of deep learning models, making them more accessible to biologists and clinicians.

Additionally, advancements in computational resources and new algorithms and model architectures will enable the handling of even larger and more complex datasets, further pushing the boundaries of what is possible in biological research. Emerging areas such as personalized medicine, synthetic biology, and systems biology stand to benefit immensely from these advancements. As AI continues to evolve, its applications in bioinformatics will undoubtedly lead to groundbreaking discoveries, fostering a deeper understanding of biological systems and accelerating the development of innovative solutions to some of the most pressing challenges in health and disease.

## REFERENCES


Anstine, D. M., & Isayev, O. (2023). Generative models as an emerging paradigm in the chemical sciences. *Journal of the American Chemical Society*, 145(16), 8736–8750. DOI: 10.1021/jacs.2c13467 PMID: 37052978

Auslander, N., Gussow, A. B., & Koonin, E. V. (2021). Incorporating machine learning into established bioinformatics frameworks. *International Journal of Molecular Sciences*, 22(6), 2903. DOI: 10.3390/ijms22062903 PMID: 33809353

Barroso, L. A., Holzle, U., & Ranganathan, P. (2019), The Datacenter as a computer. Designing warehouse-scale machines. https://link.springer.com/book/10.1007/978-3-031-01761-2

Bhukya, R., & Ashok, A. (2020). Gene expression prediction using deep neural networks. *The International Arab Journal of Information Technology*, 17(3), 422–431. DOI: 10.34028/iajit/17/3/16

Binding, D. B. (n.d.). Available at: https://www.bindingdb.org/bind/

Born, J., & Manica, M. (2021). Trends in deep learning for property-driven drug design. *Current Medicinal Chemistry*, 28(38), 7862–7886. DOI: 10.2174/0929867328666210729115728 PMID: 34325627

Brown, T. B., Mann, B., Ryder, N., Subbiah, M., Kaplan, J., Dhariwal, P., Neelakantan, A., Shyam, P., Sastry, G., Askell, A., Agarwal, S., Herbert-Voss, A., Krueger, G., Henighan, T., Child, R., Ramesh, A., Ziegler, D. M., Wu, J., Winter, C., . . . Amodei, D. (2020). Language models are few-shot learners, *NeurIPS 2020,* 33, 1877-1901. https://doi.org//arXiv.2005.14165.DOI: 10.48550

Bubeck, S., Chandrasekaran, V., Eldan, R., Gehrke, J., Horvitz, E., Kamar, E., Lee, P., Lee, Y. T., Li, Y., Lundberg, S., Nori, H., Palangi, H., Ribeiro, M. T., & Zhang, Y. (2023). Sparks of Artificial General Intelligence: Early experiments with GPT-4. *arXiv,* 2303.12712. https://doi.org//arXiv.2303.12712DOI: 10.48550

Caffe. (n.d.). Available at: https://caffe.berkeleyvision.org/

Callaway, E. (2023). AI tools are designing entirely new proteins that could transform medicine. *Nature*, 619(7969), 236–238. DOI: 10.1038/d41586-023-02227-y PMID: 37433935

Callaway, E. (2023a). After AlphaFold: Protein-folding contest seeks next big breakthrough. *Nature*, 613(7942), 13–14. DOI: 10.1038/d41586-022-04438-1 PMID: 36513827

Canatalay, P. J., & Osman, N. U. (2022). A bidirectional LSTM-RNN and GRU method to exon prediction using splice-site mapping. *Applied Sciences (Basel, Switzerland)*, 12(9), 4390. DOI: 10.3390/app12094390

Chen, H., Engkvist, O., Wang, Y., Olivecrona, M., & Blaschke, T. (2018). The rise of deep learning in drug discovery. *Drug Discovery Today*, 23(6), 1241–1250. DOI: 10.1016/j.drudis.2018.01.039 PMID: 29366762

Congressional Budget Office (CBO). (2021). Research and development in the pharmaceutical industry. https://www.cbo.gov/publication/57025

Corso, G., Stärk, H., Jing, B., Barzilay, R., & Jaakkola, T. S. (2023). DiffDock: diffusion steps, twists, and turns for molecular docking. *arXiv,* 2210.01776. https://doi.org//arXiv.2210.01776.DOI: 10.48550

CTD. (n.d.). Available at: https://ctdbase.org/downloads/

David, L., Thakkar, A., Mercado, R., & Engkvist, O. (2020). Molecular representations in AI-driven drug discovery: A review and practical guide. *Journal of Cheminformatics*, 12(1), 56. DOI: 10.1186/s13321-020-00460-5 PMID: 33431035

Dean, J., Corrado, G. S., Monga, R., Chen, K., Devin, M., Le, Q. V., & Mao, M. Z. (2012). Large-scale distributed deep networks. *NeurIPS 2012*, 25, 1223-1231. https://dl.acm.org/doi/10.5555/2999134.2999271

Defresne, M., Barbe, S., & Schiex, T. (2021). Protein design with deep learning. *International Journal of Molecular Sciences*, 22(21), 11741. DOI: 10.3390/ijms222111741 PMID: 34769173

DrugBank. (n.d.). Available at: https://go.drugbank.com/releases/latest

Fan, Y., Xiong, H., & Sun, G. (2023). DeepASDPred: A CNN-LSTM-based deep learning method for autism spectrum disorders risk RNA identification. *BMC Bioinformatics*, 24(1), 261. DOI: 10.1186/s12859-023-05378-x PMID: 37349705

Fassio, A. V., Shub, L., Ponzoni, L., McKinley, J., O'Meara, M. J., Ferreira, R. S., Keiser, M. J., & Minardi, R. C. M. (2022). Prioritizing virtual screening with interpretable interaction fingerprints. *Journal of Chemical Information and Modeling*, 62(18), 4300–4318. DOI: 10.1021/acs.jcim.2c00695 PMID: 36102784

Gao, W., Mahajan, S. P., Sulam, J., & Gray, J. J. (2020). Deep learning in protein structural modeling and design. *Patterns (New York, N.Y.)*, 1(9), 100142. DOI: 10.1016/j.patter.2020.100142 PMID: 33336200

Goodfellow, I., Bengio, Y., & Courville, A. (2016). Deep learning. MIT Press. http://www.deeplearningbook.org

Ho, J., Jain, A., & Abbeel, P. (2020). Denoising diffusion probabilistic models. *NeurIPS 2020,* 33, 6840. https://doi.org//arXiv.2006.11239DOI: 10.48550

Jiang, D., Hsieh, C.-Y., Wu, Z., Kang, Y., Wang, J., Wang, E., Liao, B., Shen, C., Xu, L., Wu, J., Cao, D., & Hou, T. (2021). InteractionGraphNet: A novel and efficient deep graph representation learning framework for accurate protein–ligand interaction predictions. *Journal of Medicinal Chemistry*, 64(24), 18209–18232. DOI: 10.1021/acs.jmedchem.1c01830 PMID: 34878785

John, G., Sahajpal, N. S., Mondal, A. K., Ananth, S., Williams, C., Chaubey, A., Rojiani, A. M., & Kolhe, R. (2021). Next-Generation sequencing (NGS) in COVID-19: A tool for SARS-CoV-2 diagnosis, monitoring new strains and phylodynamic modeling in molecular epidemiology. *Current Issues in Molecular Biology*, 43(2), 845–867. DOI: 10.3390/cimb43020061 PMID: 34449545

Jumper, J., Evans, R., Pritzel, A., Green, T., Figurnov, M., Ronneberger, O., Tunyasuvunakool, K., Bates, R., Žídek, A., Potapenko, A., Bridgland, A., Meyer, C., Kohl, S. A. A., Ballard, A. J., Cowie, A., Romera-Parades, B., Nikolov, S., Jain, R., Adler, J., & Back, T. (2021). Highly accurate protein structure prediction with AlphaFold. *Nature*, 596(7873), 583–589. DOI: 10.1038/s41586-021-03819-2 PMID: 34265844

Karim, M. R., Islam, T., Shajalal, M., Beyan, O., Lange, C., Cochez, M., Rebholz-Schuhmann, D., & Decker, S. (2023). Explainable AI for Bioinformatics: Methods, Tools and Applications. *Briefings in Bioinformatics*, 24(5), bbad236. Advance online publication. DOI: 10.1093/bib/bbad236 PMID: 37478371

Keras. (n.d.). Available at: https://keras.io/

Kim, J., Park, S., Min, D., & Kim, W. (2021). A comprehensive survey of recent drug Discovery using deep learning. *International Journal of Molecular Sciences*, 22(18), 9983. DOI: 10.3390/ijms22189983 PMID: 34576146

Kloczkowski, A., Jernigan, R., Wu, Z., Song, G., Yang, L., Kolinski, A., & Pokarowski, P. (2009). Distance matrix-based approach to protein structure prediction. *Journal of Structural and Functional Genomics*, 10(1), 67–81. DOI: 10.1007/s10969-009-9062-2 PMID: 19224393

Krishna, R., Wang, J., Ahern, W., Sturmfels, P., Venkatesh, P., Kalvet, I., Lee, G. R., Morey-Burrows, F. S., Anishchenko, I., Humphreys, I. R., McHugh, R., Vafeados, D., Li, X., Sutherland, G. A., Hitchcock, A., Hunter, C. N., Kang, A., Brackenbrough, E., Bera, A. K., & Baker, D. (2024). Generalized biomolecular modeling and design with RoseTTAFold All-Atom. *Science*, 384(6693), eadl2528. Advance online publication. DOI: 10.1126/science.adl2528 PMID: 38452047

Lander, E. S., Linton, L. M., Birren, B., Nusbaum, C., Zody, M. C., Baldwin, J., Devon, K., Dewar, K., Doyle, M., FitzHugh, W., Funke, R., Gage, D., Harris, K., Heaford, A., Howland, J., Kann, L., Lehoczky, J., LeVine, R., McEwan, P., & Morgan, M. J. (2001). Initial sequencing and analysis of the human genome. *Nature*, 409(6822), 860–921. DOI: 10.1038/35057062 PMID: 11237011

LeCun, Y., Bengio, Y., & Hinton, G. (2015). Deep learning. *Nature*, 521(7553), 436–444. DOI: 10.1038/nature14539 PMID: 26017442

Levinthal, C. (1968). Are there pathways for protein folding? *Journal de Chimie Physique*, 65, 44–45. DOI: 10.1051/jcp/1968650044

Levinthal, C. (1969). How to fold graciously. In Mossbauer Spectroscopy in Biological Systems: Proceedings of a Meeting Held at Allerton House, Monticello, Illinois (pp. 22-24). Retrieved from https://web.archive.org/web/20110523080407/http://www-miller.ch.cam.ac.uk/levinthal/levinthal.html

Lin, D., Cho, Y., Fan, H., & Zhang, Y. (2024). From explainable to interactive AI: A literature review on current trends and future directions. *Environmental Science and Ecotechnology*, 13, 100248. DOI: 10.1016/j.dsp.2024.01.001

Lin, Z., Akin, H., Rao, R., Hie, B., Zhu, Z., Lu, W., Smetanin, N., Verkuil, R., Shmueli, Y., Costa, A. S., Fazel-Zarandi, M., Sercu, T., Candido, S., & Rives, A. (2023). Evolutionary-scale prediction of atomic-level protein structure with a language model. *Science*, 379(6637), 1123–1130. DOI: 10.1126/science.ade2574 PMID: 36927031

Lipton, Z. C. (2016). The mythos of model interpretability. (*WHI 2016*) *arXiv*, 1606.03490 DOI: 10.48550/arXiv.1606.03490

Liu, J., Guo, Z., Wu, T., Roy, R. S., Chen, C., & Cheng, J. (2023). Improving AlphaFold2-based protein tertiary structure prediction with MULTICOM in CASP15. *Communications Chemistry*, 6(1), 188. DOI: 10.1038/s42004-023-00991-6 PMID: 37679431

Liu, M., Li, C., Chen, R., Cao, D., & Zeng, X. (2024). Geometric deep learning for drug discovery. *Expert Systems with Applications*, 240, 122498. DOI: 10.1016/j.eswa.2023.122498

Lo, Y.-C., Rensi, S. E., Torng, W., & Altman, R. B. (2018). Machine learning in chemoinformatics and drug discovery. *Drug Discovery Today*, 23(8), 1538–1546. DOI: 10.1016/j.drudis.2018.05.010 PMID: 29750902

Lu, W., Wu, Q., Zhang, J., Rao, J., Li, C., & Zheng, S. (2022). TANKBind: trigonometry-aware neural networKs for drug-protein binding structure prediction. *bioRxiv,* 2022.06.06.495043 DOI: 10.1101/2022.06.06.495043

McNutt, A. T., Francoeur, P., Aggarwal, R., Masuda, T., Meli, R., Ragoza, M., Sunseri, J., & Koes, D. R. (2021). GNINA 1.0: Molecular docking with deep learning. *Journal of Cheminformatics*, 13(1), 43. DOI: 10.1186/s13321-021-00522-2 PMID: 34108002

Min, S., Lee, B., & Yoon, S. (2017). Deep learning in bioinformatics. *Briefings in Bioinformatics*, 18(5), 851–869. DOI: 10.1093/bib/bbw068 PMID: 27473064

Montavon, G., Samek, W., & Müller, K. R. (2018). Methods for interpreting and understanding deep neural networks. *Digital Signal Processing*, 73, 1–15. DOI: 10.1016/j.dsp.2017.10.011

Ozden, B., Kryshtafovych, A., & Karaca, E. (2023). The impact of AI-based modeling on the accuracy of protein assembly prediction: Insights from CASP15. *Proteins*, 91(12), 1636–1657. https://doi.org/doi. DOI: 10.1002/prot.26598 PMID: 37861057

Pharm, G. K. B. (n.d.). Available at: https://www.pharmgkb.org/downloads

Pharmaceutical Research and Manufacturers of America. (2020). PhRMA annual membership survey. https://phrma.org/research-and-development/2020-phrma-annual-membership-survey

PyTorch. Available at: https://pytorch.org/

Ragoza, M., Hochuli, J., Idrobo, E., Sunseri, J., & Koes, D. R. (2017). Protein-ligand scoring with convolutional neural networks. *Journal of Chemical Information and Modeling*, 57(4), 942–957. DOI: 10.1021/acs.jcim.6b00740 PMID: 28368587

Santos, L. M. dos, Mariano, D., & Melo-Minardi, R. C. de. (2023). O impacto da inteligência artificial nas ciências da vida através da bioinformática. *Revista da UFMG*, 30. https://doi.org/DOI: 10.35699/2965-6931.2023.47996

Sapoval, N., Aghazadeh, A., Nute, M. G., Antunes, D. A., Balaji, A., Baraniuk, R., Barberan, C. J., Dannenfelser, R., Dun, C., Edrisi, M., Elworth, R. A. L., Kille, B., Kyrillidis, A., Nakhleh, L., Wolfe, C. R., Yan, Z., Yao, V., & Treangen, T. J. (2022). Current progress and open challenges for applying deep learning across the biosciences. *Nature Communications*, 13(1), 1728. DOI: 10.1038/s41467-022-29268-7 PMID: 35365602

Senior, A. W., Evans, R., Jumper, J., Kirkpatrick, J., Sifre, L., Green, T., Qin, C., Žídek, A., Nelson, A. W. R., Bridgland, A., Penedones, H., Petersen, S., Simonyan, K., Crossan, S., Kohli, P., Jones, D. T., Silver, D., Kavukcuoglu, K., & Hassabis, D. (2020). Improved protein structure prediction using potentials from deep learning. *Nature*, 577(7792), 706–710. DOI: 10.1038/s41586-019-1923-7 PMID: 31942072

Shen, C., Zhang, X., Hsieh, C.-Y., Deng, Y., Wang, D., Xu, L., Wu, J., Li, D., Kang, Y., Hou, T., & Pan, P. (2023). A generalized protein–ligand scoring framework with balanced scoring, docking, ranking and screening powers. *Chemical Science (Cambridge)*, 14(30), 8129–8146. DOI: 10.1039/D3SC02044D PMID: 37538816

Shen, X., Jiang, C., Wen, Y., Li, C., & Lu, Q. (2022). A brief review on deep learning applications in genomic studies. *Frontiers in Systems Biology*, 2, 877717. DOI: 10.3389/fsysb.2022.877717

Shrestha, A., & Mahmood, A. (2019). Review of deep learning algorithms and architectures. *IEEE Access : Practical Innovations, Open Solutions*, 7, 53040–53065. DOI: 10.1109/ACCESS.2019.2912200

Singh, J., & Banerjee, R. (2019). A study on single and multi-layer perceptron neural network. *3rd International Conference on Computing Methodologies and Communication (ICCMC)*, 35–40. DOI: 10.1109/ICCMC.2019.8819775

Sohl-Dickstein, J., Weiss, E., Maheswaranathan, N., & Ganguli, S. (2015). Deep unsupervised learning using nonequilibrium thermodynamics. In F. Bach & D. Blei (Eds.), *Proceedings of the 32nd International Conference on Machine Learning*. Lille, France: PMLR. Available at: https://proceedings.mlr.press/v37/sohl-dickstein15.html

Srivastava, A., Mahmood, A., & Srivastava, R. (2017). A comparative analysis of SVM random forest methods for protein function prediction. *2017 International Conference on Current Trends in Computer, Electrical, Electronics and Communication (CTCEEC)*, 1008-1010. DOI: 10.1109/CTCEEC.2017.8455066

Stärk, H., Ganea, O., Pattanaik, L., Barzilay, D. R., & Jaakkola, T. (2022) EquiBind: Geometric deep learning for drug binding structure prediction. *arXiv,* 2202.05146. https://doi.org//arXiv.2202.05146DOI: 10.48550

Sun, T., Zhou, B., Lai, L., & Pei, J. (2017). Sequence-based prediction of protein-protein interaction using a deep-learning algorithm. *BMC Bioinformatics*, 18(1), 277. Advance online publication. DOI: 10.1186/s12859-017-1700-2 PMID: 28545462

TensorFlow. (n.d.). Available at: https://www.tensorflow.org/

Torrisi, M., Pollastri, G., & Le, Q. (2020). Deep learning methods in protein structure prediction. *Computational and Structural Biotechnology Journal*, 18, 1301–1310. DOI: 10.1016/j.csbj.2019.12.011 PMID: 32612753

Tropsha, A., Isayev, O., Varnek, A., Schneider, G., & Cherkasov, A. (2024). Integrating QSAR modelling and deep learning in drug discovery: The emergence of deep QSAR. *Nature Reviews. Drug Discovery*, 23(2), 141–155. DOI: 10.1038/s41573-023-00832-0 PMID: 38066301

Vaswani, A., Shazeer, N., Parmar, N., Uszkoreit, J., Jones, L., Gomez, A. N., Kaiser, Ł., & Polosukhin, I. (2017). Attention is all you need. *NeurIPS 2017, 30*. https://doi.org//arXiv.1706.03762.DOI: 10.48550

Watson, J. L., Juergens, D., Bennett, N. R., Trippe, B. L., Yim, J., Eisenach, H. E., Ahern, W., Borst, A. J., Ragotte, R. J., Milles, L. F., Wicky, B. I. M., Hanikel, N., Pellock, S. J., Courbet, A., Sheffler, W., Wang, J., Venkatesh, P., Sappington, I., Torres, S. V., & Baker, D. (2023). De novo design of protein structure and function with RFdiffusion. *Nature*, 620(7976), 1089–1100. DOI: 10.1038/s41586-023-06415-8 PMID: 37433327

Weininger, D. (1988). SMILES, a chemical language and information system. 1. Introduction to methodology and encoding rules. *Journal of Chemical Information and Computer Sciences*, 28(1), 31–36. DOI: 10.1021/ci00057a005

Xia, X. (2017). Bioinformatics and drugs discovery. *Current Topics in Medicinal Chemistry*, 17(15), 1709–1726. DOI: 10.2174/1568026617666161116143440 PMID: 27848897

Xu, C., Liu, R., Huang, S., Li, W., Li, Z., & Luo, H.-B. (2023). 3D-SMGE: A pipeline for scaffold-based molecular generation and evaluation. *Briefings in Bioinformatics*, 24(6), 1–11. DOI: 10.1093/bib/bbad327 PMID: 37756591

Yang, L., Zhang, Z., Hong, S., Xu, R., Zhao, Y., Shao, Y., Zhang, W., Yang, M., & Cui, B. (2022). Diffusion models: A comprehensive survey of methods and applications. *ACM Computing Surveys*, 56, 1–39. DOI: 10.1145/3505243

Zaki, M. J., & Meira, W.Jr. (2020). *Data mining and machine learning: fundamental concepts and algorithms* (2nd ed.). Cambridge University Press. DOI: 10.1017/9781108564175

Zeng, Z., Shi, H., Wu, Y., & Hong, Z. (2015). Survey of natural language processing techniques in bioinformatics. *Computational and Mathematical Methods in Medicine*, 674296, 1–10. Advance online publication. DOI: 10.1155/2015/674296 PMID: 26525745

Zhang, S., Fan, R., Liu, Y., Chen, S., Liu, Q., & Zeng, W. (2023). Applications of transformer-based language models in bioinformatics: A survey. *Bioinformatics Advances*, 3(1), vbad001. Advance online publication. DOI: 10.1093/bioadv/vbad001 PMID: 36845200

Zhang, X., Shen, C., Zhang, H., Kang, Y., Hsieh, C.-Y., & Hou, T. (2024). Advancing Ligand Docking through Deep Learning: Challenges and Prospects in Virtual Screening. *Accounts of Chemical Research*, 57(10), 1500–1509. DOI: 10.1021/acs.accounts.4c00093 PMID: 38577892

Zhao, Y., Zheng, K., Guan, B., Guo, M., Song, L., Gao, J., Qu, H., Wang, Y., Shi, D., & Zhang, Y. (2020). DLDTI: A learning-based framework for drug-target interaction identification using neural networks and network representation. *Journal of Translational Medicine*, 18(1), 434. DOI: 10.1186/s12967-020-02602-7 PMID: 33187537

Zhou, X., Li, Y., Liu, S., Yang, Q., Su, X., Zhou, L., Tang, M., Fu, R., Li, J., & Huang, Q. (2013). Ultra-deep sequencing enables high-fidelity recovery of biodiversity for bulk arthropod samples without PCR amplification. *GigaScience*, 2(1), 4. DOI: 10.1186/2047-217X-2-4 PMID: 23587339

# Chapter 6
# Structural Bioinformatics

**Leandro Morais de Oliveira**
https://orcid.org/0009-0006-9748-4049
*Universidade Federal de Minas Gerais, Brazil*

**Luana Luiza Bastos**
https://orcid.org/0000-0002-6932-0438
*Universidade Federal de Minas Gerais, Brazil*

**Vivian Morais Paixão**
https://orcid.org/0000-0002-4154-304X
*Universidade Federal de Minas Gerais, Brazil*

**Leticia Aparecida Gontijo**
*Universidade Federal de Minas Gerais, Brazil*

**Tatiane Senna Bialves**
https://orcid.org/0000-0002-3827-2859
*Universidade Federal de Minas Gerais, Brazil*

**Diego Mariano**
https://orcid.org/0000-0002-5899-2052
*Universidade Federal de Minas Gerais, Brazil*

**Raquel Cardoso de Melo Minardi**
*Universidade Federal de Minas Gerais, Brazil*

## ABSTRACT

*Within this chapter, the authors offer a concise introduction to the concept of structural bioinformatics. In addition to proteins, DNA, and RNA, peptides are another type of biomolecule covered in this chapter. Peptides are short chains of amino acids (generally, 2-50) connected by peptide bonds. Due to their versatility, they play essential roles in several biological processes in living beings, acting as hormones, enzymes, antibiotics, structural support, etc. Additionally, the chapter will address in detail the interaction between RNA, peptides, and protein, the application of structural bioinformatics tools for studying these complexes, and their implications in research related to interactions, the development of new drugs, future prospects,*

DOI: 10.4018/979-8-3693-3192-7.ch006

*and challenges related to the topic.*

## INTRODUCTION TO STRUCTURAL BIOINFORMATICS

According to Bayat et al. (2002), bioinformatics is the application of computational tools in the analysis and interpretation of biological data, including DNA, RNA, and protein sequences. Thus, it is an interdisciplinary field that integrates computer science, mathematics, physics, and biology (biochemistry, molecular biology, genetics, among others). Within bioinformatics, there is a field that focuses on the analysis and modeling of the three-dimensional structures of biomolecules, such as proteins, nucleic acids, and protein-RNA complexes. This area is called Structural Bioinformatics.

The first studies in Structural Bioinformatics focused on determining protein structures, which are essential molecules for life. The first experimentally resolved three-dimensional structure deposited in the database was myoglobin (Figure 1, PDB ID: 1MBN) from a sperm whale (*Physeter catodon*). Deposited on April 5, 1973, it was generated through X-ray diffraction, a technique widely used in the structural area that involves passing a beam of X-rays through the crystallized molecule, allowing the determination of the location of its atoms. Since then, thousands of other structures have become available.

*Figure 1. Cartoon visualization of a Myoglobin, the first experimentally resolved three-dimensional structure. PDB ID: 1MBN*

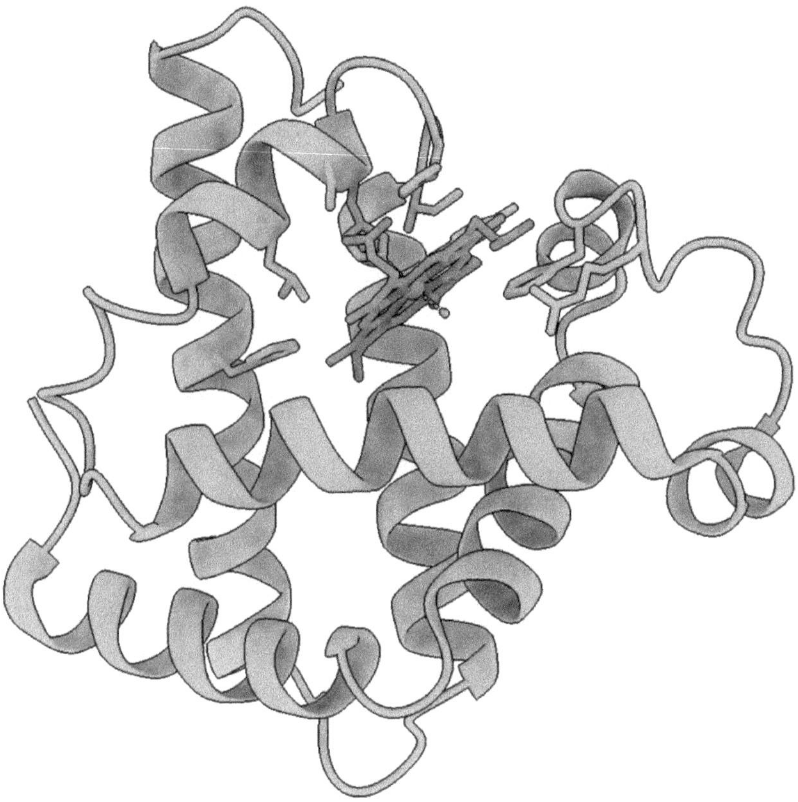

With the increasing use of DNA sequencing technology, there has been an explosive growth in biological data, making it essential to develop advanced computational methods to analyze and interpret this data, which was accumulated in various databases (Kanehisa & Bork, 2003). As Gu and Bourne (2021) observed, genetic analyses are limited to identifying the relationship between genetic sequences and their functional implications. On the other hand, structural analysis goes further, providing not only a deeper understanding of the mechanisms underlying these implications but also offering insights into how biological function is intrinsically linked to molecular structure.

Thus, Structural Bioinformatics has emerged as an important solution to this need, comprising algorithms, molecular modeling, and computational simulations to predict and analyze molecular structures. Thanks to these techniques, it is possible to predict and model the three-dimensional structure of molecules from their primary sequences, allowing a detailed understanding of their function and interactions with other molecules. Additionally, they allow the study and understanding of the molecular mechanisms underlying biological processes, such as protein-RNA recognition and ligand binding to proteins.

Some of the most commonly used experimental methods in the study of the three-dimensional structure of molecules include X-ray crystallography, nuclear magnetic resonance (NMR), and transmission electron microscopy (TEM), widely employed to determine biomolecular structures with high resolution and reliability. With the numerous data resulting from these experiments, the Protein Data Bank (PDB; http://pdb.org; Berman et al., 2000) was created, a widely used database for molecule studies, currently containing 217,705 experimentally resolved structures. From this data, computational techniques such as modeling, docking, and molecular dynamics simulation have emerged, allowing the study and prediction of structures, predicting interactions between different molecules, and studying their dynamics at the atomic scale.

Justifying its great importance, the areas of application of structural bioinformatics are considered widely comprehensive. It is widely used in the discovery of new drugs, allowing the rational design of therapeutic molecules based on the structure of specific molecular targets, for example, in the creation of HIV protease inhibitors based on their crystallized three-dimensional structure (Erickson & Kempf, 1994). In terms of clinical applications, there is a fundamental role in personalized medicine and the development of targeted therapies. For example, it can be used to predict the efficacy of drugs based on the target structure and identify genetic variants associated with diseases. Additionally, during the COVID-19 pandemic, it was an important area for the study of new variants of the SARS-CoV-2 virus, evaluating their structural changes (Durmaz et al., 2022; Yang et al., 2021). There is also the application in protein engineering for various purposes, including industrial and agricultural applications.

The advances in Structural Bioinformatics are remarkable but not without challenges and limitations. For example, despite successes in resolving molecular structures, accuracy can be limited due to the high flexibility and dynamics of some molecules (Gu & Bourne, 2011). Additionally, while protein structures tend to be more conserved than their sequences, most of the available data refers to sequences, which can undergo mutations without affecting their structure. This discrepancy can hinder the correlation between both. Another difficulty arises from the fact that docking and molecular dynamics study methods use algorithms that employ a series of mathematical functions with approximations that do not precisely consider all thermodynamic elements of binding energy, representing only an approximation of the real free binding energy (Pinzi & Rastelli, 2019). These challenges highlight the need for continuous improvement and development of Structural Bioinformatics.

Understanding the structural complexities of biomolecules, such as proteins, nucleic acids, and their complexes, is crucial for future advances in medicine, biotechnology, and many other areas. As new technologies and approaches are developed to overcome these challenges, Structural Bioinformatics will continue to play a fundamental role in the quest for deeper insights into the relationship between molecular structure and biological function.

## Structures of Proteins and Peptides

Proteins are macromolecules that play a fundamental role in a variety of biological processes, ranging from the catalysis of chemical reactions to the structural support of cells. Their functional diversity is closely linked to their three-dimensional structure, which can be described at four hierarchical levels. Understanding protein structure is essential for comprehending the mechanisms of action and biological function of these macromolecules, and an in-depth understanding of protein structures is fundamental for the further development of medical therapies and biotechnological applications (NELSON, 2008).

### Amino Acids and Peptide Chains

Proteins are polymers of amino acids linked together by covalent bonds, called peptide bonds. The general structure of an amino acid (Figure 2) features an amino group (-NH2) and a carboxyl group (-COOH), both attached to the alpha carbon, to which a hydrogen atom and a variable group called a side chain or R group are also attached. The R group can have different structures and differentiate the amino acids from each other. Amino acids line up in different quantities to form a simple peptide structure that can contain a few amino acids or a structure or thousands, forming a protein with a high degree of complexity and functionality.

*Figure 2. Amino acid structure. Amino group (-NH2), Side chain (R) and Carboxyl group (-COOH) (Wikimedia Commons (CC-BY 4.0))*

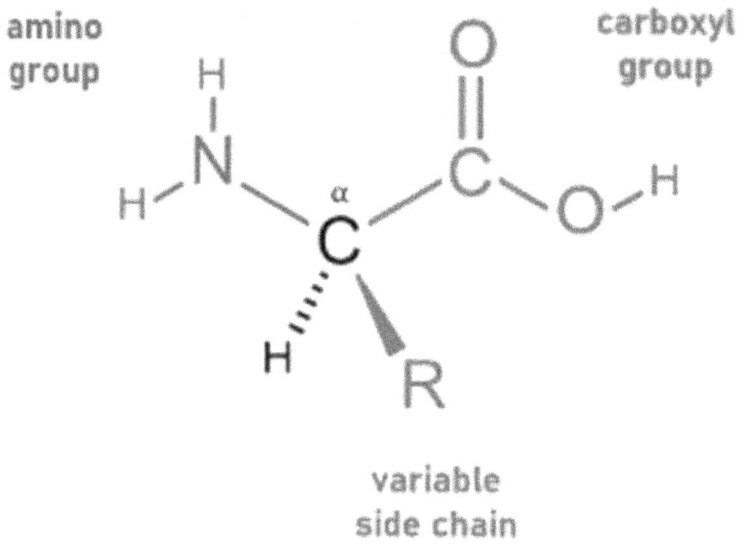


## Hierarchical Levels of Protein Structure

The complexity of protein 3D-structures is generally described at four hierarchical levels, namely primary, secondary, tertiary, and quaternary structure (Figure 3). In the synthesis of a peptide or protein, the final structure of this molecule, whatever it may be, goes through the hierarchical ladder of the structure, with no organization of a secondary structure without first having the primary structure, or the tertiary structure without first forming the secondary structure and so on. A linear sequence of amino acids constitutes the primary structure of the protein. The final three-dimensional structure and functionality of a protein is largely determined by its primary structure. The primary structure of a protein is the sequence of amino acids in the protein. The sequence of amino acids in a protein is determined by the DNA of the organism that produces it.

*Figure 3. Illustrative image of protein structure hierarchy (Wikimedia commons)*

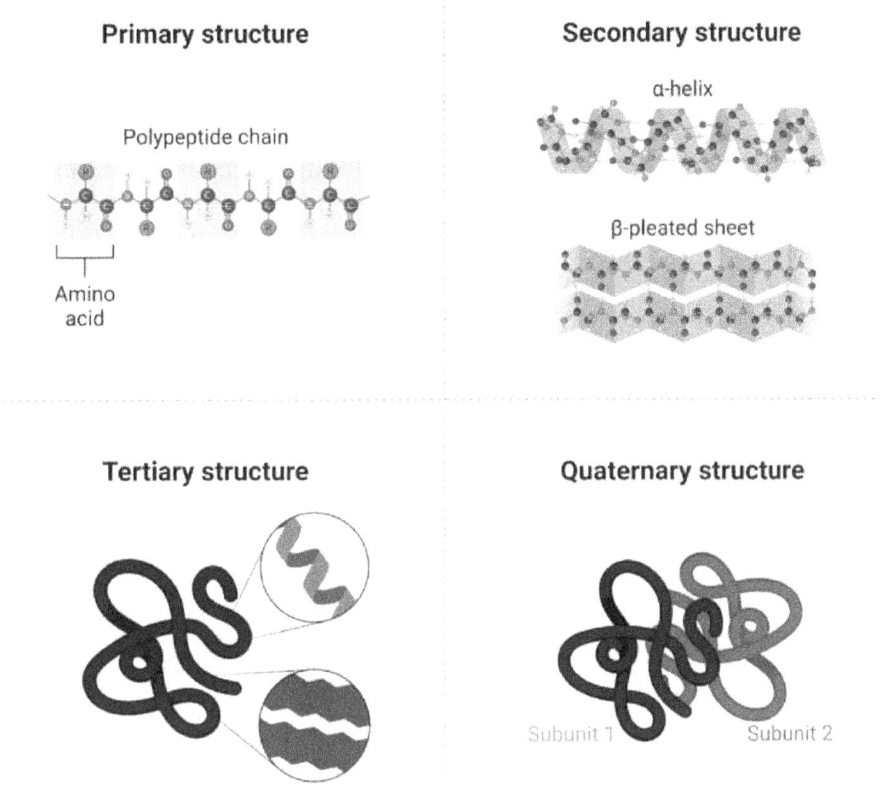


The secondary structure arises from the interaction of amino acids present in the primary structure, giving rise to known structural patterns, called alpha helices and beta sheets. The alpha helix structure is characterized by a helical structure, in which the polypeptide chain forms around an imaginary axis. The bond between the amino acids in the alpha helix is formed by hydrogen bonds between the amino group of one amino acid and the carboxyl group of the amino acid four residues ahead in the chain. This configuration allows the alpha helix to be stable and resistant to changes such as temperature and pH variations.

The tertiary structure arises from the three-dimensional rearrangement of the structure through the interaction between amino acids by hydrogen bonds, hydrophobic interactions, disulfide bridges, and electrostatic interactions. These interactions influence the spatial organization of the protein. The tertiary structure may present specific regions known as domains, which may be involved in functions such as enzyme active sites, substrate binding sites, or interaction sites with other molecules. The complex tertiary protein structure resulting from the interactions between different peptide regions can result in a globular or fibrous shape. Globular proteins are characterized by their spherical or globular shape, resulting from a compact folding, while fibrous proteins are characterized by their elongated and fibrillar structure, which results in an elongated and linear shape.

The quaternary structure of proteins is formed by the association of multiple identical or different individual peptide subunits forming a functional complex. The formation of the quaternary structure is highly specific and can present a variety of three-dimensional arrangements, including linear oligomers, rings, helices, and three-dimensional networks. The quaternary structure is often associated with the regulation of enzymatic activity, structural stability, modulation of binding affinity, and protein-protein interactions. These protein complexes are crucial for cellular homeostasis and play fundamental roles in physiological and pathological processes.

*Figure 4. Cartoon visualization of a protein (left). Surface visualization of a protein (right). PDB ID: 1EVE*

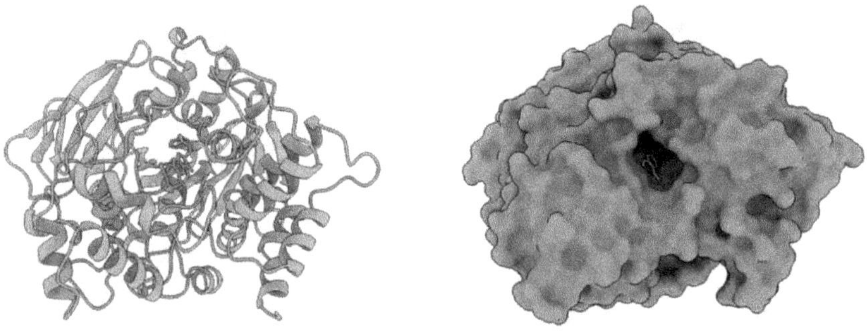

## DNA and RNA Structures

DNA (deoxyribonucleic acid) is a double-stranded helical biomolecule that stores the genetic instructions necessary for the development of living beings and some viruses. DNA is vital for the storage and transmission of genetic information. Its structure is formed by nucleotides, which are composed of a phosphate group linked to a sugar molecule (deoxyribose) and a nitrogenous base that can be adenine (A), cytosine (C), guanine (G) or thymine (T). In the case of the nucleotides in DNA, the sugar is deoxyribose attached to a single phosphate group (hence the name deoxyribonucleic acid). The two strands of DNA are held together by hydrogen bonds between pairs of complementary bases: pairs A with T and pairs C with G. The nucleotides are covalently linked in the chain through sugars and phosphates, which thus form a "backbone" of alternating sugar-phosphate-sugar-phosphate (Figure 5).

*Figure 5. Structure and composition of DNA. Nucleotide: composed of a phosphate group linked to a sugar molecule (Deoxyribonucleic Acid) and a nitrogenous base that can be adenine (A), cytosine (C), guanine (G) or thymine (T) (Created using BioRender)*

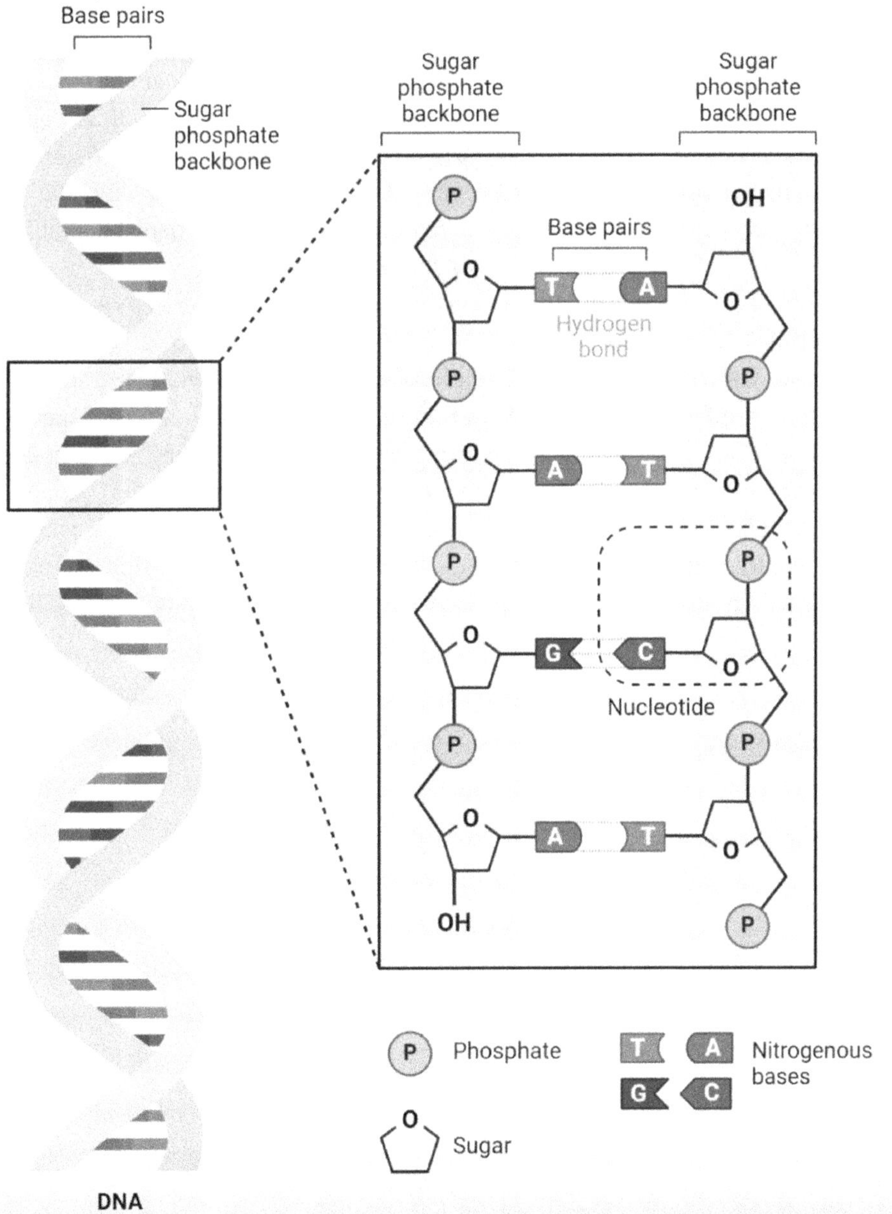

Although DNA has a relatively simple composition, only four nitrogenous bases, billions of combinations can be made along the extensive length of the molecule, and small differences in the combinations of these bases give individuals unique characteristics. The determination of the composition and structure of DNA were essential for its definition as a molecule that stores genetic information (Zaha et al., 2014).

Since DNA is a long molecule, it needed to be organized compactly in the form of chromosomes so that it could be stored in the cell nucleus. And therefore, whenever a cell needs to enter the process of cell division, it synthesizes copies of its chromosomes. And evolutionarily, organisms have developed mechanisms that keep the genetic code safe. In humans, for example, before cell division, all 23 pairs of chromosomes need to be replicated to form 46 pairs, so that after cell division each daughter cell has a full complement (23 pairs) of chromosomes. After understanding the importance of DNA for preserving genetic information, mechanisms that keep the genetic code safe were observed (Minchin & Lodge, 2019).

The flow of genetic information occurs unidirectionally, where DNA will transcribe part of its genetic code into mRNA (gene expression), we call this process transcription. Unlike what occurs in duplication, only one of the DNA strands is transcribed into mRNA (Junqueira & Carneiro, 2023; Zaha et al., 2014). Due to its single-stranded structure, the mRNA leaves the cell nucleus carrying the "message" for the expression of a specific protein in its nucleotide sequence. In the cytoplasm, the mRNA meets the ribosomes that read it and, together with the tRNA, translate its message into a sequence of amino acids. And finally, after translation, the amino acid sequence organizes itself, giving rise to a protein (Figure 6) (Crick, 1958).

*Figure 6. Gene expression. Gene expression begins by choosing and transcribing a specific gene into an mRNA in the cell nucleus. The mRNA leaves the nucleus carrying the message for the production of a protein in its nucleotides. In the cytoplasm, the ribosome subunits find the mRNA and, with the help of tRNA, the process of translating codons into amino acids begins. The translation process is composed of three parts: Initialization (meeting of Ribosome subunits), elongation of the amino acid chain (growth of amino acids and formation of peptide bonds between translated residues) and termination of translation, when the mRNA termination codon sequence. After the end of translation, the ribosome subunits turn off, and then the chain of amino acids is released, which will undergo a folding process and form the protein*

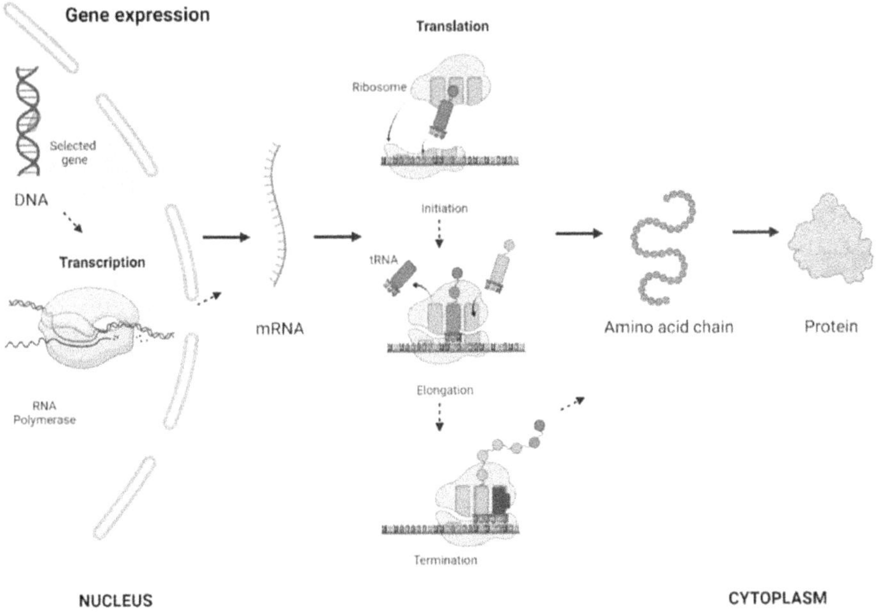


RNA, or ribonucleic acid, is a type of nucleic acid essential for protein synthesis and gene regulation in cells. It has several properties that make it essential for biological processes. Like DNA, RNA is made up of chains of nucleotides, but these are arranged in a single strand and have a different characteristic in relation to nitrogenous bases: it has the base uracil (U) instead of thymine (T). RNA is an essential molecule for life, serving a fundamental role in biological processes (Junqueira & Carneiro, 2023; Zaha et al., 2014).

RNA takes on diverse forms, including messenger RNA (mRNA), transfer RNA (tRNA), and ribosomal RNA (rRNA) (Figure 7). Their versatility is reflected in various regulatory functions, including gene expression regulation, chromatin organization, and modulation of cellular processes (Junqueira & Carneiro, 2023; Zaha et al., 2014).

*Figure 7. RNA types and their structures (Created using BioRender)*

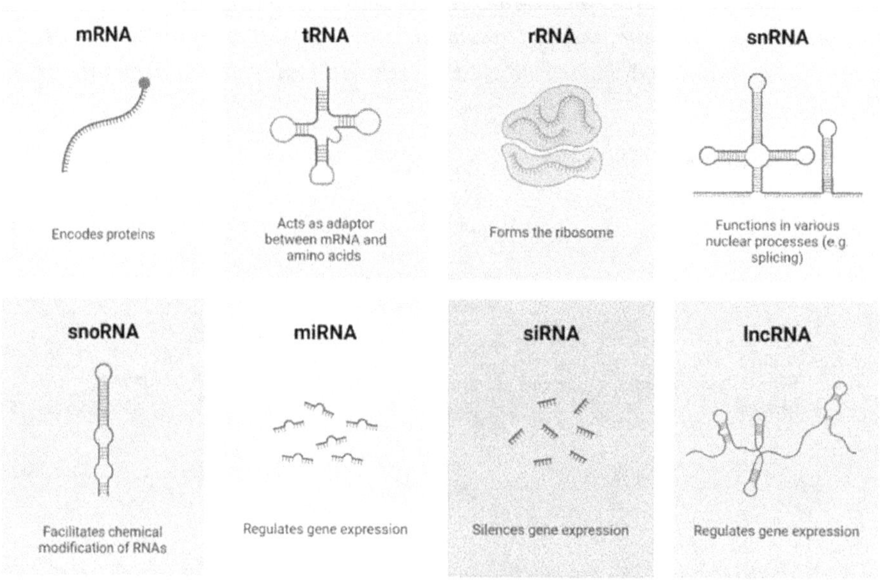


The interaction between RNA and protein is a crucial aspect of this molecular dynamics, with proteins playing diverse roles in stabilizing, modifying, and transporting RNA. This intricate interaction is vital for gene regulation, response to cellular stimuli, and other essential cellular processes. The versatility and complexity of RNA in its interaction with proteins make it a central piece in understanding biological mechanisms and in the search for innovative therapies. Structural bioinformatics can be used to study protein-RNA interactions in various ways, utilizing or creating computational tools. Among these, one can identify binding sites by analyzing RNA sequences and identifying those compatible with protein binding. Another way to analyze the complex is through predicting possible interactions and studying the three-dimensional structure of proteins and RNA. All these applications in the field of bioinformatics are important for research, disease diagnosis, and drug development related to RNA-protein interactions(Junqueira & Carneiro, 2023; Zaha et al., 2014).

Although its structure is very established in the literature, it is important to understand how other molecules interact with it and how this affects various biological processes.

## Inter- and Intra-Molecular Interactions

One of the most important topics in contemporary science is understanding the role of intra- and intermolecular interactions and their influence on the self-assembly of molecules, as well as their various characteristics observed at the macroscopic level. Intra- and intermolecular interactions play fundamental roles in the structure, function, and behavior of biological macromolecules (Wang & Farhana, 2024; Zaha et al., 2014).

Intramolecular interactions are responsible for the three-dimensional structure of biological macromolecules, such as proteins, nucleic acids, and polysaccharides. For example, hydrogen bonds between functional groups influence the conformation of DNA helices. Intermolecular interactions, such as van der Waals forces and hydrogen bonds, can stabilize the three-dimensional structure of biological macromolecules in solution, helping to maintain their form and function (*Biologia molecular da célula*, 2021; Derkach et al., 2022; Gu & Bourne, 2011; Palumbo et al., 2021; Samuel et al., 2023).

Additionally, intermolecular interactions influence many physical and chemical properties of biological macromolecules, such as melting and boiling points, solubility in different solvents, and the ability to form aggregates or organized structures. In summary, intra- and intermolecular interactions are essential for the structure, function, and behavior of biological macromolecules, playing fundamental roles in molecular and cellular biology (Nelson et al., 2014; Sanjeev et al., 2019).

Intramolecular bonds are essential for the cohesion of atoms within a specific molecule, directly influencing its chemical and physical properties, such as stability, reactivity, and molecular geometry. They encompass ionic, covalent, and metallic bonds, varying depending on the nature of the participating atoms. On the other hand, Intermolecular Forces refer to interactions between distinct molecules, arising from electrostatic attractions between partially opposite charges, permanent or momentary dipoles, and induced-dipole-induced-dipole interactions (known as London or dispersion forces) (Bruno César Feltes, 2015; Junqueira & Carneiro, 2023; Zaha et al., 2014).

These forces determine macroscopic properties in liquids and solids, such as melting point, boiling point, viscosity, and solubility. The difference in energy between Intramolecular and Intermolecular Forces is significant, with the former generally being stronger. Therefore, more energy is required to break bonds within a molecule than to separate molecules from one another. For example, the boiling point of water (due to hydrogen bonding) is higher than that of methane (influenced by London forces). These discrepancies between Intramolecular and Intermolecular Forces are crucial for understanding various chemical and physical phenomena observed in molecular systems (Munshi & Guru Row, 2006; Palumbo et al., 2021; Uversky & Finkelstein, 2019).

## Intramolecular Interactions

Intramolecular interactions refer to the forces that occur within the same molecule, playing a crucial role in determining its structure and properties. These interactions include covalent bonds, which are the strongest and involve atoms sharing electron pairs to form stable bonds, such as single, double, and triple bonds. Ionic bonds, although typically considered intermolecular, can also occur within molecules, involving complete electron transfer between atoms and resulting in the formation of positive and negative ions that attract each other electrically (Jabłoński, 2021; Palumbo et al., 2021; Uversky & Finkelstein, 2019). Hydrogen bonds, often associated with intermolecular interactions, can form within a molecule when a hydrogen atom bonded to an electronegative atom forms a bond with another electronegative atom in the same molecule (Jabłoński, 2021; Jarvis, 2023). Similarly, Van der Waals bonds, while primarily intermolecular, contribute to molecular conformation and stability in large molecules through dipole-dipole interactions and London dispersion within the molecule. These intramolecular interactions significantly influence the physical and chemical properties of substances, as well as their reactivities and binding capacities with other molecules (Jabłoński, 2021; Jarvis, 2023).

These interactions directly influence the formation of different protein structures. In secondary structure, for example, hydrogen bonds stabilize the α-helix, where each amino acid discharge forms hydrogen bonds with residues located a few turns forward or backward in the polypeptide chain, maintaining a helix in its helical form (Figure 3). Furthermore, intramolecular interactions between amino acid residues in different segments of the polypeptide chain form β-sheets, contributing to the stabilization of the planar structure of β-sheets. In tertiary structure, intramolecular interactions, including protective bonds, hydrophobic interactions, electrostatic interactions, and disulfide bonds, help stabilize the three-dimensional structure of the protein (Gu & Bourne, 2011; Munshi & Guru Row, 2006; Uversky & Finkelstein, 2019).

For example, as protective bonds between functional groups of amino acids in regions of protein folds and loops directed towards their tertiary structure. In the quaternary structure, in proteins that have multiple subunits, intramolecular interactions between different subunits ensure the stability of the quaternary structure of the protein. For example, protective bonds and hydrophobic interactions between subunits help maintain the protein's overall structure. In summary, intramolecular interactions play a crucial role in determining the three-dimensional structure of proteins, directly influencing their biological function (Jabłoński, 2021; Jarvis, 2023).

When considering RNA and DNA, intramolecular interactions are crucial for their structure and function. In RNA, hydrogen bonds are essential for forming secondary structures like hairpin loops and double-stranded RNA helices, while base pairing between adenine and uracil, and between cytosine and guanine, stabilizes these structures. The compact packaging of RNA molecules and their phosphodiester bonds determine their three-dimensional structure and stability (Bruno César Feltes, 2015; Wang & Farhana, 2024; Zaha et al., 2014).

In DNA, hydrogen bonds facilitate the formation of the double helix, enabling specific base pairing between adenine and thymine, and between cytosine and guanine, which is the primary stabilizing force. Additionally, hydrophobic interactions between stacked bases and the helical structure, along with phosphodiester bonds along the DNA chain, maintain the stability of the double helix (Gu & Bourne, 2011; Zaha et al., 2014).

## Intermolecular Interactions

Intermolecular interactions are attractive or repulsive forces that occur between separate molecules, influencing their physical, chemical, and biological properties. These forces arise due to differences in the distribution of electric charge within molecules and can be classified into several main types (Gu & Bourne, 2011; Munshi & Guru Row, 2006; Uversky & Finkelstein, 2019).

A prominent example of these interactions in biology is the binding between proteins and DNA e proteins and RNA (Figure 8). In cells, DNA, an essential molecule for storing and transmitting genetic information, interacts with proteins to regulate gene expression. These interactions are mediated by various intermolecular forces, including hydrogen bonds and dipole-dipole interactions. For instance, histones, highly basic proteins, bind to DNA through electrostatic interactions between the positively charged groups of histones and the negatively charged groups of DNA. This binding compacts DNA into a structure known as chromatin, controlling gene access and regulating vital cellular processes such as DNA replication and transcription (ALBERTS, 2017; Bruno César Feltes, 2015; Zaha et al., 2014).

*Figure 8. Cartoon visualization of a protein-RNA complex. PDB ID: 2QUX*

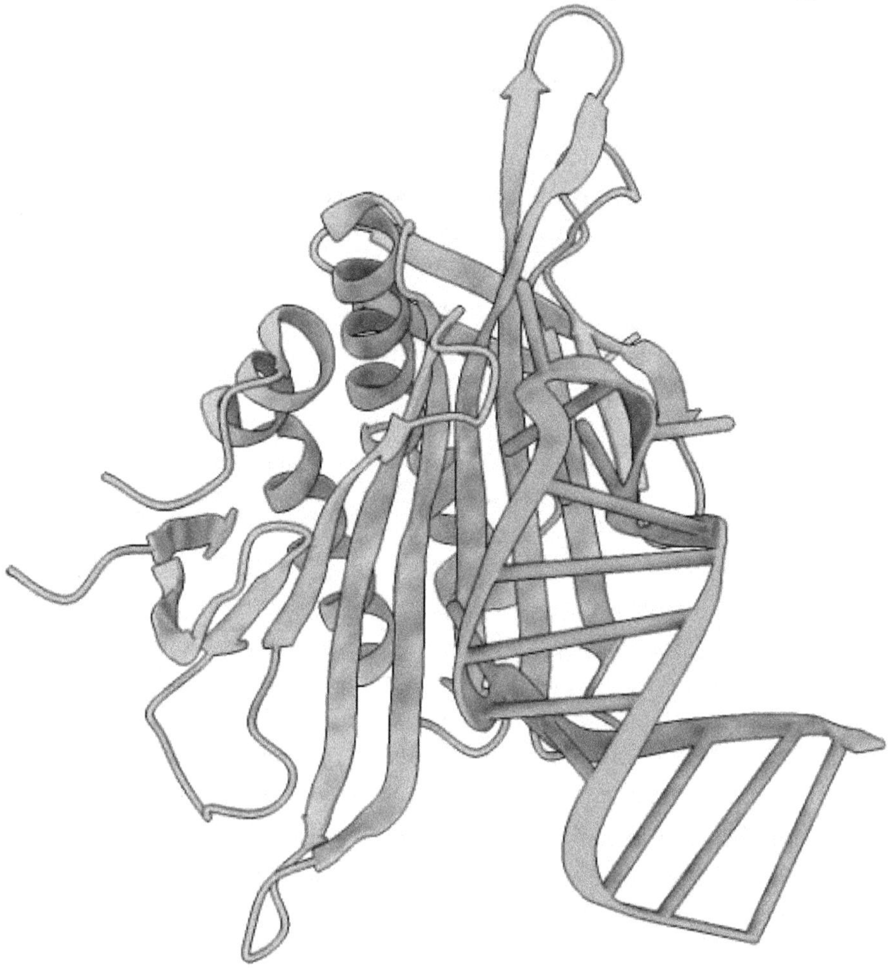

Furthermore, interactions between proteins and peptides are another important example of intermolecular interactions in biology. Proteins, composed of amino acid chains, can interact with smaller peptides in various ways, including substrate recognition by enzymes and cell signaling. For example, many hormones, such as insulin, act by binding to protein receptors on the surface of target cells, triggering specific responses within the cell that regulate metabolism and other physiological processes (Bruno César Feltes, 2015; Junqueira & Carneiro, 2023; Zaha et al., 2014).

These examples illustrate how intermolecular interactions play a vital role in biology, influencing fundamental cellular and molecular processes. Understanding these interactions is essential for advancing our knowledge in molecular biology, medicine, and biotechnology, enabling the development of new therapies and treatment approaches for a variety of medical conditions and diseases (Jabłoński, 2021; Jarvis, 2023; Munshi & Guru Row, 2006; Uversky & Finkelstein, 2019).

Among the types of intermolecular interactions, the following are highlighted:

1. **Dipole-Dipole Forces:** occur between molecules with permanent electric dipoles, resulting in attraction between the opposite poles of the molecules. Examples include HCl and HBr molecules (Jabłoński, 2021; Jarvis, 2023; Munshi & Guru Row, 2006; Uversky & Finkelstein, 2019).
2. **Dipole-Induced Dipole Forces:** also known as London dispersion forces, occur between molecules without permanent dipoles but can develop temporary dipoles due to electron movement. Examples include apolar molecules such as carbon tetrachloride and acetone (Jabłoński, 2021; Jarvis, 2023; Munshi & Guru Row, 2006; Uversky & Finkelstein, 2019).
3. **Instantaneous Dipole-Induced Dipole Forces:** are the weakest interactions between molecules and occur between completely apolar molecules. These interactions result from fluctuations in the electron clouds of molecules, leading to the formation of instantaneous dipoles and induced dipoles in neighboring molecules. Examples include bromine ($Br_2$) (Jabłoński, 2021; Jarvis, 2023; Munshi & Guru Row, 2006; Uversky & Finkelstein, 2019).
4. **Hydrogen Bonds:** are extremely strong interactions between a molecule containing a hydrogen bonded to a highly electronegative atom (such as N, O, or F) and an electronegative atom in another molecule. This bond is responsible for unique properties of water, such as its high boiling point and its ability to form network-like structures when frozen (Jabłoński, 2021; Jarvis, 2023; Munshi & Guru Row, 2006; Uversky & Finkelstein, 2019).

## Structural Databases

Structural databases store detailed information about the three-dimensional structures of molecules and materials, being crucial for various fields such as bioinformatics and chemistry. Structural data typically include atomic coordinates, bond lengths, bond angles, and other parameters describing the spatial arrangement of atoms within the molecule or material (Lesk, 2019).

Some databases are more well-known, examples include the Protein Data Bank (PDB) for biological macromolecules and other complexes, and the Cambridge Structural Database (CSD) for organic and metal-organic molecules. These databases facilitate research, modeling, and development of new therapies and materials. Standardization and improving accessibility are important for advancing the field.

The Protein Data Bank (PDB) is an important database that stores three-dimensional structures of biological macromolecules such as proteins and nucleic acids. These structures are obtained through techniques like X-ray crystallography and electron microscopy and are essential for studies in drug discovery, protein engineering, and molecular biology (Burley, 2017).

The Cambridge Structural Database (CSD) specializes in small organic and metal-organic molecules, providing details about their crystalline structures and how they are packed. On the other hand, the Inorganic Crystal Structure Database (ICSD) contains structural information of inorganic compounds, covering minerals, metals, and coordination complexes (Groom, 2016).

In addition to popular databases, we can also mention Propedia, which is a comprehensive and up-to-date tool containing over 19,000 high-resolution structures from the Protein Data Bank (PDB). It segments these structures into clusters to reduce redundancy, allowing searches and visualizations based on various criteria such as PDB ID, sequence similarity, peptide classification, among others. These clusters facilitate comparisons between sequences, interfaces, interactions, and functions, making Propedia a powerful tool for structural studies of protein-peptide recognition, dataset construction for docking approaches, and rational peptide design (Martins, 2021).

Structural databases are essential for computational studies and modeling, providing experimental models for molecular modeling, simulation, and docking studies. By comparing new structures with those in the database, scientists identify patterns that help understand the underlying principles of molecular structure. These databases are also crucial for the development and validation of computational methods for structure prediction, protein folding, and ligand binding. Comparison with known structures drives innovation in computational techniques, contributing to a better understanding of biological systems and complex materials.

There are several possible applications for databases in bioinformatics, but the increase in data also brings challenges. With the emergence of new databases, standardization becomes crucial to ensure that analyses are not affected by data heterogeneity. An important example of standardization is the Worldwide Protein Data Bank (wwPDB) collaboration, which aims to maintain a unified collection of structural data, ensuring its integrity, reliability, and interoperability with other bioinformatics resources (Berman, 2007).

To summarize, Table 1 provides an overview of the mentioned databases along with their descriptions and links:

*Table 1. Structural databases of macromolecules*

| Database | Description | Link |
|---|---|---|
| RCSB PDB (Protein Data Bank) | Central archive of 3D structures of biological macromolecules, such as proteins, RNA, and DNA, determined by X-ray crystallography and NMR. | https://www.rcsb.org/ |
| Cambridge Structural Database (CSD) | Database of organic and organometallic compounds with 3D structures determined by X-ray crystallography and NMR. | https://www.ccdc.cam.ac.uk/structures/ |
| Worldwide Protein Data Bank (wwPDB) | Global consortium that aggregates PDB data from around the world, such as RCSB PDB, PDBj, and PDB-Europe. | https://www.wwpdb.org/docs.html |
| Propedia | Database with a web interface that permits clustering, searching and visualizing of protein–peptide complexes according to varied criteria. | http://bioinfo.dcc.ufmg.br/propedia/ |

In summary, structural databases play a fundamental role in various areas of scientific research. By providing a wide range of information about molecular structures and materials, these databases not only facilitate scientific discoveries but also enable simulations and support the development of new treatments, materials, and technologies. Continuous investments in the curation, expansion, and accessibility of these databases are crucial to drive the advancement of scientific knowledge and promote innovation in the future (Velankar, 2021).

## Structural Alignments

The comparison of protein sequences and structures is crucial for understanding the nature and functionalities of these fundamental molecules (Hasegawa, 2009). Comparing protein structures has been a traditional challenge in the field of computational molecular biology for over twenty years. During this period, various programs have been developed to address this issue (Ye, 2003).

Several applications can be highlighted in both multiple and pairwise alignments, such as database searching, structure classification, phylogenetic analysis, structural organization, and functional protein prediction. The pairwise alignment method involves comparing two structures, while multiple alignment refers to the analysis of three or more structures (Ma, 2014).

There are two forms of structure alignment behind the tools: global and local alignment. Global alignment is characterized by the complete alignment of the studied protein, thus maximizing the overall similarity of the structure. Local alignment focuses only on the most similar regions of the compared structures, essential for identifying short and conserved areas in proteins (Altschul, 1990; Meng, 2016).

The task of protein structure alignment is similar to the overlay of rigid bodies, but there are significant differences, such as the size of the compared molecules and the reduced chemical diversity of proteins (Lemmen, 2000). To perform the comparison between two protein structures, three fundamental steps must be followed: first, identify their shared similarities; then, align the structures based on these similarities; finally, calculate a statistical measure to assess the similarity. Many of the frequently employed methods demonstrate efficacy in identifying obvious similarities between protein structures. However, aligning two or more structures constitutes a more complex task, whose accuracy may vary depending on the method or program used (Marti-Renom, 2009).

There is an important relationship between the conformations of structures and their sequences. When there was a limited amount of available structures, it was noted that the conservation of protein structure was more significant than the conservation of amino acid sequences. Therefore, protein structures have the ability to provide sequence alignments with an accuracy that would not be possible based solely on sequence alignments. This ability is becoming a crucial contribution to the field of structural bioinformatics and is particularly evident when considering evolution through the analysis of protein structures (Chothia, 1986).

The relevance of comparing and aligning protein structures lies in their ability to identify and measure structural and functional similarities among proteins that initially appear unrelated. Some examples include: Identifying similarity between two structures, found in colicin A and c-phycocyanin, may indicate convergent evolution through a common ancestor (HOLM, 1993); and analysis of the lectin-type C structure, which provides stable support for a wide range of sequences, representing a distinct evolutionary approach to balancing diversity with stability, differing from the immunoglobulin structure (McMahon, 2005). The methods mentioned refer to an analysis from the rigid perspective of the protein, but in a natural context, this type of perspective can lead to erroneous comparisons.

Proteins are often treated as rigid molecules due to limitations in resolution methods, although it is known that they possess flexibility. However, comparing two flexible molecules without taking these changes into account may result in the loss of identifying similarities between them. Therefore, it is necessary to use tools that consider these subtle changes in molecular conformation, allowing for comparisons that closely approximate identified reality (Ye, 2003).

In recent years, various methods have been developed to compare three-dimensional protein structures, reflecting scientists' continuous interest in this area. This is due to the availability of large structural databases and the need for rapid techniques to navigate through them. Additionally, the search for remote similarities and the extraction of biological information from new protein structures drive the development of slower methods. Aligning multiple structures and comparing different alignments are also important research focuses. However, there is still no detailed evaluation of the performance of these methods due to the lack of standardization in datasets and computational protocols. Therefore, a more comprehensive analysis of protein structure comparison methods is necessary, especially considering the growing interest in computational structural biology and bioinformatics (Carugo, 2007).

RMSD (Root Mean Square Deviation) is the most commonly used quantitative metric to assess the similarity between two superimposed atomic coordinates. RMSD values are expressed in Å (angstroms), and the calculation required to obtain the desired value is presented below:

$$RMSD = \sqrt{\frac{1}{n}\sum_{i=l}^{n} d_i^2}$$

RMSD is calculated over n pairs of equivalent atoms, where did_idi is the distance between the atoms of the i-th pair. It can be applied to any type or subset of atoms, such as Cα atoms of an entire protein, specific residues (transmembrane helices, binding sites, loops), all heavy atoms of specific residues, or all heavy atoms in small molecule ligands (Kufareva, 2012). However, it is worth noting that RMSD should not be the only metric used in structure comparison, as it has some errors. A small perturbation in one part of the protein, such as a hinge between two domains, can result in high RMSD values, giving the false impression that the two structures are significantly different, even if they are not (Carugo, 2007).

The field of protein structure comparison and alignment faces significant challenges, including the need for accurate and fast methods to align multiple structures, consider the flexibility of structures, ensure biologically relevant alignments, and optimize parameters automatically. Additionally, there is a demand for better clustering and classification methods, a more precise division of structural space, and better utilization of available structure alignments. Overcoming these challenges is essential to ensure that protein structure comparison and alignment remain active and relevant areas of research in the future (Marti-Renom, 2009).

## Molecular Modeling

Molecular modeling is an essential tool for obtaining 3D-structures of biomolecules, especially when traditional experimental approaches prove to be costly or unfeasible. Methods for predicting protein structure can be categorized into two approaches: model-based and model-free. Model-based methods can use previously determined related structures in the PDB as templates to model the target sequence when predicting the structure of a protein's amino acid sequence. Alternatively, if no related proteins are found, model-free methods attempt to predict the protein's structure directly from the protein sequence, using energy functions combined with conformational sampling (Baxevanis et al., 2020).

Comparative modeling, also known as model-based, relies on comparing unknown protein sequences with those of proteins whose three-dimensional structures are known. With the advent of next-generation sequencing (NGS) technologies, protein sequences can be obtained at low cost, which can be compared with structures of homologous proteins to infer the target protein's structure (Baxevanis et al., 2020).

In situations where no similar experimental structures are available for the target sequence, *ab initio* and de novo methods are employed. These methods are used when there is no prior information about the protein structure. Due to the absence of an initial structural model, these methods require conformational sampling and scoring criteria to identify conformations that approach the native state. These methods are based on Anfinsen's thermodynamic hypothesis, which posits that the native state of a protein is the one with the lowest free energy. Thus, they combine sampling of alternative conformations with scoring functions to evaluate and rank the sampled conformations, aiming to identify the one with the lowest energy, corresponding to the protein's native state. However, due to the vast size of the conformational space, exhaustive search becomes infeasible, necessitating the use of search algorithms to navigate the energy landscape towards near-native conformations (Baxevanis et al., 2020).

These are some of the main molecular modeling tools currently used:

- **SWISS-MODEL:** Based on homology modeling, it utilizes experimentally determined model structures that share high similarity with the target sequence. The models are aligned with the target sequence, and then a 3D protein model is generated for each selected model (Arnold et al., 2006).
- **Modeller:** Also based on homology modeling, it uses spatial restraints to perform comparative modeling of protein structures. These restraints include bond lengths and angles, distance restraints, and dihedral angles in the query sequence. (Webb & Sali, 2017)
- **I-TASSER:** Constructs 3D protein models through interactive floor assembly simulations. The target protein is compared with a set of representative protein structures to find possible folds, and the reassembly process is conducted through replica exchange Monte Carlo simulations (Yang et al., 2015; Zhang, 2008).
- **AlphaFold:** Utilizes co-evolution methods and convolutional neural networks to predict protein structures. It constructs multiple alignments of protein sequences (AlQuraishi, 2019; Jumper et al., 2021).
- **Rosetta:** It is a tool with an ab initio approach to protein modeling, which uses the assembly of small fragments taken from the PDB library. It performs

multiple rounds of Monte Carlo minimization, evaluating each move by a scoring function. The scoring function is a linear combination of physical and statistical terms, describing non-covalent interactions, residue solvation, and torsion angles. It is a well-established tool for protein model prediction, especially effective for free modeling targets in CASP experiments. Additionally, Rosetta also provides a comparative modeling approach (Leman et al., 2020).

- **QUARK**: It is an ab initio pipeline for protein structure prediction, based on continuous assembly of fragments using energy terms and physical knowledge. It utilizes high-resolution non-redundant PDB structures to generate structural fragments, and Monte Carlo simulations with replica exchange to assemble them into complete models (Xu & Zhang, 2012).

Several important techniques in structural bioinformatics are in the field of molecular modeling, including: homology (or comparative) modeling, de novo modeling, molecular docking, virtual screening, and molecular dynamics. These methods will be detailed throughout this chapter.

## Comparative Modeling

Among the methods for predicting 3D structures, homology modeling (comparative modeling) is the simplest, based on the observations that a protein's structure is determined by its amino acid sequence and that during evolution, protein structures change at a much slower rate compared to sequences. Thus, similar sequences tend to fold into identical structures, and even sequences with low similarity can adopt similar structures (Krieger, 2003). The comparative modeling process consists of several steps, which will be detailed below:

- **Model Recognition:**In homology modeling, the unknown sequence is compared with known structures in the PDB using tools like BLAST, which can be limited with sequence identities below 30%. Methods such as PSI-BLAST and Hidden Markov Models (HMM) produce more accurate alignments. Programs like ClustalW and T-Coffee, and advanced tools such as PROBCONS and COMPASS, improve alignment accuracy, especially with reliable homologs (Vyas, 2012).
- **Alignment Correction:** Alignments can be target-template or template-template when multiple templates are used. Errors in alignment can cause α-carbon shifts, and a gap in an α-helix can result in the rotation of the remaining residues. After selecting the appropriate alignments, it is necessary to correct any possible errors. Therefore, the correct alignment of sequences is essential in homology modeling (Muhammed, 2019).

- **Model Construction:** There are several approaches to generate 3D models for a target sequence: rigid body assembly, which divides the protein into conserved regions, loops, and side chains, using tools such as 3D-JIGSAW and SWISS-MODEL; segment matching, which utilizes segments from known structures to assemble the complete model, as in SEGMOD/ENCAD; spatial restraints, which applies constraints derived from the template structure with tools like MODELLER; and artificial evolution, which combines rigid body assembly and evolutionary mutations to adjust the template sequence to the target, using programs like NEST (Muhammed, 2019).
- **Model Refinement:** Model refinement is crucial, requiring efficient sampling of conformational space and precise methods to identify near-native structures (Vyas, 2012). Issues such as insertions, deletions, and misalignments are common, especially with distant relationships between the query sequence and the model, resulting in errors such as incorrect packing of side chains, poorly defined loop conformations, and distortions in secondary structure elements (SSEs). In addition to alignment errors, inevitable differences between the query sequence and the model also lead to errors. Refinement should be divided into two stages: first, local structural errors are corrected while keeping the backbone structure fixed; second, the global protein structure is adjusted to improve overall conformation (Zhu, 2008).
- **Model Optimization:** Model optimization begins with energy minimization using molecular mechanics force fields, which eliminates large errors but may introduce small errors that accumulate. To reduce these errors, it is possible to restrict atomic positions, perform energy minimization with several hundred steps, and use more accurate force fields, such as quantum and self-parametrized ones. Additional methods, such as molecular dynamics and Monte Carlo, can also be employed for more advanced model optimization (Muhammed, 2019).
- **Model Validation:** For a generated model, it is essential to have favorable stereochemistry, and scoring function plays a crucial role in its assessment. During validation, we compare the location of each amino acid residue in the model with its predicted position in the high-resolution X-ray structure. Methods used to analyze errors in X-ray structures can also evaluate alignment failures in the homology model (KHAN, 2022). Validation programs fall into two categories: (1) primary checks, such as PROCHECK and WHATIF, which assess protein stereochemistry, including symmetry, geometry, solvation, and structural packing quality; (2) secondary checks, such as VERIFY3D and PROSAII, which assess the suitability of the sequence to the structure, assigning a score to each residue based on its environment (Vyas, 2012).

## De Novo Modeling

The consistent determination of structure from sequence is one of the greatest unresolved challenges in science, recently surpassing the 50-year mark. Accurately predicting the three-dimensional structure of a protein involves several steps applied to an amino acid sequence: secondary structure prediction (identification of local segments such as helices, beta sheets, or loops), structural alignment with candidate models, conformational sampling, and selection (Khoury, 2014). Before the late 1980s, protein design seemed impossible due to the small thermodynamic stability of the native fold and the vast number of possible sequences, greater than the number of atoms in the universe. The random search for a specific sequence and the large number of possible conformations made it impractical to predict or test all options (Korendovych, 2020).

"De novo" protein design creates new proteins using the physical principles of molecular interactions. While the primary focus is on new structures, there is a growing emphasis on developing new biological functions. Designed proteins are impacting biomedical and synthetic biology research, with applications that include viral inhibitors, immunomodulators, self-assembling biomaterials, and signaling systems (Pan, 2021).

The evolution of "de novo" design has occurred in three distinct periods. In the first period, from the late 1970s to the early 1980s, manual protein design predominated, limited by peptide synthesis. In the second phase, from the mid-1980s to the early 2000s, the focus shifted to computational design, guided by fundamental physicochemical principles. Finally, the third phase, from the early 2000s onward, has taken advantage of the growing database of crystallographic structures for fragment-based and bioinformatics-driven design (Korendovych, 2020).

In a scenario where both the sequence and structure are unknown, computational techniques are employed to efficiently sample backbones. This generally involves generating a large set of alternative conformations using short peptide fragments or algebraic equations. Sequence optimization calculations are then applied to find the lowest energy sequence for each designed backbone conformation. Sequence-independent constraints, such as chain connectivity and limited flexibility, are considered to reduce the space of possible backbone structures. The goal is to generate robust and stable protein structures, although many natural proteins exhibit irregular features that complicate structural analysis and reduce the folding free energy (Huang, 2016).

In summary, structure prediction, including protein design, is extremely important for filling existing gaps in structural bioinformatics. Databases such as SWISS-MODEL, Genome3D, and ModBase, along with tools like AlphaFold2, which uses artificial intelligence to predict conformations, have been essential for increasing prediction accuracy and expanding the field. All these changes have contributed to the expansion of applications for these artificially developed structures (Tunyasuvunakool, 2021).

## Molecular Docking

The molecular docking consists of finding the best orientation for the formation of complexes between molecules. The main components of docking are the sampling algorithm and the scoring function. The sampling algorithm generates different conformations of the ligand within a simulation box. There are two main categories: systematic search, which explores the entire allowed space, and stochastic search, which refines the search through genetic algorithms, for example. The effectiveness of docking is influenced by the flexibility of the molecules and the simulation space. The scoring function ranks the generated conformations and predicts the affinity between the components. There are three main types of scoring functions: empirical, force field-based, and knowledge-based (Bhagat et al., 2021; Sunny & Jayaraj, 2022; Tripathi & Bankaitis, 2017; X. Xu et al., 2018). Molecular docking is a crucial technique in understanding interactions between macromolecules, such as proteins and ligands, proteins, and peptides, or even between proteins and nucleic acids. However, the efficient application of this technique requires careful consideration of the characteristics of the involved macromolecules, including their nature, flexibility, and the most suitable search strategy (Bhagat et al., 2021; Sunny & Jayaraj, 2022; Tripathi & Bankaitis, 2017; X. Xu et al., 2018).

In general, the types of macromolecules involved can be classified into different groups: protein-ligand, protein-peptide, protein-nucleic acid, and protein-protein. Each interaction presents unique challenges in terms of flexibility and binding specificity. In protein-ligand docking, the ligand is usually a smaller and more flexible molecule, while the protein provides the binding site. On the other hand, in protein-peptide docking, the peptide structure can vary in terms of conformation and flexibility, necessitating the use of search methods that can capture this variability. In protein-nucleic acid docking, the structure of nucleic acids can be highly flexible and dynamic, requiring specific molecular modeling approaches to capture specific interactions (Bhagat et al., 2021; Sunny & Jayaraj, 2022; Tripathi & Bankaitis, 2017; X. Xu et al., 2018).

The flexibility of macromolecules plays a crucial role in choosing the docking method. Ligands, peptides, and nucleic acids tend to be more flexible than proteins, able to adopt various conformations during interaction. Therefore, for these types of docking, it is essential to use search methods that can handle this flexibility, allowing the exploration of different conformations to find the best interaction with the protein (Bhagat et al., 2021; Sunny & Jayaraj, 2022; Tripathi & Bankaitis, 2017; X. Xu et al., 2018).

Regarding the search method, both systematic search and stochastic search can be applied, depending on the specific context of the interaction and the analysis objective. Systematic search is useful for exhaustively exploring the search space, while stochastic search may be more efficient in finding solutions in complex and high-dimensional search spaces. In summary, the success of molecular docking depends on the careful selection of the search method and specific tool, considering the characteristics of the involved macromolecules, their flexibility, and the analysis objective. This adaptive approach is essential for obtaining accurate predictions and meaningful insights into molecular interactions in complex biological systems (Bhagat et al., 2021; Sunny & Jayaraj, 2022; Tripathi & Bankaitis, 2017; X. Xu et al., 2018).

Docking is a crucial step in drug discovery, and several computational tools are available. Here is a summary of the main platforms:

- **GOLD (Genetic Optimization for Ligand Docking)**: One of the pioneers in protein-ligand docking, it uses genetic algorithms to adjust the position, orientation, and conformation of the ligand in the protein's active site. It offers an enhanced scoring function that considers ligand energy parameters. Commercially available through the CCDC (Verdonk et al., 2003).
- **AutoDock Vina:** A fast and accurate docking program that automatically calculates grid maps. It uses an iterated local-global optimization algorithm along with gradient methods. Available free for academic purposes (Trott & Olson, 2010)
- **SwissDock:** A web service with a user-friendly interface that utilizes the EADock DSS engine to perform protein-ligand docking. It offers the advantage of docking processing and result visualization on the server side (Grosdidier et al., 2011).
- **ClusPro:** A web server for protein-protein docking that performs docking in three stages, including rigid-body docking, clustering, and energy refinement. Recognized by CAPRI, but its accuracy may be limited compared to flexible methods (Kozakov et al., 2017).
- **pepATTRACT**: A web service for blind peptide-protein docking that uses a rigid docking approach followed by two stages of local docking. It offers great flexibility to the peptide but may take considerable time to execute (de Vries et al., 2017).
- **HDOCK:** A web server for protein-protein docking that offers the option of inputting either a three-dimensional structure or amino acid sequence. It uses a hybrid docking strategy and can predict protein-RNA/DNA interactions (Yan et al., 2020).

- **ZDOCK:** An easy-to-use web platform for protein-protein docking, with options to edit scoring parameters and visualize results. The process is divided into three stages and fast execution time (Pierce et al., 2014).
- **HADDOCK (High Ambiguity Driven Biomolecular DOCKing):** An open web server for protein-protein and protein-ligand docking. It offers various interfaces for different levels of control over simulation parameters, from basic to advanced (de Vries et al., 2010)

## Virtual Screening

In structural bioinformatics, virtual drug screening is a computational process used in the discovery and development of new medications. Unlike traditional laboratory testing methods with real chemical compounds, virtual screening is conducted through computational simulations that evaluate the interaction between molecules and biological targets, such as proteins or enzymes associated with diseases. In this process, molecular modeling and bioinformatics techniques predict the affinity and activity of potential candidate compounds against the desired therapeutic target. This allows for a quick and cost-effective analysis of many molecules, helping to identify the most promising candidates for further laboratory studies (Kitchen et al., 2004; X. Xu et al., 2018).

Virtual drug screening is crucial in accelerating the drug discovery process, helping scientists select the most promising candidates more efficiently and economically. This may eventually lead to the development of safer and more effective therapies for various medical conditions (Cheng et al., 2012). There are several types of virtual screening, each with specific techniques. Below are some of the main types:

- **Molecular docking:** One of the most common techniques in virtual screening, it involves predicting the interaction between a ligand (the chemical compound) and a receptor (the target protein) through computational calculations. The goal is to identify compounds that fit ideally into the active site of the target protein, potentially inhibiting or activating its activity (Schneider & Böhm, 2002).
- **Structure-based screening:** Where the three-dimensional structure of the biological target (such as a protein) is used as a basis for selecting candidate compounds. The analysis is done considering the target's structural characteristics and the candidate compounds' complementarity.
- **Ligand-based virtual screening:** Instead of using the structure of the biological target, screening is performed based on the properties of ligands (known compounds that bind to the target). Methods such as molecular similarity, pharmacophore, and QSAR (Quantitative Structure-Activity Relationship)

are commonly employed to identify new candidates based on known compounds (Cheng et al., 2012b; Lyne, 2002).

In summary, virtual screening stands out as a valuable and efficient approach in the discovery of new bioactive compounds, driven by the wide range of available tools. However, it is essential to recognize that its effectiveness can be further enhanced when combined with complementary techniques, such as molecular dynamics and experimental validation techniques. By integrating these methods, it is possible not only to optimize the accuracy and reliability of the results but also to explore the chemical space more comprehensively in search of promising therapeutic candidates. Therefore, to maximize the potential of virtual screening and advance in the discovery of new medications, it is essential to adopt a holistic approach that incorporates a variety of techniques and multidisciplinary expertise (Kimber et al., 2021; Vyas et al., 2008).

## Molecular Dynamics Simulations

Molecular Dynamics (MD) Simulation is a computational technique that simulates the behavior and interaction of atoms and molecules over time. It relies on applying classical physics laws, such as Newton's laws, to model the movement of particles in a molecular system. This modeling is done at the atomic and molecular scale, allowing for an understanding of how molecules behave under different conditions and environments. In molecular dynamics, molecules are represented as a collection of interacting atoms, where the forces between atoms are described using mathematical functions called force fields. These force fields assign potential energies to different molecular configurations, reflecting interactions between atoms such as chemical bonds, electrostatic repulsion, and van der Waals interactions (Case et al., 2005; Lindahl, 2008).

By applying the equations of motion and the forces calculated by the force field, it's possible to simulate how molecules move over time in response to the forces acting on them. These simulations can study various phenomena, such as protein folding, drug-protein interactions, liquid dynamics, and more. There are three main types of molecular dynamics simulations: conventional simulations, used in various studies like protein folding and complex stability analysis; QM/MM simulations, which simulate chemical reactions; and metadynamics, which predicts the free energy of systems. All these simulations employ algorithms to calculate the integration of motion and energy equations, allowing for the physical properties of the system to be obtained over time (Kondratyuk et al., 2021; E. R. Lindahl, 2008).

Molecular dynamics emerges as a highly relevant technique, widely used to deepen our understanding of the formation of complexes between macromolecules and ligands, as well as cellular phenomena on a larger scale. Now, let's move on to the summaries of the mentioned tools:

- **Amber:** Developed in the 1970s and continuously maintained by the developer community, Amber is a molecular dynamics simulation tool consisting of programs for setting up, running, and analyzing molecular simulations. It offers three main tools: Sander, Pmemd, and Pmemd.cuda. Amber supports various force fields and is available for free on its website (Case et al., 2005).
- **GROMACS:** Widely used in the chemical field for molecular dynamics simulations, GROMACS is known for its efficiency in building non-bonded interactions. It offers a variety of integrated force fields and is available for free on its website (E. Lindahl et al., 2001; Van Der Spoel et al., 2005).
- **NAMD:** An efficient parallel implementation molecular dynamics program, NAMD supports multiple force fields and offers simulations in different thermodynamic ensembles. It's open-source and used in biomedical research, available for free on its website (Nelson et al., 1996; Phillips et al., 2005).

In conclusion, structural bioinformatics emerges as an indispensable tool in the era of biotechnology and personalized medicine. By integrating advanced techniques such as molecular modeling, docking, virtual screening, and molecular dynamics, it allows for a deeper understanding of molecular interactions fundamental to life. These approaches not only provide valuable insights into the structure and function of biomolecules but also pave the way for the rational design of new drugs and more effective therapies. Therefore, structural bioinformatics plays a vital role in accelerating drug discovery, driving significant advancements towards more personalized and precisely targeted healthcare (Baxevanis et al., 2020; E. R. Lindahl, 2008; Webb & Sali, 2017).

## FINAL REMARKS

In this chapter, we addressed structural bioinformatics and the important applications of this area that has been undergoing constant development for over 50 years. Initially, we introduced the topic in an introductory manner, outlining the theoretical basis in general, including essential concepts of molecular biology and computational techniques to study structures, compare them, and identify the functions of macromolecules.

We highlighted how the combination of advanced techniques, such as molecular modeling, docking, virtual screening, and molecular dynamics, has provided an unprecedented understanding of molecular interactions at the atomic level. We exemplified this through studies of proteins, nucleic acids, and protein-ligand complexes, demonstrating how these approaches can provide valuable insights into crucial biological processes.

Furthermore, we explored the applications of structural bioinformatics in diverse areas using different methods. Various tools (AlphaFold and Rosetta), databases (PDB, CSD, wwPDB, Propedia), and webservers (SWISS-MODEL) have been used in many studies that have presented structural analyses in their results. Some practical applications are more common, such as intelligent drug design.

However, it should be recognized that there are important obstacles to be faced. Improving the accuracy of computational methods, better integration of multi-omic data, and application in a variety of emerging areas represent significant research opportunities.

Ultimately, structural bioinformatics plays a crucial role in drug design, precision medicine, and understanding biomolecules and their functions. We hope that this chapter has provided a comprehensive and stimulating overview of the applications and prospects of such an important research area in the field of bioinformatics.

## REFERENCES


Adelusi, T. I., Oyedele, A.-Q. K., Boyenle, I. D., Ogunlana, A. T., Adeyemi, R. O., Ukachi, C. D., Idris, M. O., Olaoba, O. T., Adedotun, I. O., Kolawole, O. E., Xiaoxing, Y., & Abdul-Hammed, M. (2022). Molecular modeling in drug discovery. *Informatics in Medicine Unlocked*, 29, 100880. DOI: 10.1016/j.imu.2022.100880

Alberts. (2017). *Biologia molecular da célula*. Artmed Editora.

Alberts, B., Bray, D., Hopkin, K., Johnson, A., Lewis, J., Raff, M., Roberts, K., Walter, P., Andrade, A. E. B., & Renard, G. (2017). *Fundamentos da Biologia Celular* (4ª edição). *The Art of Medication*.

AlQuraishi, M. (2019). AlphaFold at CASP13. *Bioinformatics (Oxford, England)*, 35(22), 4862–4865. DOI: 10.1093/bioinformatics/btz422 PMID: 31116374

Altschul. (1990). Basic local alignment search tool. Journal of Molecular Biology, 215(3), 403-410.

Arnold, K., Bordoli, L., Kopp, J., & Schwede, T. (2006). The SWISS-MODEL workspace: A web-based environment for protein structure homology modelling. *Bioinformatics (Oxford, England)*, 22(2), 195–201. DOI: 10.1093/bioinformatics/bti770 PMID: 16301204

Baxevanis, A. D., Bader, G. D., & Wishart, D. S. (2020). *Bioinformatics*. John Wiley & Sons.

Bayat, A. (2002). Bioinformatics. *BMJ (Clinical Research Ed.)*, 324(7344), 1018–1022. DOI: 10.1136/bmj.324.7344.1018 PMID: 11976246

Bhagat, R. T., Butle, S. R., Khobragade, D. S., Wankhede, S. B., Prasad, C. C., Mahure, D. S., & Armarkar, A. V. (2021). Molecular Docking in Drug Discovery. *Journal of Pharmaceutical Research International*, 46–58. DOI: 10.9734/jpri/2021/v33i30B31639

Burley. (2017). *Protein Data Bank* (PDB): the single global macromolecular structure archive. Protein crystallography: methods and protocols, 627-641.

Carugo, O. (2007). Oliviero. Recent progress in measuring structural similarity between proteins. *Current Protein & Peptide Science*, 8(3), 219–241. DOI: 10.2174/138920307780831839 PMID: 17584118

Case, D. A., Cheatham, T. E.III, Darden, T., Gohlke, H., Luo, R., Merz, K. M.Jr, Onufriev, A., Simmerling, C., Wang, B., & Woods, R. J. (2005a). The Amber biomolecular simulation programs. *Journal of Computational Chemistry*, 26(16), 1668–1688. DOI: 10.1002/jcc.20290 PMID: 16200636

Cheng, T., Li, Q., Zhou, Z., Wang, Y., & Bryant, S. H. (2012a). Structure-Based Virtual Screening for Drug Discovery: A Problem-Centric Review. *The AAPS Journal*, 14(1), 133–141. DOI: 10.1208/s12248-012-9322-0 PMID: 22281989

Chothia, C., & Lesk, A. M. (1986). Cyrus; LESK, Arthur M. The relation between the divergence of sequence and structure in proteins. *The EMBO Journal*, 5(4), 823–826. DOI: 10.1002/j.1460-2075.1986.tb04288.x PMID: 3709526

Crick, F. H. (1958). On protein synthesis. *Symposia of the Society for Experimental Biology*, 12, 138–163. PMID: 13580867

de Vries, S. J., Rey, J., Schindler, C. E. M., Zacharias, M., & Tuffery, P. (2017). The pepATTRACT web server for blind, large-scale peptide–protein docking. *Nucleic Acids Research*, 45(W1), W361–W364. DOI: 10.1093/nar/gkx335 PMID: 28460116

de Vries, S. J., van Dijk, M., & Bonvin, A. M. J. J. (2010). The HADDOCK web server for data-driven biomolecular docking. *Nature Protocols*, 5(5), 883–897. DOI: 10.1038/nprot.2010.32 PMID: 20431534

Derkach, S. R., Voron'ko, N. G., & Kuchina, Y. A. (2022). Intermolecular Interactions in the Formation of Polysaccharide-Gelatin Complexes: A Spectroscopic Study. *Polymers*, 14(14), 2777. DOI: 10.3390/polym14142777 PMID: 35890554

Durmaz, V., Köchl, K., Krassnigg, A., Parigger, L., Hetmann, M., Singh, A., Nutz, D., Korsunsky, A., Kahler, U., König, C., Chang, L., Krebs, M., Bassetto, R., Pavkov-Keller, T., Resch, V., Gruber, K., Steinkellner, G., & Gruber, C. C. (2022). Structural bioinformatics analysis of SARS-CoV-2 variants reveals higher hACE2 receptor binding affinity for Omicron B.1.1.529 spike RBD compared to wild type reference. *Scientific Reports*, 12(1), 14534. DOI: 10.1038/s41598-022-18507-y PMID: 36008461

Erickson, J., & Kempf, D. (1994). Structure-based design of symmetric inhibitors of HIV-1 protease. *Archives of Virology. Supplementum*, 9, 19–29. DOI: 10.1007/978-3-7091-9326-6_3 PMID: 8032250

Feltes. (2015). *Bioinformática: Da biologia à flexibilidade molecular*. SBBq.

Groom. (2016). The Cambridge structural database. Acta Crystallographica Section B: Structural Science. Crystal Engineering and Materials, 72(2), 171-179.

Grosdidier, A., Zoete, V., & Michielin, O. (2011). SwissDock, a protein-small molecule docking web service based on EADock DSS. *Nucleic Acids Research*, 39(suppl_2), W270–W277. DOI: 10.1093/nar/gkr366 PMID: 21624888

Gu, J., & Bourne, P. E. (2011). *Structural Bioinformatics*. John Wiley & Sons.

Hasegawa, H., & Holm, L. (2009). Advances and pitfalls of protein structural alignment. *Current Opinion in Structural Biology*, 19(3), 341–348. DOI: 10.1016/j.sbi.2009.04.003 PMID: 19481444

Holm, L., & Sander, C. (1993). Structural alignment of globins, phycocyanins and colicin A. *FEBS Letters*, 315(3), 301–306. DOI: 10.1016/0014-5793(93)81183-Z PMID: 8422921

Huang, P.-S., Boyken, S. E., & Baker, D. (2016). The coming of age of de novo protein design. *Nature*, 537(7620), 320–327. DOI: 10.1038/nature19946 PMID: 27629638

Jabłoński, M. (2021). Intramolecular Hydrogen Bonding 2021. *Molecules (Basel, Switzerland)*, 26(20), 6319. DOI: 10.3390/molecules26206319 PMID: 34684899

Jarvis, M. C. (2023). Hydrogen bonding and other non-covalent interactions at the surfaces of cellulose microfibrils. *Cellulose (London, England)*, 30(2), 667–687. DOI: 10.1007/s10570-022-04954-3

Jiang, Z. H. U. (2008). Refining homology models by combining replica-exchange molecular dynamics and statistical potentials. *Proteins*, 72(4), 1171–1188. DOI: 10.1002/prot.22005 PMID: 18338384

Jumper, J., Evans, R., Pritzel, A., Green, T., Figurnov, M., Ronneberger, O., Tunyasuvunakool, K., Bates, R., Žídek, A., Potapenko, A., Bridgland, A., Meyer, C., Kohl, S. A. A., Ballard, A. J., Cowie, A., Romera-Paredes, B., Nikolov, S., Jain, R., Adler, J., & Hassabis, D. (2021). Highly accurate protein structure prediction with AlphaFold. *Nature*, 596(7873), 583–589. DOI: 10.1038/s41586-021-03819-2 PMID: 34265844

Junqueira, L. C., & Carneiro, J. (2023). *Biologia Celular e Molecular* (10ª edição). Guanabara Koogan.

Kanehisa, M., & Bork, P. (2003). Bioinformatics in the post-sequence era. *Nature Genetics*, 33(3), 305–310. DOI: 10.1038/ng1109 PMID: 12610540

Kathryn, T. (2021). Highly accurate protein structure prediction for the human proteome. *Nature*, 596(7873), 590–596. DOI: 10.1038/s41586-021-03828-1 PMID: 34293799

Khan, S. U., Ahemad, N., Chuah, L.-H., Naidu, R., & Htar, T. T. (2022). G protein-coupled estrogen receptor-1: Homology modeling approaches and application in screening new GPER-1 modulators. *Journal of Biomolecular Structure & Dynamics*, 40(7), 3325–3335. DOI: 10.1080/07391102.2020.1844059 PMID: 33164654

Khoury. (2014). Protein folding and de novo protein design for biotechnological applications. Trends in Biotechnology, 32(2), 99-109.

Kimber, T. B., Chen, Y., & Volkamer, A. (2021). Deep Learning in Virtual Screening: Recent Applications and Developments. *International Journal of Molecular Sciences, 22*(9). *International Journal of Molecular Sciences*, 9(9), 4435. Advance online publication. DOI: 10.3390/ijms22094435 PMID: 33922714

Kitchen, D. B., Decornez, H., Furr, J. R., & Bajorath, J. (2004). Docking and scoring in virtual screening for drug discovery: Methods and applications. *Nature Reviews. Drug Discovery*, 3(11), 935–949. DOI: 10.1038/nrd1549 PMID: 15520816

Kondratyuk, N., Nikolskiy, V., Pavlov, D., & Stegailov, V. (2021). GPU-accelerated molecular dynamics: State-of-art software performance and porting from Nvidia CUDA to AMD HIP. *International Journal of High Performance Computing Applications*, 35(4), 312–324. DOI: 10.1177/10943420211008288

Korendovych, I. V., & DeGrado, W. F. (2020). De novo protein design, a retrospective. *Quarterly Reviews of Biophysics*, 53, e3. DOI: 10.1017/S0033583519000131

Kozakov, D., Hall, D. R., Xia, B., Porter, K. A., Padhorny, D., Yueh, C., Beglov, D., & Vajda, S. (2017). The ClusPro web server for protein–protein docking. *Nature Protocols*, 12(2), 255–278. DOI: 10.1038/nprot.2016.169 PMID: 28079879

Krieger. (2003). Homology modeling. Structural Bioinformatics, 44, 509-523.

Leman, J. K., Weitzner, B. D., Lewis, S. M., Adolf-Bryfogle, J., Alam, N., Alford, R. F., Aprahamian, M., Baker, D., Barlow, K. A., Barth, P., Basanta, B., Bender, B. J., Blacklock, K., Bonet, J., Boyken, S. E., Bradley, P., Bystroff, C., Conway, P., Cooper, S., & Bonneau, R. (2020). Macromolecular modeling and design in Rosetta: Recent methods and frameworks. *Nature Methods*, 17(7), 665–680. DOI: 10.1038/s41592-020-0848-2 PMID: 32483333

Lemmen, C., & Lengauer, T. (2000). Computational methods for the structural alignment of molecules. *Journal of Computer-Aided Molecular Design*, 14(3), 215–232. DOI: 10.1023/A:1008194019144 PMID: 10756477

Lesk. (2019). *Introduction to bioinformatics*. Oxford University Press.

Lindahl, E., Hess, B., & van der Spoel, D. (2001). GROMACS 3.0: A package for molecular simulation and trajectory analysis. *Journal of Molecular Modeling*, 7(8), 306–317. DOI: 10.1007/s008940100045

Lindahl, E. R. (2008). Molecular Dynamics Simulations. In A. Kukol (Org.), *Molecular Modeling of Proteins* (p. 3–23). Humana Press. DOI: 10.1007/978-1-59745-177-2_1

Lyne, P. D. (2002). Structure-based virtual screening: An overview. *Drug Discovery Today*, 7(20), 1047–1055. DOI: 10.1016/S1359-6446(02)02483-2 PMID: 12546894

Ma, J., & Wang, S. (2014). Algorithms, applications, and challenges of protein structure alignment. *Advances in Protein Chemistry and Structural Biology*, 94, 121–175. DOI: 10.1016/B978-0-12-800168-4.00005-6 PMID: 24629187

Marti-Renom. (2009). Structure comparison and alignment. Structural Bioinformatics, 2.

Martins. (2021). Propedia: a database for protein–peptide identification based on a hybrid clustering algorithm. BMC Bioinformatics, 22, 1-20.

McMahon. (2005). The C-type lectin fold as an evolutionary solution for massive sequence variation. Nature Structural & Molecular Biology, 12(10), 886-892.

Meng, L., Striegel, A., & Milenković, T. (2016). Local versus global biological network alignment. *Bioinformatics (Oxford, England)*, 32(20), 3155–3164. DOI: 10.1093/bioinformatics/btw348 PMID: 27357169

Minchin, S., & Lodge, J. (2019). Understanding biochemistry: Structure and function of nucleic acids. *Essays in Biochemistry*, 63(4), 433–456. DOI: 10.1042/EBC20180038 PMID: 31652314

*Molecular Dynamics*. (n.d.). Recuperado 5 de abril de 2024, de https://link.springer.com/book/10.1007/978-3-319-16375-8

Muhammed, M. T., & Aki-Yalcin, E. (2019). Homology modeling in drug discovery: Overview, current applications, and future perspectives. *Chemical Biology & Drug Design*, 93(1), 12–20. DOI: 10.1111/cbdd.13388 PMID: 30187647

Munshi, P., & Guru Row, T. N. (2006). Intra- and intermolecular interactions in small bioactive molecules: Cooperative features from experimental and theoretical charge-density analysis. *Acta Crystallographica. Section B, Structural Science*, 62(4), 612–626. DOI: 10.1107/S0108768106017393 PMID: 16840811

Nelson, D. L., Cox, M. M., Termignoni, C., Renard, G., Pereira, M. L. S., & Farias, S. E. (2014). *Lehninger principles of biochemistry* (6ª edição). *The Art of Medication*.

Nelson, M. T., Humphrey, W., Gursoy, A., Dalke, A., Kalé, L. V., Skeel, R. D., & Schulten, K. (1996). NAMD: A Parallel, Object-Oriented Molecular Dynamics Program. *The International Journal of Supercomputer Applications and High Performance Computing*, 10(4), 251–268. DOI: 10.1177/109434209601000401

Okazaki, R., Okazaki, T., Sakabe, K., Sugimoto, K., & Sugino, A. (1968). Mechanism of DNA chain growth. I. Possible discontinuity and unusual secondary structure of newly synthesized chains. *Proceedings of the National Academy of Sciences of the United States of America*, 59(2), 598–605. DOI: 10.1073/pnas.59.2.598 PMID: 4967086

Palumbo, O., Trequattrini, F., Cimini, A., Tsurumaki, A., Navarra, M. A., & Paolone, A. (2021). Inter- and Intramolecular Interactions in Ether-Functionalized Ionic Liquids. *The Journal of Physical Chemistry B*, 125(9), 2380–2388. DOI: 10.1021/acs.jpcb.0c11429 PMID: 33625218

Pan. (2021). Recent advances in de novo protein design: Principles, methods, and applications. Journal of Biological Chemistry, 296.

Phillips, J. C., Braun, R., Wang, W., Gumbart, J., Tajkhorshid, E., Villa, E., Chipot, C., Skeel, R. D., Kalé, L., & Schulten, K. (2005). Scalable molecular dynamics with NAMD. *Journal of Computational Chemistry*, 26(16), 1781–1802. DOI: 10.1002/jcc.20289 PMID: 16222654

Pierce, B. G., Wiehe, K., Hwang, H., Kim, B.-H., Vreven, T., & Weng, Z. (2014). ZDOCK server: Interactive docking prediction of protein–protein complexes and symmetric multimers. *Bioinformatics (Oxford, England)*, 30(12), 1771–1773. DOI: 10.1093/bioinformatics/btu097 PMID: 24532726

Pinzi, L., & Rastelli, G. (2019). Molecular Docking: Shifting Paradigms in Drug Discovery. *International Journal of Molecular Sciences*, 20(18), 4331. DOI: 10.3390/ijms20184331 PMID: 31487867

Saleh. (2017). Design and development of some viral protease inhibitors by QSAR and molecular modeling studies. In *Viral proteases and their inhibitors* (pp. 25–58). Academic Press. DOI: 10.1016/B978-0-12-809712-0.00002-2

Samuel, H., Nweke-Maraizu, U., & Etim, E. (2023). Understanding Intermolecular and Intramolecular Hydrogen Bonds. *Spectroscopic and Computational Approaches.*, 5, 439–465. DOI: 10.48309/JCR.2023.407989.1235

Sanjeev, A., Mattaparthi, V. S. K., & Kaushik, S. (2019). Nucleic-Acid Structure Database. In S. Ranganathan, M. Gribskov, K. Nakai, & C. Schönbach (Orgs.), *Encyclopedia of Bioinformatics and Computational Biology* (p. 567–574). Academic Press. DOI: 10.1016/B978-0-12-809633-8.20285-9

Schneider, G., & Böhm, H.-J. (2002). Virtual screening and fast automated docking methods. *Drug Discovery Today*, 7(1), 64–70. DOI: 10.1016/S1359-6446(01)02091-8 PMID: 11790605

Sunny, S., & Jayaraj, P. B. (2022). Protein–Protein Docking: Past, Present, and Future. *The Protein Journal*, 41(1), 1–26. DOI: 10.1007/s10930-021-10031-8 PMID: 34787783

Tripathi, A., & Bankaitis, V. A. (2017). Molecular Docking: From Lock and Key to Combination Lock. *Journal of Molecular Medicine and Clinical Applications*, 2(1). Advance online publication. DOI: 10.16966/2575-0305.106 PMID: 29333532

Trott, O., & Olson, A. J. (2010). AutoDock Vina: Improving the speed and accuracy of docking with a new scoring function, efficient optimization, and multithreading. *Journal of Computational Chemistry*, 31(2), 455–461. DOI: 10.1002/jcc.21334 PMID: 19499576

Uversky, V. N., & Finkelstein, A. V. (2019). Life in Phases: Intra- and Inter- Molecular Phase Transitions in Protein Solutions. *Biomolecules*, 9(12), 842. DOI: 10.3390/biom9120842 PMID: 31817975

Van De Waterbeemd. (1997). Glossary of terms used in computational drug design (IUPAC Recommendations 1997). Pure and Applied Chemistry, 69(5), 1137-1152.

Van Der Spoel, D., Lindahl, E., Hess, B., Groenhof, G., Mark, A. E., & Berendsen, H. J. C. (2005). GROMACS: Fast, flexible, and free. *Journal of Computational Chemistry*, 26(16), 1701–1718. DOI: 10.1002/jcc.20291 PMID: 16211538

Verdonk, M. L., Cole, J. C., Hartshorn, M. J., Murray, C. W., & Taylor, R. D. (2003). Improved protein–ligand docking using GOLD. *Proteins*, 52(4), 609–623. DOI: 10.1002/prot.10465 PMID: 12910460

Vyas, V., Jain, A., Jain, A., & Gupta, A. (2008). Virtual Screening: A Fast Tool for Drug Design. *Scientia Pharmaceutica, 76*(3). *Scientia Pharmaceutica*, 3(3), 333–360. Advance online publication. DOI: 10.3797/scipharm.0803-03

Vyas, V. K., Ukawala, R. D., Chintha, C., & Ghate, M. (2012). Homology modeling a fast tool for drug discovery: Current perspectives. *Indian Journal of Pharmaceutical Sciences*, 74(1), 1. DOI: 10.4103/0250-474X.102537 PMID: 23204616

Wang, D., & Farhana, A. (2024). Biochemistry, RNA Structure. In *StatPearls*. StatPearls Publishing. https://www.ncbi.nlm.nih.gov/books/NBK558999/

Webb, B., & Sali, A. (2017). Protein Structure Modeling with MODELLER. In M. Kaufmann, C. Klinger, & A. Savelsbergh (Orgs.), *Functional Genomics: Methods and Protocols* (p. 39–54). Springer. DOI: 10.1007/978-1-4939-7231-9_4

Xu, D., & Zhang, Y. (2012). Ab initio protein structure assembly using continuous structure fragments and optimized knowledge-based force field. *Proteins*, 80(7), 1715–1735. DOI: 10.1002/prot.24065 PMID: 22411565

Xu, X., Huang, M., & Zou, X. (2018). Docking-based inverse virtual screening: Methods, applications, and challenges. *Biophysics Reports*, 4(1), 1–16. DOI: 10.1007/s41048-017-0045-8 PMID: 29577065

Yan, Y., Tao, H., He, J., & Huang, S.-Y. (2020). The HDOCK server for integrated protein–protein docking. *Nature Protocols*, 15(5), 1829–1852. DOI: 10.1038/s41596-020-0312-x PMID: 32269383

Yang, J., Yan, R., Roy, A., Xu, D., Poisson, J., & Zhang, Y. (2015). The I-TASSER Suite: Protein structure and function prediction. *Nature Methods*, 12(1), 7–8. DOI: 10.1038/nmeth.3213 PMID: 25549265

Yang, T.-J., Yu, P.-Y., Chang, Y.-C., Liang, K.-H., Tso, H.-C., Ho, M.-R., Chen, W.-Y., Lin, H.-T., Wu, H.-C., & Hsu, S.-T. D. (2021). Effect of SARS-CoV-2 B.1.1.7 mutations on spike protein structure and function. *Nature Structural & Molecular Biology*, 28(9), 731–739. DOI: 10.1038/s41594-021-00652-z PMID: 34385690

Ye. (2003). Flexible structure alignment by chaining aligned fragment pairs allowing twists. Bioinformatics, 19(2), ii246-ii255.

Zaha, A., Ferreira, H. B., & Passaglia, L. M. P. (2014). *Biologia Molecular Básica—5. ed.* Artmed.

Zhang, Y. (2008). I-TASSER server for protein 3D structure prediction. *BMC Bioinformatics*, 9(1), 40. DOI: 10.1186/1471-2105-9-40 PMID: 18215316

# Chapter 7
# Virtual Screening: Introduction and Importance in Drug Discovery

**Angie Atoche Puelles**
*Universidade Federal de Minas Gerais, Brazil*

**Luana Luiza Bastos**
https://orcid.org/0000-0002-6932-0438
*Universidade Federal de Minas Gerais, Brazil*

**Vivian Morais Paixão**
https://orcid.org/0000-0002-4154-304X
*Universidade Federal de Minas Gerais, Brazil*

**Sheila Cruz Araujo**
https://orcid.org/0000-0001-7413-0893
*Universidade Federal de Minas Gerais, Brazil*

**Raquel Cardoso de Melo Minardi**
*Universidade Federal de Minas Gerais, Brazil*

## ABSTRACT

*Advancements in computational methods have significantly transformed the traditional drug discovery process, which has been historically lengthy and costly, often taking over a decade and billions of dollars. Virtual screening (VS) is a computational technique used in drug discovery to identify potential drug candidates by screening large chemical libraries. This technique has transformed drug discovery by enabling rapid and cost-effective screening of chemical libraries, reducing time and resources. This leverages algorithms, such as similarity-based, quantitative, and machine learning, to prioritize compounds binding to target proteins. Additionally, databases*

DOI: 10.4018/979-8-3693-3192-7.ch007

*provide vast compound collections for screening. However, experimental validation is essential to ensure accuracy. In this chapter, the authors delve into the concepts and methodologies of this technique, focusing on its crucial role in drug discovery.*

## INTRODUCTION AND IMPORTANCE IN DRUG DISCOVERY

In recent years, computational methods have revolutionized the conventional drug discovery process. These processes were considered lengthy and costly, often taking over a decade and billions of dollars. Recently, they have been optimized by computational techniques. One of these techniques is Virtual Screening (VS), a term coined in the late 1990s that describes the application of algorithms and computational models to identify new bioactive molecules (Schneider, 2010). VS technique allows for a quick and cost-effective selection from vast chemical libraries, significantly reducing the time and resources needed to identify promising compounds. Applying new strategies and computational methods, such as Genetic Algorithms Optimization for virtual screening processes, effectively discovers new drug candidates. (Morales-Helguera et al., 2015)

Virtual screening is beneficial in the initial phase of identifying potential compounds from large chemical libraries. According to Bajorath (2002), this method is designed to search large in silico compound databases and select a limited number of candidate molecules for testing to identify new chemical entities with the desired biological activity. It uses comparison and filtering algorithms, identifying compounds with desired characteristics such as target affinity, favorable pharmacokinetic properties, and low toxicity. This significantly reduces the number of compounds to be experimentally tested, saving time and precious resources. It is important to note that virtual screening expands the possibilities for molecules that may not necessarily exist physically in a researcher's collection but can be easily obtained through purchase or synthesis (McInnes, 2007).

The methodology involves initial preparation, filtering, and experimental analysis (Figure 1). In the first step, preparing a library with compound structures, generating conformations, and correcting possible stereochemical errors is necessary (Santana et al., 2021). Databases can be obtained from different collections, such as bioactivity databases, commercial databases, and natural products. Then, different techniques are applied to filter the desired compounds. Various filters such as pharmacophore, similarity, and other computational tools can be used. Finally, it may be necessary to optimize the structure of the chosen ligands to increase their affinity with the target or improve some pharmacological properties. After these steps, the chosen molecules are tested experimentally *in vitro* and *in vivo*.

*Figure 1. Scheme representing the virtual screening methodology (created by CANVA)*

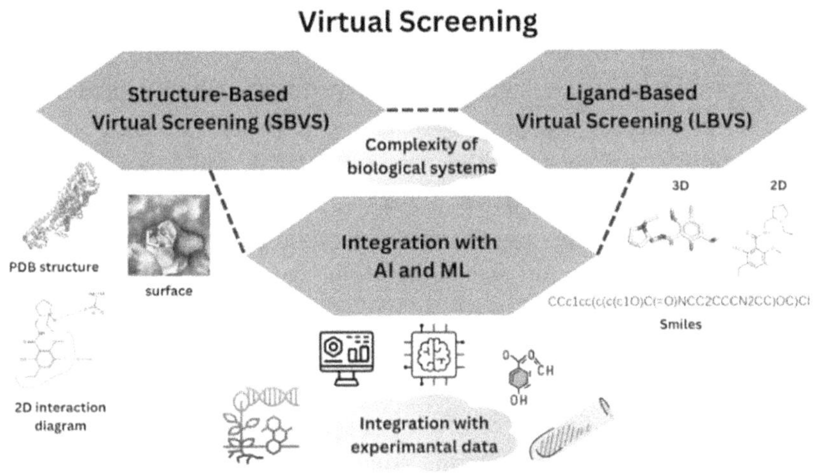


Among the methodologies applied in Virtual Screening, the main ones are Ligand-Based Virtual Screening (LBVS) and Structure-Based Virtual Screening (SBVS). In Ligand-Based Virtual Screening, the characteristics of the compound's structure are used, including electronic, topological, physicochemical, and structural properties related to its molecular activity, using as a starting point a set of compounds with experimentally proven properties (Santana et al., 2021). On the other hand, in Structure-Based Virtual Screening, screening is based on considering the structure of the receptor, which is the molecular target of the active ligands under investigation (Kitchen et al., 2004a). For example, Peng et al. (2003) described the identification of new BCR-ABL tyrosine kinase inhibitors for the treatment of chronic myeloid leukemia (CML) through this technique. This approach allowed researchers to identify promising candidate compounds that were subsequently tested experimentally, resulting in the discovery of new inhibitors.

Thus, in the context of Drug Discovery, virtual screening is an essential computational tool that plays a crucial role in various stages of the process. Its techniques enable efficient screening of large compound libraries to identify promising candidates with the potential to interact with therapeutic target proteins. Furthermore, it is an essential tool for drug metabolism analysis, a crucial aspect in drug development as it influences the efficacy, safety, and toxicity of compounds. For example, VS allows us to predict how compounds interact with cytochrome P450 (CYP), a family of liver enzymes responsible for metabolizing various compounds (Park; Harris, 2003). Analyzing the interaction between drugs and CYP isoforms is essential to predict how compounds will be metabolized in the body.

Virtual screening (VS) has emerged as a pivotal technique in drug discovery, providing a streamlined and economical method for identifying promising candidate molecules. This approach utilizes methodologies such as Ligand-Based Virtual Screening (LBVS) and Structure-Based Virtual Screening (SBVS), which significantly propel the research process and expedite the development of more effective and safer therapies. The primary objective of virtual screening is to pinpoint potential candidates that merit further experimental investigation (Bhagat et al., 2021; Bleicher et al., 2003; Kitchen et al., 2004b). In conclusion, virtual screening constitutes a breakthrough in drug discovery, facilitating the quicker identification of novel drugs and substantially enhancing global health outcomes.

## Structure and Ligand-Based Filtering

One of the most essential stages of virtual screening is filtering the compound library to select the best ones. Knowing which types are most commonly used is essential. This filtering stage aims to reduce the number of compounds to be experimentally tested, selecting the most promising ones to advance to the subsequent stages of the drug discovery process. In this section, we discuss the types that are most used.

Regarding Ligand-Based Virtual Screening, various compounds characteristics are evaluated in relation to their structure's intrinsic properties. The filters used in this stage aim to identify compounds similar to those known to be active against a specific target (Yang et al., 2019). This includes analyses of electronic, topological, physicochemical, and structural properties related to the molecular activity of the compounds (Santana et al., 2021). Additionally, filters can be designed to identify compounds with specific desired characteristics, such as affinity for the biological target, favorable pharmacokinetic properties, and low toxicity. This filtering is performed through computational algorithms, including:

1. Molecular Fingerprints: Based on the premise that molecules with similar structures have similar biological activities. They are numerical representations of molecular structures that capture information about the presence or absence of different structural features (Yang et al., 2022). Typically, each specific bit position represents the presence (1) or absence (0) of predefined functional groups, substructure motifs, or fragments (Rácz; Bajusz; Héberger, 2018) (Figure 2). These fingerprints are used to calculate the similarity between the ligand of interest and the compounds in the selected library, allowing the identification of candidates with similar structures.

*Figure 2. Scheme representing the molecular fingerprint of Nicotinamide ($C_6H_6N_2O$). Each bit position corresponds to a specific substructure, so if it is "1," it means that the substructure is present and "0" is absent (Wikimedia Commons (CC-BY 4.0))*

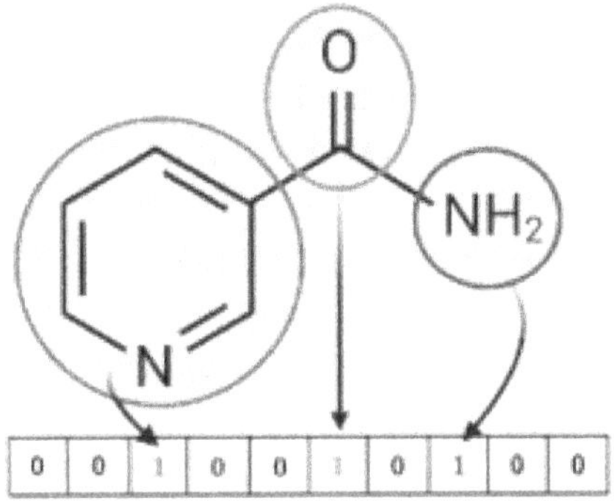


2. 3D shape similarity search: The concept behind this method is based on comparing the three-dimensional shapes of molecules, relying on the premise that molecules with similar three-dimensional structures tend to exhibit similar biological properties and can be recognized by the same receptor (Koes; Camacho, 2014). Thus, compounds are selected based on their overlap with known active molecules and subsequent analysis of the relative positions of all atoms, selecting those that are most similar (Seddon et al., 2019). Calculations can be performed, for example, using the Tanimoto coefficient or Euclidean distance (Shin et al., 2015), and then the compounds are ranked according to the similarity score.
3. Ligand-based pharmacophore modeling: According to the official IUPAC definition (Wermuth et al., 1998). A pharmacophore is the ensemble of steric and electronic features necessary to ensure the optimal supra-molecular interactions with a specific biological target structure and trigger (or block) its biological response. Thus, LBVS and SBVS use pharmacophore modeling to obtain the ideal ligand. The method involves selecting experimentally validated active compounds, analyzing their chemical and structural characteristics, and generating a pharmacophore model, which will be used as a reference in screening the ligands to be tested (Santana et al., 2021).
4. Machine learning algorithms: According to Nichols et al. (2018), Machine Learning refers to a broad range of algorithms that perform intelligent predictions based on a data set. This approach is used for various prediction purposes since the tool can be trained with different types of data sets. For example, it can be used to predict some pharmacokinetic properties (Qiang et al., 2020) or

molecular targets (Jeon et al., 2014), among various others. For this, the training data set consists of compounds previously known as active for the target, and thus, the algorithm can predict similar properties of compound sets for testing. The types of algorithms that can be used are mentioned in the subsequent topics.

Regarding the filtering linked to Structure-Based Virtual Screening, the approach is mainly based on the target receptor's three-dimensional (3D) structure and the molecular interactions between the receptor and the ligands (Da; Kireev, 2014). Thus, the conformation of the complex, the binding affinity, the intermolecular interactions between the ligand and the receptor, and the composition of the binding site are commonly analyzed. For this purpose, some of the methods used include:

1. Molecular docking: This is a widely used computational technique for predicting and analyzing the ligand-receptor interaction. Typically, it first predicts the molecular orientation of a ligand within a receptor and then estimates its complementarity using a scoring function (Kitchen et al., 2004a).
2. Molecular dynamics simulation: According to Hollingsworth and Dror (2018), these simulations predict how every atom in a protein or other molecular system will move over time based on a general physics model governing interatomic interactions. Thus, this methodology analyzes the ligand and target interaction over time, simulating a natural environment.
3. Structure-based pharmacophore modeling: Similarly to LBVS, it is possible to model pharmacophores based on the target structure. For this purpose, experimentally resolved structures of molecular targets complexed with an active ligand are used, and then the structures are employed in screening the ligands to be tested (Jiang et al., 2020).

LBVS evaluates various compound characteristics to identify potential ligands with properties similar to those of known active compounds. On the other hand, SBVS relies on the three-dimensional structure of the target receptor and the interactions between the receptor and ligands. These filtering approaches are essential for reducing the number of compounds to be experimentally tested and selecting the most promising candidates for subsequent stages of drug discovery.

## Docking and Virtual Screening

Molecular Docking (MD) and Virtual Screening are two fundamental techniques in drug discovery, harnessing computational power to predict and assess the interaction between molecules and target proteins. Molecular docking, also known as virtual docking or only docking, is a computational method that models the interaction between a small molecule (ligand) and a macromolecule (typically a protein) target. Using advanced algorithms and techniques, docking simulates how the ligand fits into the protein's binding site, evaluating aspects such as structural complementarity and molecular interactions. This process allows for the prediction of binding affinity and geometry, identifying promising candidates for pharmaceutical development (Agrawal et al., 2019a; Feltes, 2015; Kitchen et al., 2004b; Ma; Shiu-Hin Chan; Leung, 2011a; Xu; Huang; Zou, 2018).

These techniques are highly complementary. Molecular docking is often used after virtual screening for detailed analysis of identified candidates, refining selection, and providing more accurate information on molecular interactions. Together, these approaches significantly accelerate the drug discovery process, allowing scientists to quickly identify promising compounds and optimize their efficacy and safety before clinical testing (Cheng et al., 2012a; Kitchen et al., 2004b; Schneider; Böhm, 2002; Xu; Huang; Zou, 2018).

In addition to its general applications, molecular docking can be adapted for different types of biomolecular interactions, further expanding its scope of application.

1. Protein-Protein Docking: This type of docking is used to study the interaction between two proteins. It can be valuable for understanding complex biological processes involving protein complex formation, such as gene expression regulation, cell signaling, and assembly of supramolecular structures (Sunny; Jayaraj, 2022).
2. Protein-Peptide Docking: Focused on the interaction between proteins and peptides, this type of docking is valuable for studying specific interactions, such as the binding between a peptide ligand and the active site of an enzyme. It's relevant for understanding physiological processes and developing therapeutic peptides (Agrawal et al., 2019b).
3. Protein-Small Molecule Docking: This methodology is widely employed in drug discovery, where the interaction between a small molecule (ligand) and a target protein is meticulously analyzed. The objective of this approach is to predict the conformation and binding affinity of the ligand to the active site of the protein. This prediction aids in the rational design of new pharmaceutical compounds, providing insights essential for developing effective therapeutic agents. (Kitchen et al., 2004c)
4. Nucleic Acid-Protein Docking is a specialized form of docking that investigates the interactions between nucleic acids (DNA or RNA) and proteins. This type of docking is crucial for elucidating key biological processes such as DNA replication, transcription, gene expression regulation, and ribonucleoprotein complexes. Understanding these interactions is fundamental to advancing our

knowledge of cellular mechanisms and can lead to innovations in genetic and molecular therapies. (Tuszynska et al., 2015)

Furthermore, molecular docking can be classified based on the flexibility of the molecules involved in the docking process. This classification includes:

1. Rigid-Flexible Docking: One component (usually the receptor or the protein) is treated as a rigid structure, while the other component (generally the ligand or the smaller molecule) exhibits flexibility. This flexibility may include torsional movements around bond angles or exploring different conformations that a molecule can adopt when not bound to the protein. This hybrid approach within molecular docking integrates elements of rigid and flexible docking techniques to balance computational efficiency with accuracy. This method is designed to handle the complexity of molecular interactions while managing the computational load effectively. (Dnyandev et al., 2021; Fan; Fu; Zhang, 2019a; Kaur et al., 2019; Ma; Shiu-Hin Chan; Leung, 2011a; Schneider; Böhm, 2002; Tripathi; Bankaitis, 2017a; XU; Huang; Zou, 2018).
2. Fully Flexible Docking: In this approach, both the receptor and the ligand are treated as completely flexible entities throughout the docking process. This method permits dynamic adjustment of their conformations, thereby more accurately mirroring the inherent flexibility and adaptability of biomolecular interactions. Although it is computationally demanding, fully flexible docking yields crucial insights into molecular interactions under conditions that closely resemble real-world scenarios. This is particularly advantageous for analyzing large molecules or instances where binding induces substantial conformational changes. (Dnyandev et al., 2021; Fan; Fu; Zhang, 2019a; Kaur et al., 2019; Ma; Shiu-Hin Chan; Leung, 2011a; Schneider; Böhm, 2002; Tripathi; Bankaitis, 2017a; Xu; Huang; Zou, 2018).
3. Partially Flexible Docking: This technique treats the receptor and the ligand as partially flexible during the docking process. It offers an intermediate level of flexibility that falls between the rigid and fully flexible approaches, capturing essential conformational changes more effectively while avoiding the extensive computational demands of fully flexible docking. Partially flexible docking is commonly used to strike a balance between precision and computational efficiency, making it highly suitable for studies of biomolecular interactions that require a nuanced understanding of molecular dynamics without excessive resource utilization (Dnyandev et al., 2021; Fan; Fu; Zhang, 2019a; Kaur et al., 2019; Ma; Shiu-Hin Chan; Leung, 2011a; Schneider; Böhm, 2002; Tripathi; Bankaitis, 2017a; Xu; Huang; Zou, 2018).

Furthermore, we can perform Molecular Docking with Prior Knowledge of the Binding Site. In Molecular Docking, with prior knowledge of the binding site, detailed information about the structure and characteristics of the target protein's binding site is provided. This includes the identification of key residues and structural features that are essential for ligand-protein interaction. With this data, the docking process can be specifically directed to explore and optimize interactions at the known binding site. This increases the accuracy of predictions and facilitates the selection of ligand candidates with high affinity and selectivity (Fan; Fu; Zhang, 2019b; Ma; Shiu-Hin Chan; Leung, 2011b; Tripathi; Bankaitis, 2017a).

Another method known is Molecular Docking without prior knowledge of the binding site, also known as "blind docking". In this method, the structural details and characteristics of the binding site on the target protein either need to be discovered or better understood. This scenario necessitates a more exploratory approach to docking, wherein a variety of potential binding sites across the protein's surface are evaluated to identify possible locations for ligand interaction. This method involves scanning the entire surface of the protein, systematically investigating each region to discover suitable sites that the ligand might feasibly bind to. This comprehensive surface scanning is essential because, without a predefined binding site, the algorithm must consider a broader range of possibilities, which can significantly increase the computational complexity and time required for the docking process. This kind of method involves some challenges and considerations:

1. Computational Demand: Blind docking is computationally more intensive than targeted docking because it explores many potential interactions across the entire protein surface.
2. Accuracy and Precision: The lack of specific target information can lead to less precise and potentially less accurate predictions. This is particularly true for proteins with complex or flexible structures, where defining potential binding sites without explicit data is challenging.
3. Utility: Despite these challenges, blind docking is invaluable when the binding sites on a protein are not well-characterized. It can provide novel insights into protein functionality and interaction dynamics, potentially identifying unexpected sites of activity that are not evident from the protein structure alone.

Blind docking is, therefore, a crucial tool in structural bioinformatics, particularly for novel or poorly characterized proteins. It can uncover new avenues for therapeutic intervention and enhance our understanding of protein-ligand interactions. (Fan; Fu; Zhang, 2019b; Ma; Shiu-Hin Chan; Leung, 2011b; Tripathi; Bankaitis, 2017a).

## Key Algorithms and Techniques

As we already know, Virtual Screening (VS) is a critical step in discovering new chemical compounds with pharmacological potential. It uses sophisticated algorithms and computational techniques to navigate through extensive databases of molecules. The goal is to identify those most likely to bind successfully to specific biological targets, such as proteins implicated in various diseases. (Lyne, 2002a; Malvezzi, 2008; Tripathi; Bankaitis, 2017b). The filtering methods employed in virtual screening can be divided into three main categories:

1. One- and two-dimensional (1D and 2D) filters: These filters rely on compounds' physicochemical properties and connectivity, seeking characteristics often associated with bioactive molecules or "drug-like" compounds. They examine properties such as molecular size, molecular weight, functional groups, polarity, and solubility, among others. These filters help discard compounds that do not possess the desired characteristics to efficiently interact with biological targets (Lyne, 2002a; Malvezzi, 2008; Tripathi; Bankaitis, 2017b).
2. Three-dimensional (3D) filters: Unlike 1D and 2D filters, 3D filters consider the spatial distribution of specific molecular features. They are based on three-dimensional models that represent molecular structure and look for patterns that fit into the binding sites of relevant biological targets. These filters, known as pharmacophoric filters, are especially useful for identifying molecules with appropriate three-dimensional conformations to bind to the biological targets of interest.
3. Molecular interaction filters: These filters evaluate the affinity between the compound and its biological target. They are often based on molecular docking techniques, which simulate the interaction between the candidate molecule and the active site of the biological target. Docking filters consider aspects such as binding energy, structural complementarity, and specific interactions such as hydrogen bonding and hydrophobic interactions. They are used to predict the ability of a molecule to bind to the biological target selectively and effectively (Lyne, 2002a; Malvezzi, 2008; Tripathi; Bankaitis, 2017b).

In summary, the filters applied in virtual data screening are essential for identifying and selecting candidate molecules with pharmacological potential. They streamline the drug discovery process by reducing the number of compounds that require experimental testing, thus saving time and resources and enhancing overall efficiency (Lyne, 2002a; Malvezzi, 2008; Tripathi; Bankaitis, 2017b).

Drug-like filters offer a significant advantage in rapidly processing vast amounts of data, enabling the screening of thousands of compounds per second, even on personal computers. They play a crucial role in reducing the number of compounds in databases, making identifying promising candidates more efficient. However, it's crucial to acknowledge their subjectivity and limitations. The criteria defining a "drug-like" compound are arbitrary. They may vary depending on the context, potentially leading to the erroneous exclusion of therapeutically valuable compounds and retention of less valuable ones. To address this, regular review and adjustment of filter criteria are necessary.

A prominent example is Lipinski's "rule of 5," a guideline formulated by pharmaceutical scientist Christopher A. Lipinski in 1997 to determine whether a chemical compound will likely be an orally active drug in humans. Despite its limitations, the rule remains widely used in virtual screening to identify potential drug candidates. However, it should be applied judiciously alongside other considerations to comprehensively evaluate a molecule's therapeutic potential (Lipinski et al., 2001; Malvezzi, 2008). The rules are called the "Rule of 5" because they involve four critical properties of a molecule, each of which is related to the number five:

1. Molecular weight: The compound's molecular weight should be less than 500 daltons.
2. LogP (octanol-water partition coefficient): This measures the compound's lipophilicity, which should be less than 5.
3. Hydrogen bond donors: The number of hydrogen bonds that a compound can donate should not exceed 5.
4. Hydrogen bond acceptors: The number of hydrogen bonds that a compound can accept should be no more than 10.

If a compound violates more than one of these criteria, it typically has lower chances of being effectively absorbed when administered orally. The "Rule of 5" serves as a quick screening tool in drug discovery to evaluate the oral bioavailability potential of drug candidates. While highly helpful, these guidelines are not infallible. Some orally effective drugs do not conform to these rules, often because they are absorbed through mechanisms other than passive diffusion. Then, it was designed specifically to predict oral bioavailability and does not account for drugs administered through different routes such as intravenous, subcutaneous, or topical. This limitation excludes many potentially effective drugs that do not fit the "rule of 5" criteria but are still therapeutically viable. Nonetheless, the "Rule of 5" provides a foundational basis for assessing drug-likeness in the early stages of the drug development process. (Pollastri, 2010)

## Pharmacophoric Filters in Drug Discovery

Pharmacophoric filters are essential tools in virtual compound screening for drug discovery. They utilize pharmacophoric models to represent the spatial arrangement of steric and electronic properties crucial for optimal ligand-target interactions. These models aid in identifying compounds with structural features resembling bioactive ligands, indicating potential pharmacological activity. A pharmacophore represents an abstract description of molecular features necessary for the optimal interaction with a biological target. These features typically include hydrogen bond acceptors, hydrogen bond donors, hydrophobic regions, aromatic rings, and positive or negative ionizable areas (Giordano et al., 2022; Malvezzi, 2008).

Constructing pharmacophoric models involves defining a compound's bioactive conformation, which is often challenging for flexible compounds. Methods like using minimum energy conformations of potent analogs can be employed, but model validity hinges on proper conformation selection. These models can be developed from active compound series (ligand-based), receptor structures (structure-based), or both. Ligand-based models offer independence from prior target knowledge but rely on the quality and representativeness of the compound series used. Challenges in model construction include aligning compounds to a reference and determining interaction points, especially for poorly active compounds. Despite challenges, pharmacophoric models are invaluable for identifying drug candidates. Validated models predict compound activity, optimize ligand efficacy and selectivity, and streamline drug discovery processes (Giordano et al., 2022; Malvezzi, 2008).

The anchoring filters, essential for molecular docking practice, play a crucial role in efficiently exploring and evaluating the binding space of a ligand. The scoring function, a central component of these filters, not only guides the docking program in searching for the optimal binding pose but also assesses the binding affinity once the correct pose is identified. Consequently, the predictability and accuracy of scoring functions have a direct impact on the productivity of Data-Based Virtual Screening (DBVS). A wide variety of scoring functions have been developed over the past decades, with new approaches continuously emerging. These functions can be categorized into three main types: force field-based, empirical, and knowledge-based (Charifson; Walters, 2002; Cheng et al., 2012b; Lyne, 2002b; Ma; Shiu-Hin Chan; Leung, 2011a; Malvezzi, 2008).

## Force Field-Based Scoring Functions in Drug Discovery

Force field-based scoring functions are critical in computational drug discovery for evaluating the interactions between proteins and ligands. These functions use physical chemistry principles to predict binding affinities and help identify potential drug candidates by simulating molecular interactions. It employs principles of molecular physics to calculate non-covalent ligand-target interactions, such as van der Waals and electrostatic energies. (Guedes et al., 2021). Solvation terms often complement them to capture the effects of the solvent environment. On the other hand, empirical scoring functions derive the free energy of binding from various energetic terms, including hydrogen bonding and hydrophobic interactions, by adjusting the weights of each term based on known experimental data sets. Knowledge-based scoring functions calculate ligand-target interactions using distance-dependent statistical potentials. (Charifson; Walters, 2002; Cheng et al., 2012b; Lyne, 2002b; Ma; Shiu-Hin Chan; Leung, 2011a; Malvezzi, 2008).

The performance of these scoring functions has been widely investigated through comparative studies, addressing their ability to reproduce known binding poses, predict binding affinities, and rank compound libraries. Although modern scoring functions can correctly identify the binding position of a ligand in many cases, the precise prediction of binding affinity remains a significant challenge. Various strategies have been adopted to improve the performance of existing scoring functions. These include adding additional factors to capture effects such as solvation and entropy, refining energy terms through high-level quantum calculations, and implementing consensus scores that combine multiple scoring functions (Charifson; Walters, 2002; Cheng et al., 2012b; Lyne, 2002b; Ma; Shiu-Hin Chan; Leung, 2011a; Malvezzi, 2008).

Recently, there has been a growing interest in developing target-specific scoring functions. These can be obtained by recalibrating existing generic scoring functions or adjusting their parameters to fit specific target classes. Alternatively, machine learning-based approaches have been explored, leveraging algorithms such as neural networks and support vector machines to build more accurate and robust predictive models. These advancements represent a continuous evolution in the field of scoring functions in molecular docking, driving the search for better tools and techniques for the discovery and development of new bioactive compounds (Charifson; Walters, 2002; Cheng et al., 2012b; Lyne, 2002b; Ma; Shiu-Hin Chan; Leung, 2011a; Malvezzi, 2008).

Inside VS, a variety of algorithms can be employed, each with its own advantages and limitations, such as: Artificial Intelligence Algorithms, Evolutionary Algorithms, Similarity-Based Methods, and Quality Evaluation Methods for Simulation.

## Artificial Intelligence Algorithm

1. Artificial Neural Networks (ANNs): ANNs are computational models inspired by the functioning of the human brain. They consist of multiple layers of interconnected neurons, each of which performs a simple mathematical operation.

In virtual screening, ANNs can be trained to recognize complex patterns in ligand and protein data. Once trained, they can be used to predict the interaction between ligands and proteins based on their structural characteristics (E. Graff; I. Shakhnovich; W. Coley, 2021; Gentile et al., 2022; Kimber; Chen; Volkamer, 2021; Patel et al., 2020; Sadybekov; Katritch, 2023).

2. Support Vector Machines (SVM): SVMs are supervised learning algorithms that find the optimal separating hyperplane between different classes of data. In virtual screening, SVMs can be used to classify ligands as active or inactive based on their structural characteristics, allowing the identification of potential compounds (E. Graff; I. Shakhnovich; W. Coley, 2021; Gentile et al., 2022; Kimber; Chen; Volkamer, 2021; Patel et al., 2020; Sadybekov; Katritch, 2023).
3. Bayesian Techniques: Bayesian techniques use Bayes' theorem to model the probability of a true hypothesis based on observed data. In virtual screening, Bayesian techniques can calculate the probability of a particular ligand binding to the target protein based on structural data and prior information (E. Graff; I. Shakhnovich; W. Coley, 2021; Gentile et al., 2022; Kimber; Chen; Volkamer, 2021; Patel et al., 2020; Sadybekov; Katritch, 2023).
4. Decision Trees: Decision trees are machine learning models that recursively split data into subsets based on specific features. In virtual screening, decision trees can be used to classify ligands as active or inactive based on criteria such as chemical structure, physicochemical properties, among others (E. Graff; I. Shakhnovich; W. Coley, 2021; Gentile et al., 2022; Kimber; Chen; Volkamer, 2021; Patel et al., 2020; Sadybekov; Katritch, 2023).
5. k-Nearest Neighbors (kNN): The kNN algorithm classifies a data point based on the majority class of its nearest neighbors in the feature space. In virtual screening, kNN can be used to classify ligands based on structural characteristics similar to other known ligands (E. Graff; I. Shakhnovich; W. Coley, 2021; Gentile et al., 2022; Kimber; Chen; Volkamer, 2021; Patel et al., 2020; Sadybekov; Katritch, 2023).

These algorithms are employed in virtual screening to identify potential ligands that may interact with a specific target protein. Each algorithm has its own advantages and limitations, and the choice of the appropriate algorithm depends on the data characteristics and screening objectives. In many cases, it is expected to use a combination of different algorithms to improve the accuracy and effectiveness of virtual screening (E. Graff; I. Shakhnovich; W. Coley, 2021; Gentile et al., 2022; Kimber; Chen; Volkamer, 2021; Patel et al., 2020; Sadybekov; Katritch, 2023).

## Evolutionary Algorithms

1. Genetic Algorithms (GA): Inspired by Darwin's theory of evolution, GAs are optimization algorithms that simulate the process of natural selection. In virtual screening, GAs operate on a population of candidate solutions (ligands) and apply genetic operators such as selection, crossover (recombination), and mutation to create new solutions. These solutions are evaluated based on their fitness (how well they bind to the target protein), and the process is repeated for several generations until a satisfactory solution is found (Atilgan; Hu, 2011; Opo et al., 2021).
2. Differential Evolution (DE): DE is a population-based optimization algorithm that operates on a set of candidate solutions. In virtual screening, DE generates new solutions by randomly modifying the parameter vectors of existing solutions. These new solutions are evaluated, and if they have better fitness than the current solutions, they are kept in the population. This process is repeated over several iterations to converge to an optimal solution (Atilgan; Hu, 2011; Opo et al., 2021)
3. Local and Exhaustive Search Algorithms: These are local search algorithms that explore the ligand conformational space to find favorable binding configurations. In virtual screening, these algorithms perform a local search around the initial position of the ligand, adjusting its orientation and conformation to optimize the interaction with the target protein. They use efficient heuristic methods to explore the search space and are generally faster than exhaustive methods (Holliday; Sani; Willett, 2018; Maia et al., 2020).

## Similarity-Based Methods

Substructure-Based, Pharmacophoric, Overlapping Volumes, Molecular Interaction Fields (MIFs): These methods use measures of similarity between molecules to identify candidates with characteristics similar to known ligands. In virtual screening, these methods compare the structure or properties of candidate ligands with known ligands using similarity comparison algorithms. Candidates with higher similarity are selected as potential ligands for the target protein. These algorithms, combined with evaluation techniques such as Root Mean Square Deviation (RMSD), Receiver Operating Characteristic (ROC) Curve and Area under the ROC Curve (AUC-ROC), Enrichment Factors (EFs), and Boltzmann-Enhanced Discrimination of ROC (BEDROC), provide a comprehensive approach to identifying potential ligands (Cheng et al., 2012b; Kitchen et al., 2004b; Kröger; Seidel; Gauglitz, 2003; Lyne, 2002b; Schneider; Böhm, 2002).

### Quality Evaluation Methods for Simulation

After conducting a virtual screening simulation, it is crucial to evaluate the quality of the protein-ligand complexes generated. Some of the most common methods for this evaluation include Root Mean Square Deviation (RMSD), Receiver Operating Characteristic (ROC) Curve and Area under the ROC Curve (AUC-ROC), Enrichment Factors (EFs), and Boltzmann-Enhanced Discrimination of ROC (BEDROC) (Cheng et al., 2012b; Kitchen et al., 2004b; Kröger; Seidel; Gauglitz, 2003; Lyne, 2002b; Schneider; Böhm, 2002).

RMSD is widely used to assess the accuracy of the geometry generated during molecular docking, allowing comparison of the ligand-receptor complex conformation obtained computationally with known crystallographic structures. RMSD values below 2.0 Å are generally considered acceptable. On the other hand, ROC Curve and AUC-ROC are used to evaluate the ability of VS to distinguish true positive compounds (TPCs) from false positive compounds (FPCs). The higher the area under the curve, the better the ability to separate TPCs from FPCs. AUC-ROC values above 0.7 are generally considered acceptable. Furthermore, Enrichment Factors (EFs) measure the ability of VS to retrieve active compounds compared to a random search, while BEDROC exponentially weights initial rankings of active compounds. These metrics provide a comprehensive assessment of the effectiveness of VS in identifying potential ligands (Cheng et al., 2012b; Kitchen et al., 2004b; Kröger; Seidel; Gauglitz, 2003; Lyne, 2002b; Schneider; Böhm, 2002).

## Virtual Screening on Research Application

### Integration With Machine Learning and AI

At this point, virtual screening has become a crucial technique in drug discovery, even more so by applying new computational techniques such as machine learning (ML) and artificial intelligence (AI) to continue supporting to identify potential drug candidates as efficiently and accurately as possible. These kinds of methods have demonstrated the acceleration of the identification of new drug targets, enhance the efficiency of drug development processes, and enable personalized medical therapies. (Nguyen & Wei, 2019)

The integration of machine learning and artificial intelligence uses algorithms and models to enable systems to learn from data, recognize patterns, and make decisions. This integration aims to improve automation, optimization, and prediction processes in various fields such as language processing and computer vision (Banaee; Zeidi; Faggio, 2024; Field et al., 2021; Kolluri et al., 2022; Myszczynska et al., 2020). The overview of this methodology consists of selecting and preprocessing data, choosing suitable algorithms, training models, and testing their performance. Collaboration between scientists is crucial for successful integration.

While traditional biological methods are important for assessing the potential risks of chemicals to health and the environment, they also have some limitations. For example, new findings such as biotechnological approaches can offer advantages over traditional methods. Some successful examples include computational toxicology as a complementary tool that can replace some animal testing methods. Researchers perform several validations to ensure that in silico methods are considered reliable, as Mahdi et al. showed in computational and aquatic toxicology (Banaee; Zeidi; Faggio, 2024).

For example, the African Research group with resource-limited products produced a tool based on AI and ML for quantitative structure-activity/property relationship: Quantitative Structure-Activity Relationship (QSAR) modeling. From a data set house, they developed a virtual screening cascade for malaria and tuberculosis drug discovery which can enable assays ranging from whole-cell phenotypic screening and cytotoxicity to aqueous solubility, permeability, microsomal metabolic stability, cytochrome inhibition, and cardiotoxicity. This work showed the possibility of computational profiling of compounds before synthesis and testing, which can inform the progression of novel compounds (Turon et al., 2023).

Other approaches introduce a methodology for a data-driven strategy to speed up two-dimensional (2D) virtual screening. They generated the prediction properties for the final hits using ML (Sorkun et al., 2020).

## QSAR-Based Virtual Screening: Advances and Applications in Drug Discovery

Quantitative Structure-Activity Relationship (QSAR) models utilize statistical and machine learning techniques to establish correlations between the chemical structures of molecules and their biological activities. These models are pivotal in predicting new compounds' activity and designing novel drug candidates (Neves et al., 2018). To develop QSAR models, chemical descriptors that represent molecular properties are calculated and utilized. Machine learning methods are then employed to construct models that can predict biological activities based on these descriptors.

The accuracy and reliability of these models are further confirmed through validation with experimental data. The integration of Virtual Screening (VS) with QSAR methodologies significantly augments the predictive accuracy and efficiency of the drug discovery process. For instance, applying QSAR models to pre-filter compounds prior to docking can drastically cut computational demands and enhance the success rate of identifying promising candidates (Rudrapal & Chetia, 2020). This combined approach streamlines the workflow and optimizes resource utilization in the quest for new therapeutic agents.

According to several studies, recent machine learning applications to virtual screening have been increasingly utilized across various fields, such as organic photovoltaic devices and clinical medicine. Numerous research efforts have concentrated on enhancing the precision and effectiveness of virtual screening by employing a range of machine learning techniques, such as support vector machines, neural networks, and decision trees. Successfully deploying these approaches requires meticulous data preparation, validation, and optimization procedures.

Challenges and some considerations are data privacy and security because there is a lot of collection and analysis of vast data with significant privacy and security concerns. There are further ethical implications about bias in AI, job displacement, and the precision-making AI systems process. Faster advancement of AI technology poses challenges for regulation, as laws and guidelines struggle to keep pace with innovation (Paul et al., 2021). Integrating ML and AI is about reimagining how work can make decisions and interact with the world. All these technologies continue to evolve, and their integration into various domains will undoubtedly deepen, promising to unlock new potentials and challenges.

## Application in Personalized Medicine and Multitarget Drug Design

Personalized medicine and multitarget drug design are two interrelated areas revolutionizing healthcare and pharmaceutical development. Both aim to enhance the efficacy and safety of treatments but approach this from slightly different angles. These concepts can be applied in both areas (Sorkun et al., 2020).

Precision medicine or personalized medicine adapted clinical treatment to the specific characteristics of each patient. Consider agents such as genetics, environment, and lifestyle into account to predict more precisely which treatment strategy for a particular pathology will serve which patient group (Sorkun et al., 2020). This compares to a unique approach in which disease treatment strategies are developed for the average person, with little or no regard for individual differences (Sorkun et al., 2020). The applications in personalized medicine are in genomic medicine, pharmacogenomics, biomarker discovery, and wearable technologies (Sorkun et al., 2020).

Multitarget Drug Design, an interrelated area with personalized medicine, focuses on designing and developing drugs that operate as multiple targets for complex diseases such as cancer, neurodegenerative, and cardiovascular diseases (Sorkun et al., 2020). Thus, driving various pathways can be significantly more efficient and minimize the chance of drug resistance (Sorkun et al., 2020). These approaches are applied in polypharmacology, combination therapies, computer-aided drug design

(CADD), and systems biochemicals (Sorkun et al., 2020). One recent example is the Nanomedicine for Targeted Therapy which integrates nanotechnology to enhance drug delivery, diagnosis, and treatment efficacy, particularly in cancer therapy. By engineering nanoparticles to target specific cells or tissues, nanomedicine can improve the precision and effectiveness of treatments while minimizing side effects. It ensures that the appropriate drug is delivered precisely to the correct location within the right patient, optimizing therapeutic outcomes (Gu et al., 2021) (Irvine; Dane, 2020).

Some cases report anticancer drug discovery given innovative treatment options and best-selling implementations, including personalized medicine in clinical approaches. A Nature review presented studies depicting therapeutic strategies exploiting targets conferring oncogene or non-oncogene addiction. For tumors that are oncogenic changes, genomics-guides target therapy can be an effective approach (Makhoba et al., 2020).

Target therapy can generate new vulnerabilities for non-oncogene tumors. There are strategies as dual biomarkers guide personal treatment for anticancer therapy. Biomarkers can be measured by molecular alterations associated with the drug response during the treatment and can indicate the development of acquired resistance and promote the selection of alternative therapies integrating different therapies (Makhoba et al., 2020) (Cabral; Kinoh; Kataoka, 2020)

The combination of histone deacetylase (HDAC)c and JAK (Janus Kinases) signaling inhibitors is efficient against breast cancer due to HDAC inhibition-induced feedback activation of the leukemia inhibitory factor receptor, yet it is only effective in a subset of cancer models. Studies of enhancers of zeste homolog 2 (EZH2) inhibitors allowed the development of a personalized combination strategy (Sorkun et al., 2020).

Personalized medicine and the development of multitarget drugs hold significant promise, yet they introduce complex challenges. Ethical concerns related to genetic data, the management and interpretation of vast amounts of information, and the necessity for regulations that reflect these innovative approaches are pivotal issues. Additionally, the costs associated with developing personalized therapies and ensuring their accessibility remain considerable hurdles. Despite their potential, these advances in healthcare demand a comprehensive regulatory framework and careful consideration of ethical implications. The successful advancement of these fields will require ongoing research and effective interdisciplinary collaboration, leveraging personalized nanomedicine, intelligent drug delivery systems, and advanced computational methods to transform the future of healthcare.

Personalized medicine and multitarget drug design are setting the stage for a future in which healthcare is more precise, effective, and tailored to the individual, promising significant improvements in public health. By leveraging technological advancements and integrating multidisciplinary approaches, these fields promise to revolutionize how we treat complex diseases, offering hope for better patient outcomes.

## Emerging Technologies and Methodological Improvements

The progress in machine intelligence has significantly transformed virtual screening, enabling rapid identification of hit molecules from the vast amounts of data in extensive chemical and biological databases. This leap forward is attributed to deploying sophisticated machine-learning algorithms and the surge in available computational power.

For example, according to Shen et al., 2019, the transition from machine learning to deep learning has shown promise in enhancing scoring functions for protein-ligand docking, with deep learning approaches surpassing classical methods. These new methods offer new perspectives on virtual screening by focusing on structural properties rather than partial structural templates like pharmacophores. On the other hand, the QSAR analysis has been enhanced after collecting genomics data; then chemical descriptors might be calculated and correlated with biological properties of novel compounds. (Neves; et al, 2018).

# CONCLUSION

In this chapter, we have explored the intricate realm of Virtual Screening (VS), a fundamental computational technique in the field of drug discovery and development. VS not only accelerates the identification of potential drug candidates from extensive libraries of virtual molecules but also saves significant time and resources by reducing the need for the physical testing of each compound.

Among the key strategies employed in virtual screening, Molecular Docking stands out for its ability to simulate interactions between a target molecule, such as a protein, and a pool of candidate compounds. This method leverages sophisticated algorithms to predict the optimal three-dimensional configuration of protein-ligand complexes, categorizing candidates based on their affinity and structural alignment with the target's active site. Another pivotal approach is Structure-Based Filtering, which utilizes the structural data of the target molecule to refine the search for suitable candidates. This includes techniques like pharmacophore modeling, which identifies essential structural motifs for biological activity, and molecular superposition, which assesses the geometric congruence of molecules.

Ligand-Based Filtering also plays a critical role by utilizing data from known ligands to scout for new candidates using methodologies such as Quantitative Structure-Activity Relationship (QSAR) and Molecular Similarity, aiming to find molecules with properties similar to established compounds.

The efficiency of Virtual Screening is further enhanced by its capacity to rapidly and accurately screen large compound libraries, significantly reducing the costs associated with drug discovery. By integrating principles of chemistry and molecular biology, VS boosts the likelihood of identifying novel drugs, spanning diverse areas of drug discovery from enzyme inhibitors to receptor ligands. Moreover, it is routinely paired with experimental techniques to confirm computationally derived results.

Advancements in technology have brought about the integration of machine learning and artificial intelligence with virtual screening, particularly as vast datasets become increasingly available. This synergy has shown great promise in identifying viable drug candidates swiftly, marking a significant evolution in the drug discovery process. Innovations such as machine learning-powered tools and web servers have streamlined these processes, enhancing the efficacy and reach of drug discovery efforts (Yang, et al., 2022) (Liu; et al., 2019).

As we conclude this chapter, it is clear that the VS, with cutting-edge artificial intelligence techniques, is set to revolutionize drug discovery and development, opening the way for faster, more efficient, and increasingly personalized therapeutic solutions. This promising frontier holds immense potential and promises to reshape our approach to healthcare and medicine in profound ways.

## REFERENCES


Agrawal, P., Singh, H., Srivastava, H. K., Singh, S., Kishore, G., & Raghava, G. P. S. (2019a, February 4). Benchmarking of different molecular docking methods for protein-peptide docking. *BMC Bioinformatics*, 19(13), 426. DOI: 10.1186/s12859-018-2449-y PMID: 30717654

Atilgan, E., & Hu, J. (2011, January 1). Improving Protein Docking Using Sustainable Genetic Algorithms. *International Journal of Computer Information Systems and Industrial Management Applications*, 3, 8–8.

Bajorath, J. (2002, November). Integration of virtual and high-throughput screening. *Nature Reviews. Drug Discovery*, 1(11), 882–894. DOI: 10.1038/nrd941 PMID: 12415248

Banaee, M., Zeidi, A., & Faggio, C. (2024). Application of Artificial Intelligence and Machine Learning in Computational Toxicology in Aquatic Toxicology. *Pollution*, 10(1), 210–235.

Cabral, H., Kinoh, H., & Kataoka, K. (2020, December 15). Tumor-Targeted Nanomedicine for Immunotherapy. *Accounts of Chemical Research*, 53(12), 2765–2776. DOI: 10.1021/acs.accounts.0c00518 PMID: 33161717

Charifson, P. S., & Walters, W. P. (2002). Filtering databases and chemical libraries. *Molecular Diversity*, 5(4), 185–197. DOI: 10.1023/A:1021316023707 PMID: 12549671

Cheng, T., Li, Q., Zhou, Z., Wang, Y., & Bryant, S. H. (2012b, March 1). Structure-Based Virtual Screening for Drug Discovery: A Problem-Centric Review. *The AAPS Journal*, 14(1), 133–141. DOI: 10.1208/s12248-012-9322-0 PMID: 22281989

Da, C., & Kireev, D. (2014, September 22). Structural Protein–Ligand Interaction Fingerprints (SPLIF) for Structure-Based Virtual Screening: Method and Benchmark Study. *Journal of Chemical Information and Modeling*, 54(9), 2555–2561. DOI: 10.1021/ci500319f PMID: 25116840

Dnyandev, K. M., Babasaheb, G. V., Chandrashekhar, K. V., Chandrakant, M. A., & Vasant, O. K. (2021, May). A Review on Molecular Docking. *International Research Journal of Pure and Applied Chemistry*, 20, 60–68. DOI: 10.9734/irjpac/2021/v22i330396

Fan, J., Fu, A., & Zhang, L. (2019a, June 1). Progress in molecular docking. *Quantitative Biology*, 7(2), 83–89. DOI: 10.1007/s40484-019-0172-y

Field, M., Hardcastle, N., Jameson, M., Aherne, N., & Holloway, L. (2021, July). Machine learning applications in radiation oncology. *Physics and Imaging in Radiation Oncology*, 19, 13–24. DOI: 10.1016/j.phro.2021.05.007 PMID: 34307915

Gentile, F., Yaacoub, J. C., Gleave, J., Fernandez, M., Ton, A.-T., Ban, F., Stern, A., & Cherkasov, A. (2022, March). Artificial intelligence–enabled virtual screening of ultra-large chemical libraries with deep docking. *Nature Protocols*, 17(3), 672–697. DOI: 10.1038/s41596-021-00659-2 PMID: 35121854

Giordano, D., Biancaniello, C., Argenio, M. A., & Facchiano, A. (2022, May). Drug Design by Pharmacophore and Virtual Screening Approach. *Pharmaceuticals (Basel, Switzerland)*, 15(5), 646. DOI: 10.3390/ph15050646 PMID: 35631472

Graff, E., Shakhnovich, E. I., & Coley, C. W. (2021). Accelerating high-throughput virtual screening through molecular pool-based active learning. *Chemical Science (Cambridge)*, 12(22), 7866–7881. DOI: 10.1039/D0SC06805E PMID: 34168840

Gu, W., Meng, F., Haag, R., & Zhong, Z. (2021, January). Actively targeted nanomedicines for precision cancer therapy: Concept, construction, challenges and clinical translation. *Journal of Controlled Release*, 329, 676–695. DOI: 10.1016/j.jconrel.2020.10.003 PMID: 33022328

Guedes, I. A., Barreto, A. M. S., Marinho, D., Krempser, E., Kuenemann, M. A., Sperandio, O., Dardenne, L. E., & Miteva, M. A. (2021, February 4). New machine learning and physics-based scoring functions for drug discovery. *Scientific Reports*, 11(1), 3198. DOI: 10.1038/s41598-021-82410-1 PMID: 33542326

Holliday, J., Sani, N., & Willett, P. (2018, June). Ligand-based virtual screening using a genetic algorithm with data fusion. *Match (Mülheim an der Ruhr, Germany)*, 80(3, 6).

Hollingsworth, S. A., & Dror, R. O. (2018, September 19). Molecular dynamics simulation for all. *Neuron*, 99(6), 1129–1143. DOI: 10.1016/j.neuron.2018.08.011 PMID: 30236283

Irvine, D. J., & Dane, E. L. (2020, May). Enhancing cancer immunotherapy with nanomedicine. *Nature Reviews. Immunology*, 20(5), 321–334. DOI: 10.1038/s41577-019-0269-6 PMID: 32005979

Jeon, J., Nim, S., Teyra, J., Datti, A., Wrana, J. L., Sidhu, S. S., Moffat, J., & Kim, P. M. (2014, July 30). A systematic approach to identify novel cancer drug targets using machine learning, inhibitor design and high-throughput screening. *Genome Medicine*, 6(7), 57. DOI: 10.1186/s13073-014-0057-7 PMID: 25165489

Jiang, S., Feher, M., Williams, C., Cole, B., & Shaw, D. E. (2020, September 28). AutoPH4: An Automated Method for Generating Pharmacophore Models from Protein Binding Pockets. *Journal of Chemical Information and Modeling*, 60(9), 4326–4338. DOI: 10.1021/acs.jcim.0c00121 PMID: 32639159

Kaur, T., Madgulkar, A., Bhalekar, M., & Asgaonkar, K. (2019, March 1). Molecular Docking in Formulation and Development. *Current Drug Discovery Technologies*, 16(1), 30–39. DOI: 10.2174/1570163815666180219112421 PMID: 29468973

Kimber, T. B., Chen, Y., & Volkamer, A. (2021, January). Deep Learning in Virtual Screening: Recent Applications and Developments. *International Journal of Molecular Sciences*, 22(9), 4435. DOI: 10.3390/ijms22094435 PMID: 33922714

Kitchen, D. B., Decornez, H., Furr, J. R., & Bajorath, J. (2004a, November). Docking and scoring in virtual screening for drug discovery: Methods and applications. *Nature Reviews. Drug Discovery*, 3(11), 935–949. DOI: 10.1038/nrd1549 PMID: 15520816

Koes, D. R., & Camacho, C. J. (2014, September 30). Shape-Based Virtual Screening with Volumetric Aligned Molecular Shapes. *Journal of Computational Chemistry*, 35(25), 1824–1834. DOI: 10.1002/jcc.23690 PMID: 25049193

Kolluri, S., Lin, J., Liu, R., Zhang, Y., & Zhang, W. (2022, January). Machine Learning and Artificial Intelligence in Pharmaceutical Research and Development: A Review. *The AAPS Journal*, 24(1, 4), 19. DOI: 10.1208/s12248-021-00644-3 PMID: 34984579

Kröger, K., Seidel, M., & Gauglitz, G. (2003). Optical Sensing Methods in High-Throughput Screening. In Potyrailo, R. A., & Amis, E. J. (Eds.), *High-Throughput Analysis: A Tool for Combinatorial Materials Science* (pp. 261–288). Springer US. DOI: 10.1007/978-1-4419-8989-5_13

Lipinski, C. A., Lombardo, F., Dominy, B. W., & Feeney, P. J. (2001, March 1). Experimental and computational approaches to estimate solubility and permeability in drug discovery and development settings. *Advanced Drug Delivery Reviews*, 46(1–3), 3–26. DOI: 10.1016/S0169-409X(00)00129-0 PMID: 11259830

Liu, Z., Du, J., Fang, J., Yin, Y., Xu, G., & Xie, L. (2019). Deepscreening: A deep learning-based screening web server for accelerating drug discovery. *Database (Oxford)*, 2019, baz104. Advance online publication. DOI: 10.1093/database/baz104 PMID: 31608949

Lyne, P. D. (2002a, October 15). Structure-based virtual screening: An overview. *Drug Discovery Today*, 7(20), 1047–1055. DOI: 10.1016/S1359-6446(02)02483-2 PMID: 12546894

Ma, D.-L., Shiu-Hin Chan, D., & Leung, C.-H. (2011a). Molecular docking for virtual screening of natural product databases. *Chemical Science (Cambridge)*, 2(9), 1656–1665. DOI: 10.1039/C1SC00152C

Ma, D.-L., Shiu-Hin Chan, D., & Leung, C.-H. (2011b). Molecular docking for virtual screening of natural product databases. *Chemical Science (Cambridge)*, 2(9), 1656–1665. DOI: 10.1039/C1SC00152C

Maia, E. H. B., Assis, L. C., de Oliveira, T. A., da Silva, A. M., & Taranto, A. G. (2020, April). Structure-Based Virtual Screening: From Classical to Artificial Intelligence. *Frontiers in Chemistry*, 8, 28. DOI: 10.3389/fchem.2020.00343 PMID: 32411671

Makhoba, X. H. (2020). Potential Impact of the Multi-Target Drug Approach in the Treatment of Some Complex Diseases. Drug Design, Development and Therapy, 14, 3235–3249.

Malvezzi, A. (2008). Modelos de virtual Screening de inibidores da cruzaína: desenvolvimento e validação experimental. Universidade de São Paulo.

McInnes, C. (2007). Virtual screening strategies in drug discovery. Current Opinion in Chemical Biology, Analytical Techniques / Mechanisms, 11(5), 494–502.

Morales-Helguera, A. (2015). Virtual screening tailored ensembles of QSAR models for the discovery of dual $A_{2A}$ Adenosine Receptor Antagonists / Monoamine Oxidase B Inhibitors. Proceedings of MOL2NET, International Conference on Multidisciplinary Sciences. https://sciforum.net/conference/MOL2NET-1/paper/3361

Myszczynska, M. (2020). Applications of machine learning to diagnosis and treatment of neurodegenerative diseases. Nature Reviews Neurology, 16(8), 440–456.

Neves, B. J., Braga, R. C., Melo-Filho, C. C., Moreira-Filho, J. T., Muratov, E. N., & Andrade, C. H. (2018, November 13). QSAR-Based Virtual Screening: Advances and Applications in Drug Discovery. *Frontiers in Pharmacology*, 9, 1275. DOI: 10.3389/fphar.2018.01275 PMID: 30524275

Nichols, J. A., Herbert Chan, H. W., & Baker, M. A. B. (2018, September 4). Machine learning: Applications of artificial intelligence to imaging and diagnosis. *Biophysical Reviews*, 11(1), 111–118. DOI: 10.1007/s12551-018-0449-9 PMID: 30182201

Opo, F. A. D. M., Rahman, M. M., Ahammad, F., Ahmed, I., Bhuiyan, M. A., & Asiri, A. M. (2021, February 18). Structure based pharmacophore modeling, virtual screening, molecular docking and ADMET approaches for identification of natural anti-cancer agents targeting XIAP protein. *Scientific Reports*, 11(1), 4049. DOI: 10.1038/s41598-021-83626-x PMID: 33603068

Park, J.-Y., & Harris, D. (2003, April 24). Construction and assessment of models of CYP2E1: Predictions of metabolism from docking, molecular dynamics, and density functional theoretical calculations. *Journal of Medicinal Chemistry*, 46(9), 1645–1660. DOI: 10.1021/jm020538a PMID: 12699383

Patel, L., Shukla, T., Huang, X., Ussery, D. W., & Wang, S. (2020, January). Machine Learning Methods in Drug Discovery. *Molecules (Basel, Switzerland)*, 25(22), 5277. DOI: 10.3390/molecules25225277 PMID: 33198233

Paul, D., Sanap, G., Shenoy, S., Kalyane, D., Kalia, K., & Tekade, R. K. (2021, January). TEKADE RK. Artificial intelligence in drug discovery and development. *Drug Discovery Today*, 26(1), 80–93. DOI: 10.1016/j.drudis.2020.10.010 PMID: 33099022

Peng, H., Huang, N., Qi, J., Xie, P., Xu, C., Wang, J., & Yang, C. (2003, November 3). Identification of novel inhibitors of BCR-ABL tyrosine kinase via virtual screening. *Bioorganic & Medicinal Chemistry Letters*, 13(21), 3693–3699. DOI: 10.1016/j.bmcl.2003.08.014 PMID: 14552760

Pollastri, M. P. (2010, June). Overview on the Rule of Five. *Current Protocols in Pharmacology*, 49(1). DOI: 10.1002/0471141755.ph0912s49 PMID: 22294375

Qiang, X. (2020, January 17). CPPred-FL: A sequence-based predictor for large-scale identification of cell-penetrating peptides by feature representation learning. *Briefings in Bioinformatics*, 21(1), 11–23. PMID: 30239616

Rácz, A., Bajusz, D., & Héberger, K. (2018, October 4). Life beyond the Tanimoto coefficient: Similarity measures for interaction fingerprints. *Journal of Cheminformatics*, 10(1), 48. DOI: 10.1186/s13321-018-0302-y PMID: 30288626

Rudrapal, M., & Chetia, D. (2020, July 15). Virtual Screening, Molecular Docking and QSAR Studies in Drug Discovery and Development Programme. *Journal of Drug Delivery and Therapeutics*, 10(4), 225–233. DOI: 10.22270/jddt.v10i4.4218

Sadybekov, A. V., & Katritch, V. (2023, April). Computational approaches streamlining drug discovery. *Nature*, 616(7958), 673–685. DOI: 10.1038/s41586-023-05905-z PMID: 37100941

Santana, K., do Nascimento, L. D., Lima e Lima, A., Damasceno, V., Nahum, C., Braga, R. C., & Lameira, J. (2021, April 29). Applications of Virtual Screening in Bioprospecting: Facts, Shifts, and Perspectives to Explore the Chemo-Structural Diversity of Natural Products. *Frontiers in Chemistry*, 9, 662688. DOI: 10.3389/fchem.2021.662688 PMID: 33996755

Schneider, G. (2010, April). Virtual screening: An endless staircase? *Nature Reviews. Drug Discovery*, 9(4), 273–276. DOI: 10.1038/nrd3139 PMID: 20357802

Schneider, G., & Böhm, H.-J. (2002, January 1). Virtual screening and fast automated docking methods. *Drug Discovery Today*, 7(1), 64–70. DOI: 10.1016/S1359-6446(01)02091-8 PMID: 11790605

Seddon, M. P., Cosgrove, D. A., Packer, M. J., & Gillet, V. J. (2019, January 28). Alignment-Free Molecular Shape Comparison Using Spectral Geometry: The Framework. *Journal of Chemical Information and Modeling*, 59(1), 98–116. DOI: 10.1021/acs.jcim.8b00676 PMID: 30462505

Shen, C., Ding, J., Wang, Z., Cao, D., Ding, X., & Hou, T. (2019). From machine learning to deep learning: Advances in scoring functions for protein–ligand docking. *Wiley Interdisciplinary Reviews. Computational Molecular Science*, 10(1), e1429. Advance online publication. DOI: 10.1002/wcms.1429

Shin, W.-H., Zhu, X., Bures, M., & Kihara, D. (2015, July 16). Three-Dimensional Compound Comparison Methods and Their Application in Drug Discovery. *Molecules (Basel, Switzerland)*, 20(7), 12841–12862. DOI: 10.3390/molecules200712841 PMID: 26193243

Sorkun, M. C. (2020). An artificial intelligence-aided virtual screening recipe for two-dimensional materials discovery. NPJ Computational Materials, 6(1), 106.

Sunny, S., & Jayaraj, P. B. (2022, February 1). Protein–Protein Docking: Past, Present, and Future. *The Protein Journal*, 41(1), 1–26. DOI: 10.1007/s10930-021-10031-8 PMID: 34787783

Tripathi, A., & Bankaitis, V. A. (2017). Molecular Docking: From Lock and Key to Combination Lock. Journal of Molecular Medicine and Clinical Applications, 2(1). DOI: 10.16966/2575-0305.106

Turon, G., Hlozek, J., Woodland, J. G., Kumar, A., Chibale, K., & Duran-Frigola, M. (2023, September 15). First fully-automated AI/ML virtual screening cascade implemented at a drug discovery centre in Africa. *Nature Communications*, 14(1), 5736. DOI: 10.1038/s41467-023-41512-2 PMID: 37714843

Tuszynska, I., Magnus, M., Jonak, K., Dawson, W., & Bujnicki, J. M. (2015, July 1). NPDock: A web server for protein–nucleic acid docking. *Nucleic Acids Research*, 43(W1, n. W1), W425–W430. DOI: 10.1093/nar/gkv493 PMID: 25977296

Walters, W. P., Stahl, M. T., & Murcko, M. A. (1998, April 1). Virtual screening—An overview. *Drug Discovery Today*, 3(4), 160–178. DOI: 10.1016/S1359-6446(97)01163-X

Wermuth, C. G., Ganellin, C. R., Lindberg, P., & Mitscher, L. A. (1998, January 1). Glossary of terms used in medicinal chemistry (IUPAC Recommendations 1998). *Pure and Applied Chemistry*, 70(5), 1129–1143. DOI: 10.1351/pac199870051129

Xu, X., Huang, M., & Zou, X. (2018, February 1). Docking-based inverse virtual screening: Methods, applications, and challenges. *Biophysics Reports*, 4(1), 1–16. DOI: 10.1007/s41048-017-0045-8 PMID: 29577065

Yang, J., Cai, Y., Zhao, K., Xie, H., & Chen, X. (2022, November 1). Concepts and applications of chemical fingerprint for hit and lead screening. *Drug Discovery Today*, 27(11), 103356. DOI: 10.1016/j.drudis.2022.103356 PMID: 36113834

Yang, R., Xia, Y., Jin, X., Yu, H., Yan, B., & Cheng, B. (2022). Identification of potential dual farnesol x receptor/retinoid x receptor α agonists based on machine learning models, admet prediction and molecular docking. *ChemistrySelect*, 7(28), e202200715. Advance online publication. DOI: 10.1002/slct.202200715

Yang, X., Wang, Y., Byrne, R., Schneider, G., & Yang, S. (2019, September 25). Concepts of Artificial Intelligence for Computer-Assisted Drug Discovery. *Chemical Reviews*, 119(18), 10520–10594. DOI: 10.1021/acs.chemrev.8b00728 PMID: 31294972

# Chapter 8
# An Introductory View of Bioinformatics in Food and Nutritional Science

**Vinay Kabara**
https://orcid.org/0009-0002-9021-0230
*University Institute of Chemical Technology, Kavayitri Bahinabai Chaudhari North Maharashtra University, India*

**Samruddhi Shrikant Vaidya**
*University Institute of Chemical Technology, Kavayitri Bahinabai Chaudhari North Maharashtra University, India*

**Sunil Jayant Kulkarni**
https://orcid.org/0000-0002-5988-3448
*Gharda Institute of Technology, India*

**Ajaygiri K. Goswami**
*University Institute of Chemical Technology, Kavayitri Bahinabai Chaudhari North Maharashtra University, India*

## ABSTRACT

*The multidisciplinary field of bioinformatics integrates information technology, biology, and computer science to analyze and comprehend biological data. It entails the creation and use of computational methods and tools to comprehend biological relationships and processes. The incorporation of bioinformatics into the field of food and nutritional science is examined in this chapter, with a focus on how it has revolutionized both research and practical applications. A thorough review of food science and nutritional research is given in the introduction, laying the groundwork for understanding how bioinformatics advances both disciplines. The foundations*

DOI: 10.4018/979-8-3693-3192-7.ch008

*of bioinformatics are examined, together with significant databases and computing instruments required for data processing and interpretation. This chapter explores the various uses of bioinformatics in food and nutritional research, highlighting developments that have advanced these fields.*

## INTRODUCTION

Food and Nutritional Science is a multidisciplinary field that explores the relationships between food, nutrients, and human health. It covers a wide range of topics, including the physiological, psychological, and social effects of eating and nutrition as well as the biological, chemical, and physical characteristics of food. Here's a detailed breakdown of the major components:

### Food Science

Food Science focuses on understanding the composition, processing, safety, and sensory characteristics of food. Food Chemistry studies the chemical makeup of food, including carbohydrates, proteins, fats, vitamins, and minerals, and how these components interact during processing and storage. Food Microbiology examines the microorganisms that inhabit, create, or contaminate food, including beneficial microbes (like probiotics) and pathogens that can cause foodborne illnesses. Food Processing and Engineering investigates the methods used to transform raw ingredients into consumable food products, focusing on techniques that enhance safety, shelf-life, and nutritional value. Food Safety and Quality Control ensures that food products are safe for consumption and meet quality standards through regulatory practices, hazard analysis, and critical control points (HACCP).Sensory Science Studies how food's sensory attributes (taste, smell, texture) affect consumer preferences and eating behaviours.

### Nutritional Science

Nutritional Science delves into how nutrients and other food components affect the body's health and function. *Nutrient Metabolism* investigates how the body processes and utilizes different nutrients, including macronutrients (carbohydrates, proteins, fats) and micronutrients (vitamins and minerals). *Diet and Disease Prevention* studies the role of diet in preventing and managing diseases such as obesity, diabetes, cardiovascular diseases, and cancer. *Nutritional Biochemistry* explores the biochemical pathways through which nutrients impact cellular and physiological functions.*Clinical Nutrition* applies nutritional knowledge to treat and manage illnesses, often involving personalized nutrition plans for patients.*Public Health Nutrition* focuses on promoting health and preventing disease within populations through diet and nutrition policies, education, and community programs.A basic definition of bioinformatics is an approach that manages, visualizes, and analyzes data using computer science, engineering, and mathematical methodologies in order to find new patterns and develop hypotheses and models. This covers the creation of databases, the management of data, the development of software (algorithms), simulation modeling, and quantitative analysis. (Rhee, S. Y. 2005).

The term bioinformatics refers to the study, creation, or utilization of computational methods and tools to increase the utilization of biological, medical, behavioural, health, and nutrition data. This includes gathering, organizing, storing, archiving, analyzing, visualizing, or developing biological knowledge from vast and conventionally unconnected sources. It has the potential to transform biological research and, more significantly, apply it to human problems. Since the human genome has been made available, the rice genome has been completed, other major crop plants have been mapped and sequenced, and more and more microorganisms have had their entire genomes made publicly available, bioinformatics has inevitably become an essential part of life science research and food research. The primary goal of bioinformatics is to understand the genetic and molecular mechanisms that underlie biological processes and diseases (Desiere et al., 2001). Here's a detailed explanation of the major components and applications of bioinformatics:

## CORE COMPONENTS OF BIOINFORMATICS

### Biological Data

a. Genomic Data: Includes DNA sequences, whole-genome sequencing, and genomic variations like SNPs (single nucleotide polymorphisms).
b. Proteomic Data: Involves the study of the entire set of proteins (proteomes) expressed by a genome, cell, tissue, or organism.
c. Transcriptomic Data: Refers to the study of RNA transcripts produced by the genome, including mRNA, rRNA, tRNA, and non-coding RNA.
d. Metabolomic Data: Focuses on the comprehensive study of metabolites, the small molecule intermediates, and products of metabolism.

## Data Management and Storage

a. Databases: Essential for storing and organizing large datasets. Examples include GenBank (nucleotide sequences), Protein Data Bank (PDB for 3D protein structures), and the Gene Expression Omnibus (GEO for gene expression data).
b. Data Standards: Ensuring consistency and interoperability of data formats across different databases and research studies.

## Data Analysis

a. Sequence Alignment: Comparing sequences to identify regions of similarity that may indicate functional, structural, or evolutionary relationships. Tools include BLAST (Basic Local Alignment Search Tool) and ClustalW.
b. Gene and Protein Annotation: Identifying the locations of genes and the functional elements within genomes. Tools include Ensembl and UCSC Genome Browser.
c. Structural Bioinformatics: Predicting and analyzing the 3D structures of biological molecules. Techniques include homology modeling, molecular dynamics simulations, and protein docking.
d. Phylogenetics: Studying the evolutionary relationships among various biological species or entities based on genetic information.

## Algorithm Development

a. Machine Learning: Applying algorithms to identify patterns in biological data, including classification, clustering, and regression methods.
b. Statistical Methods: Using statistical tools to validate hypotheses and analyze the significance of biological findings.

## Software Tools and Platforms

a. Bioinformatics Software: Tools like Bioconductor for statistical analysis of genomic data, and Cytoscape for network analysis.
b. Web-based Platforms: Accessible platforms such as NCBI (National Center for Biotechnology Information) and EMBL-EBI (European Molecular Biology Laboratory - European Bioinformatics Institute).

Bioinformatics aids in the comprehension of the intricate relationships that exist between genetic variables, human health, and food components in the context of food and nutritional science. This multidisciplinary area uses a variety of approaches and instruments to process vast amounts of biological data and produce insightful findings.

A major scientific field, bioinformatics has revolutionized a number of fields, including drug discovery, microbial genome applications, molecular medicine, biotechnology, comparative genomics and molecular evolution. Nevertheless, despite its acknowledged potential to make a substantial contribution to the field of food science, it is not as well-liked in this domain. Food sciences may benefit from the application of bioinformatics techniques, as evidenced by the numerous success stories linked to their use. (Saif et al.,2022)

## COMPUTATIONAL TOOLS AND DATABASES

Bioinformatic tools, which can be standalone programs or integrated into databases, enable users to search for specific sequences using homology or ontological features, align multiple sequences, view sequences in connection to other genomic features, and categorize conserved features like tandem repeats, open reading frames, coding regions, SNPs, etc. The program term BLAST (Basic Local Alignment Search Tool) is frequently used to characterize the function of this standardized tool for homologous sequence identification. Graphs showing sequence data in connection to other genomic properties are presented by genomic browsers. The available evidence from gene prediction programs, cDNAs, and visual analysis of protein sequences can be used to determine whether a gene is appropriately modelled and whether alternative splicing occurs inside a particular coding region. These browsers offer visualization ranging from megabases of sequence information to the level of individual nucleotides. (Saif et al.,2022). Table 1 lists tools chosen for initial sequence analysis.

*Table 1. Tools chosen for initial sequence analysis*

| Tool | Basic Information |
|---|---|
| Blast | This search tool looks for certain identities in DNA or protein sequences. |
| Clustal omega | With this program, multiple sequence alignments can be carried out. |
| Softberry tools | Numerous methods are dedicated to the annotation of bacterial, plant, and animal genomes as well as the prediction of RNA and protein structure and function. |

continued on following page

*Table 1. Continued*

| Tool | Basic Information |
|---|---|
| Ppp | Utilized to forecast the promoter sequences upstream of the gene, the prokaryotic promoter prediction tool |
| Hmmer | This program allows you to search for homologous protein sequences in the corresponding databases. |
| JIGSAW | Identifying genes and splicing site predictions in particular DNA sequences |
| novoSNP | Used to determine whether the DNA sequence contains a single nucleotide change. |

(Mehmood et al.,2014)

Databases are categorized as primary or secondary according to the degree of focus and breadth of their goals and scope. Primary databases are the first locations where information about protein or nucleotide sequences for various organisms is gathered and disseminated. These databases are a part of the International Nucleotide Sequence Database Collaboration (INSDC), which enables content consistency across sites and information transfers, along with Gen-Bank. Unique niches, such as distinct plant species or clades, unique gene families, or particular genomic traits, are catered to by secondary databases. A couple of examples include SOL-Solanaceae Genomics Network (SGN) and Genome Database for Rosaceae (GDR). Secondary databases might be anything from a single research lab's specific resource to a multi-institutional resource that feeds data to hundreds of labs throughout the globe. (Saif et al.,2022)

## APPLICATIONS OF BIOINFORMATICS IN FOOD AND NUTRITIONAL SCIENCE

The field of biotechnology began to suffer from data overload due to recent technological advancements and the creation of ultra-high-throughput sequencing research methods. This resulted in the growth of the rapidly expanding scientific subject of bioinformatics, which combines information technology and biology. One of the main tasks in bioinformatics is creating software tools that produce biological knowledge or information that is valuable. Thus, to put it simply, bioinformatics is the study of biology through computer use. Increasing knowledge of biological processes is the main objective of bioinformatics. The field of bioinformatics today encompasses the development of databases, algorithms, computational and statistical methods, and theories to address theoretical and practical issues related to the handling and examination of biological data. It helps with genome sequencing and annotation, as well as the annotation of detected mutations, in the fields of genetics and genomics. It is useful for organizing and querying biological data in the creation of biological and gene ontologies as well as textual mining of biological literature. (Ogbe et al.,2016).

## Advancements in Food and Nutritional Science

Considerable endeavours are undertaken to promote and enhance the nutritional worth and calibre of dietary supplies. Food science has advanced remarkably, embracing a variety of scientific approaches including as the "omics" suite. Owing to this diversification, it has become more difficult to interpret and combine different types of data, so it's critical to create and maintain a database that can hold and update the enormous amount of biological data and resources that are crucial to the nutritional and food sciences. (Sujatha et al 2022)

### Bioinformatics in Food Taste and Flavour

Food science and nutrition are becoming increasingly directly impacted by bioinformatics, which is having an effect on things like flavour and taste. Bioinformatics has played a key role in assessing the evolutionary history of taste receptors within the framework of molecular evolution. Studies on taste receptors using GWAS have also been carried out, and a connection between the regulation of glucose and bitter taste receptors has been found. More practically, electronic databases describing the chemical characteristics associated with the flavour and taste of different compounds have recently been established. These databases draw on techniques such as structural bioinformatics and docking strategies to infer the mechanisms underlying agonist binding to taste receptors. (Saif et al.,2022)

### Uncover Bioactive Peptides

The finding of bioactive peptides found in dietary proteins has recently been made possible using bioinformatics. Bioinformatics has the potential to significantly alter the field of bioactive peptide research because of its ability to evaluate various dietary proteins and proteolytic enzymes, which aid in protein digestion. Utilizing data from many databases, such as BIOPEP, is part of the strategy. (Sujatha et al 2022)

### Allergen Detection

For research pertaining to food allergens, bioinformatics is not anticipated to be the most important resource. Many databases, such as SDAP, Aller Match, Informal, FARRP Allergen database, and Aller Match are largely focused on food allergens and demonstrate the application of bioinformatics in this field. The effectiveness of these approaches has led the World Health Organization to include sequence similarity search in the guidelines for determining if genetically modified food is allergenic. Lately, machine learning has been employed to predict peanut allergy through the use of diverse bioinformatics tools for allergen diagnostic development. (Sujatha et al 2022)

## Food Safety and Quality Assurance

Food product safety and quality are greatly enhanced by the application of bioinformatics. By using genomic analysis, it makes it easier to quickly identify and characterize foodborne pathogens, which helps with outbreak investigations and the application of preventive measures. Furthermore, bioinformatics tools are used for traceability and authentication in order to identify food fraud and guarantee adherence to regulations.

Pathogenic bacteria can be sequenced to provide previously unheard-of levels of information about the relatedness of isolates, the presence of virulence and other marker genes relevant to pathogen identification and risk characterization, and other pertinent factors. This information can be obtained during outbreak investigations or information gathering for research purposes. These developments are leading to the incorporation of WGS techniques into regulatory decision making related to food microbiology; nevertheless, for effective application, a framework for quickly gathering, analysing, and reporting data from dispersed sources is needed. an illustration of how genetic and bioinformatics analysis are used to the varied global prevalence of food-borne disease linked to contamination by the important food-borne pathogen E. coli 0157. Based on molecular typing and the correlation between Shiga toxin genes (stx) subtypes and total Shiga toxin production, the phylogenetic relationships of a group of isolates from geographically disparate regions—Argentina, the United States, and Australia—have been further resolved. Strong geographical segregation of Single Nucleotide Polymorphisms (SNPs) lineages was found in the isolates upon examination, and various linkages, regionally different lineages were discovered. (Healy et al., 2016)

## Food Processing

Grain grinding, the processing of vegetable oil and sugar, meat processing and slaughter, the processing of aquatic goods, and the processing of vegetables, fruits, and nuts that come directly from forests, fisheries, agriculture, and animal husbandry are all considered forms of food processing. A subset of the general agricultural product processing business is processing activity. The food processing industry has successfully ensured the food's quality, flavour, and nutrition through the use of bioinformatics. (Xu, W. 2018)

The quality, flavour, and nutrition of food have been effectively ensured by the use of bioinformatics in the food processing industry. When baking bread, a number of physiochemical and biological changes occur at the same time, making it difficult to produce bread of a high calibre. For physiochemical and biological activities, temperature and water content are critical. These days, lasers are used in manufacturing as practical instruments for creating high power density laser beams that can be utilized for material processing, drilling, welding, and cutting. A mathematical approach that lowers the likelihood of microbial contamination has been created to estimate the depth of cutting in potato tubers. (Xu, W. 2018) and (Tyagi et al., 2024)

Processing food requires combining biological intellect with biomaterials expertise. Traditional food processing methods required a lot of energy to reassemble biomaterials from previously utilized organisms into simpler macrostructures that could produce stable food. However, when some molecules' dangerous qualities are eliminated, the majority of the time the resultant substance loses its inherent biological properties. The intelligence based on bioinformatics demonstrates the distinctive characteristics of biomolecules' complexity and their functions. This knowledge creates conditions that lead to accurate and appropriate evaluation of the biomaterial properties of individual molecules in complicated mixtures. (Waidha et al., 2015)

## Microorganism Detection

Nowadays, there is a growing need for wholesome, secure, and nutrient-dense food items with low levels of artificial preservatives. According to WHO estimates, 420 000 fatalities globally are attributed to diarrhoea illnesses, which affect over 600 million people worldwide.

Before food is served to customers, it passes through multiple manufacturing processes. Microbes, or microbiomes, contribute significantly to the fermentation and spoiling processes that occur in any nutritional matrix. Together with significant advancements in food science and nutrition, bioinformatics—which applies mathematics to the analysis of biological data has been used in recent years. The successful application of these techniques will yield knowledge that will aid in the assessment and characterization of functional genes, proteins, and metabolite-related responses associated to a given molecule's biological activity. (Chelliah et al., 2022)

While the majority of bacteria and viruses can be destroyed by heating or preparing food properly, toxins proteins that are created or retained during food preparation may still pose a risk to public health. Microbiological testing is necessary to guarantee food safety and stop food poisoning epidemics. Based on the distinct DNA sequences of the targeted viruses and diseases, PCR is typically a rapid assay to identify them.

A good method for characterizing and identifying microorganisms is mass spectrometry (MS). The combination of mass spectrometry and high-performance liquid chromatography has emerged as a primary technique for identifying toxin proteins. Targeted proteins extracted from sample solutions can be concentrated using MS in conjunction with affinity-based approaches to significantly increase the detection limits. Numerous techniques for data analysis and bioinformatics approaches with high fidelity and efficiency have been developed for the identification of microbes. Clinical microbiology and individual medical care rely heavily on completed microorganism databases and vital protein or DNA sequence data. (Xu, W. 2018).

## Nutritional Labelling and Analysis

Bioinformatics tools are employed to assess the nutritional composition of foods accurately. By integrating genomic, proteomic, and metabolomic data, bioinformatics enables comprehensive nutritional profiling, helping in the formulation of evidence-based dietary guidelines and the development of functional foods with specific health benefits.

Food analysis is a science-based field that has developed from conventional methods to sophisticated approaches as a result of advances in analytical chemistry and contemporary biotechnology. Given the increasing demand for food that is safe, nourishing, and sustainable on a global scale, omics techniques—which include genomics, metagenomics, metabolomics, transcriptomics, proteomics and genomics —have become essential to guaranteeing the authenticity and safety of food. These analytical techniques provide a thorough understanding of the composition and functionality of food. (Okoye et al., 2024)

Targeted proteomics based on mass spectrometry (MS) is an excellent technique for quantifying peptides selected from the allergen protein sequence. In this sense, the peptide marker from a precursor protein allergen can be monitored with great specificity and sensitivity using the Selected Reaction Monitoring mode (SRM), also referred to as Multiple Reaction Monitoring (MRM). The peptide marker should have been carefully selected in a prior step with the goal of being stable during processing and allergen-specific. Peptides from milk's β-lactoglobulin and α-S1-casein were used in a different investigation to successfully search for caseinates in samples of pâté, sausages, and hamburgers. Bioinformatic technologies play a crucial role in the development of novel food products by identifying new possible allergens that require additional validation. (López-Pedrouso et al., 2023).

## Food Products supply and Food Consumption

Optimizing the quantitative and compositional aspects of traditional processes in food processing will require the application of bioinformatics toolkit and intelligence. Food items undergo processing to improve safety and stability during storage. (Waidha et al., 2015)

Almost majority of the food and feed produced in the globe comes from agricultural crops. In addition to being helpful for long-term food processing, bioinformatics-based modeling also provides some assistance in predicting food production and consumption. The effect of climate change on maize and soybean production has been evaluated by combining crop modelling with plant science. Additionally, a metabolomics-driven method might be used to forecast the consumption of cocoa products. More importantly, although attempts have been attempted with disappointing outcomes, many significant nutritional and medicinal species are challenging to farm and supply adequately. The molecular mechanisms and useful proteins underlying edible mushrooms' antioxidant, antidiabetic anticancer and antibiotic properties as well as their environmental habits have been the subject of recent proteome analysis research. It is obvious that proteomic analysis of valuable dietary ingredients would essentially increase consumption and productivity. Furthermore, the production and supply of food are directly impacted by the stability and safety of food storage, and biological knowledge gained from genomes and proteomics could efficiently optimize food quality and prevent food corruption. (Xu, W. 2018).

## Agroinformatics

The most precious word is agriculture, which turns into a human necessity. Agriculture is a type of production that involves using specific plants to grow food, feed, fiber, and other products; but, in a larger sense, agriculture also refers to raising domestic animals. In recent times, there has been a swift transformation in the cultivation and agricultural process, with numerous equipment and technology being integrated to promote a healthy farming method. Information and communication technology applications drove the development of sophisticated agricultural systems. And we refer to these as "smart agriculture." Agro Informatics is the generic term for the use of IT and computing in agriculture.(Paul et al., 2020)

Combining the two fields or disciplines of "Agriculture" and "Informatics" results in Agricultural Informatics. Agroinformatics, the broad application of digital technology in agriculture and the food business, is posing a growing challenge to farmers, particularly when it comes to utilizing new instruments for gathering and interpreting intricate data. The universal demand for more effective and efficient agricultural product production while minimizing ecological damage is what propels this phenomenon. (István et al.,2019)

### Objective

a. To learn about the fundamental characteristics and operations of agro-informatics.
b. To get knowledgeable about the technologies and stakeholders in agro-informatics and the development of smart agriculture.
c. To become knowledgeable about the fundamentals, resources, and development of smart agriculture in general.
d. To acquire information about the Internet of Things about its use in agriculture and related industries.
e. To become knowledgeable about the prospects and obstacles facing the development of smart agriculture and related industries. (Paul et al., 2020).

## ADVANTAGES AND LIMITATIONS

Bioinformatics plays a pivotal role in advancing food and nutritional science through several key advantages. bioinformatics enhances our understanding of the intricate relationship between food, nutrition, and health, driving innovation and advancements in food science for better public health outcomes.

## Advantages of Bioinformatics

### Deep learning of multimodel

Future deep learning research appears to be headed in the direction of multimodal deep learning, which makes use of data from several input sources. The field of bioinformatics is likely to profit the most, since it is one where different kinds of data may be naturally assimilated. For instance, omics data, pictures, signals, medication reactions, and electronic health records are all accessible as input data, and a single image can also yield X-ray, CT, MRI, and PET forms. Many deep learning models have previously been applied in a few bioinformatics studies. For instance, researchers used brain images from MRIs and PET scans as well as cerebrospinal fluid to explore the classification of Alzheimer's disease, and Soleymani et al. used facial image and EEG signal data to study emotion recognition. (Min *et al.*, 2017).

## Developing Personalized Medicines

The application of proteomic methods to drug discovery and development is known as pharmacoproteomics. In addition to pharmacogenomics and pharmacogenetics, pharmacoproteomics will be crucial to the creation of customized medications in a number of ways. Molecular diagnostics, the cornerstone of individualized medicine, is being enhanced by proteomic technology. Compared to genotyping, pharmacoproteomics provides a more useful depiction of patient-to-patient variance. A specific target-based therapy may be matched to a specific marker in a patient subgroup with the aid of proteomics-based characterization of multifactorial disorders. Differential protein expression could serve as the basis for customized treatment instead of a genetic variation. (Jain *et al.*, 2004)

## Accelerate Deep Learning

There are three main approaches to accelerating deep learning: advanced optimization techniques, distributed and parallel computing, and specialized tools. Since the main problem for training time is the extremely long parameter optimization method using simple SGD, many studies have focused on advanced optimization algorithms. For this purpose, the most used algorithms are Adagrad, Adam, batch optimization and non-Hessian optimization. Many deep learning studies can be done with parallel and distributed computing, which can reduce execution times. These strategies use both scaling methods that use large clusters of systems in a distributed environment and scaling methods that use graphics processing units. Many deep learning frameworks, such as the one recently published, provide distributed and parallel computing capabilities. Although the development of specialized tools for deep learning is still in its infancy, significant advances are expected to make it more significant. Currently, gate-based programmable computers are being developed, and neuromorphic chips that mimic the structure of the brain are highly anticipated technologies of the future. The application of proteomics techniques to drug discovery and development is called pharmacoproteomics. In addition to pharmacogenomics and pharmacogenetics, pharmacoproteomics is important in the development of personalized medicine in several ways. Molecular diagnostics, the cornerstone of personalized medicine, is being enhanced by proteomic technologies. Compared to genetic profiling, pharmacoproteomics better describes the differences between patients. Comparisons with specific markers in subgroups of patients can be made using proteomic profiling of multiple diseases. Differential protein expression may be the basis for personalized treatment rather than genetic variation. (Min *et al.*, 2017)

## Drug Development

To screen huge compound libraries, major pharmaceutical corporations have invested in automation and built high throughput screening facilities. Bioinformatics advances have made genome-wide analysis possible for many different types of research. Researchers can examine the trillion bytes of data generated by the Human Genome Project because to bioinformatics technology. Finding new and more effective treatment targets is facilitated by the use of gene sequence databases, gene expression databases, protein sequence databases, and associated analysis tools, which all help to ascertain whether and how a specific molecule is directly engaged in a disease process. The time and expense required to develop effective pharmacological drugs could be decreased with a dependable and successful drug design methodology. The prediction of drug-likeness, or the identification and removal of candidate compounds that are unlikely to make it through the next stages of discovery and development, is done using computational approaches. Neural network-based methods and genetic algorithms could be used to predict drug-likeness. The traditional sequence of events in the pharmaceutical discovery process is as follows: drug target, lead compound, and drug. The first crucial stage in this process is the capacity to identify new therapeutic targets for additional investigation. Nearly all pharmaceuticals currently available on the market are said to have 483 pharmacological targets (45% receptors, 28% enzymes, 5% ion channels, and 2% nuclear receptors). (Gill et al., 2016)

## Open Access to Bioinformatics Data and Tools

Openness to bioinformatics data and tools is a huge benefit to research and progress. The speed of discovery in genomics, proteomics, and other fields will increase rapidly by enabling researchers around the world to work more efficiently through the open sharing of datasets and resources. Access fosters understanding and reproducibility, both of which are important for confirming results and extending previous research. Furthermore, regardless of industry affiliations and funding constraints, open access to science is vital in allowing scientists from diverse backgrounds and geographic regions to participate in research. In addition to promoting innovation, this collaborative environment improves the quality of research and favors the creation of standards and references. Ultimately, open access to bioinformatics paves the way to benefit society at large, helping the scientific community to solve complex biological problems better and more efficiently.

## Limitations and Challenges of Bioinformatics

Bioinformatics in food and nutritional science is a rapidly growing field that leverages computational tools to analyse biological data and gain insights into food composition, nutrition, and health. However, several challenges and limitations impede its progress. Here's a detailed examination of these issues:

### Standardization and Data Quality

Food and nutritional science involve heterogeneous data sources, including genomic, proteomic, metabolomic, and clinical data. Integrating these diverse datasets is challenging due to differences in data formats, quality, and standards. Ensuring the accuracy and completeness of data is crucial for reliable analysis and interpretation.

Extracting more biological knowledge from numerous datasets is a goal of any data methodology this data cannot be obtained from a single dataset alone is the primary objective of any data integration methodology. Data integration techniques must overcome numerous computational obstacles in order to achieve this goal. These difficulties result from the disparate shapes, sizes, and dimensionalities of the data that need to be combined, in addition to their complexity, noise, information content, and mutual concordance. While most existing data integration techniques barely address any of these difficulties, a number of them do so to some degree. One explanation for this is because a lot of data integration strategies are built upon techniques intended for analyzing a single type of data and then extended to handle several data types. Therefore, when used with different types of data, these approaches frequently have a number of drawbacks. For instance, when it comes to network integration, conventional approaches to network analysis are unable to capture the biological knowledge present in various networks while also accounting for their connectivity structure (topology). (Gligorijević, V et al., 2015).

### Database Limitations

While there are numerous databases for genomic, proteomic, and metabolic data, they often lack comprehensive coverage, especially for lesser-known organisms or rare variants relevant to nutrition. This limits the scope and accuracy of bioinformatic analyses.

Nowadays, a specialized set of people, including computational biologists, database curators, and software and database engineers, carry out bioinformatics. The collaborative groups of mathematicians, computer scientists, biologists, and mechanical or electrical engineers (bioengineers) are on the periphery of this. On the other hand, most biologists are at the other end of the spectrum, using the most basic bioinformatical tools. This is currently the main drawback of bioinformatics. (Rhee et al., 2005).

## Issues With Computation

High Throughput Data Processing: Complex algorithms and substantial computer resources are needed to handle and process large-scale omics data.

Algorithm Development: It can be difficult to create reliable algorithms that can recognize biomarkers from complicated datasets or forecast nutritional effects with accuracy.

The enormous amount and complexity of biological data is the main cause of the major computational constraints in bioinformatics. Large-scale dataset analyses, like those of protein structures or genome sequences, require powerful computers and effective algorithms that can handle a variety of data formats. The issue of developing algorithms that accurately mimic biological processes while maintaining computational feasibility gives rise to algorithmic challenges. Computational tasks in bioinformatics are further complicated by the need to integrate data from several sources, protect data privacy, and handle ethical issues. For the purpose of developing analytical methods and placing computational discoveries in biologically relevant contexts, it is imperative to close the transdisciplinary knowledge gaps between computer sciences and biology. To overcome these obstacles and fully utilize bioinformatics to comprehend biological systems, more computer capacity, ongoing algorithmic innovation, and cross-disciplinary cooperation are needed.

## Complexity of Biology

Complex Interactions is challenging to make definitive findings due to the complexity and incomplete understanding of the interactions between nutrients, genes, and metabolic pathways. The creation of broad dietary guidelines is made more difficult by individual differences in genetics, metabolism, and microbiome makeup, which influence nutritional requirements and responses.

From the perspectives of bioinformatics and systems biology, the problem is how to transform data into knowledge and information that can enhance the patient's delivery of suited therapy. There have been amazing developments. (Ram et al., 2012).

## Moral and Law Concerns

Techniques involving computer software or inventions connected to software ought to be covered by patents. International copyright rules are, nevertheless, supported by a different conceptual foundation. Practically speaking, software providers employ a variety of protection strategies, including contractual obligations to users, trade secret rights, disclosing "objects code," and a growing pursuit of patent protection.

Patent protection refers to the exclusive rights granted to an invention, which can be a product or a method that offers a fresh approach to an old problem or a new technical solution. Although many nations have accepted the patentability of software-related inventions to some degree, the laws pertaining to this topic are still not unified internationally. The "technical effects" doctrine, initially published by the European Patent Office (EPO), is the most commonly accepted theory affecting the extent of patent protection for software-related discoveries. According to this theory, software can usually be patent protected if its use has a "technical effect". (Duardo-Sanchez, A., & Gonzalez-Diaz, H. 2013)

## Practical and Technical Restraints

### Data Annotation

It takes a lot of manual labour and experience, as well as resources, to annotate and curate big datasets. Tool Accessibility: Because they are expensive and require specialized knowledge and training, advanced bioinformatics tools and software may not be readily available to all researchers.

### Application-Specific Challenges

Food Safety is a highly sensitive and specific technologies, which are currently under development, are needed to identify and track foodborne pathogens using bioinformatics. Nutrigenomics is research on how individual genetic diversity affects response to diet is still in its infancy, and it is still challenging to translate scientific discoveries into useful dietary recommendations.

### Resource and Economic Restraints

Funding Restrictions can be difficult to obtain funding for bioinformatics research on food and nutrition, which puts a cap on the size and scope of possible investigations. Allocation of sufficient resources, such as data storage options and high-performance computing centre, are required for infrastructure but aren't always accessible.

### Multidisciplinary Cooperation

Need for Multidisciplinary Approaches might be challenging to organize the cooperation of biologists, nutritionists, computer scientists, and statisticians for efficient bioinformatics research in food and nutrition.

### Technological Evolution

Quick Technical Advancements need constant learning and adaptability to keep up with the quick speed of technology improvements in data analysis, computing techniques, and sequencing.

# RECENT INVESTIGATIONS AND FUTURE DIRECTIONS AND INNOVATIONS

## Recent Investigations

Enhancing food safety through bioinformatics is one significant area of advancement. For instance, in an effort to reduce the risk of allergic reactions, researchers have created techniques to forecast a food's possible allergenicity. Bioinformatics algorithms are used by programs to evaluate an allergenic potential according to predetermined criteria. (Jan *et al.*, 2023)

### Foodomics

The vital significance that omics technologies play in food and nutritional science has been brought to light by recent bioinformatics investigations. To optimize human health and well-being, foodomics blends transcriptomics, metabolomics, proteomics, and genomes to conduct a thorough analysis of food and nutrition. In order to handle the massive amounts of data resulting from these methods and help researchers tackle challenging issues related to functional foods, nutritional genomics, and food safety, advanced bioinformatics systems are necessary. These developments make it possible to comprehend and improve food safety, nutritional value, and quality. (Capozzi. F., & Bordoni, A. 2013)

## Microbial informatics

The use of bioinformatics to comprehend the microbial components of food is another area of advancement. This includes using multi-platform applications for the numerical identification of microbes, which is essential for food safety and quality management, as well as genome-scale metabolic modeling of food bacteria like Lactococcus lactis to explore taste production. (Jan *et al.*, 2023)

## Nutritional Metabolomics

The use of bioinformatics to understand the microbial components of food is another area of advancement. This includes using multi-platform applications for the numerical identification of microbes, which is essential for food safety and quality management, as well as genome-scale metabolic modeling of food bacteria like Lactococcus lactis to explore taste production. These latest developments open the door for metabolome-wide association studies (MWAS) and nutritional metabolomics to work along with epigenomics, health phenotyping and genomes. to give the integrated models required for personalized diet and nutrition projections. (Jones *et al.*, 2012)

## Predictive Modeling and Personalized Nutrition

The fast rise in the rate of diabetes and impaired glucose tolerance, which is projected to affect 37% of adult U.S. population alone, is evidence that blood glucose levels are rapidly rising in the population. Type II diabetes mellitus (TIIDM) is largely predisposed to prediabetes, which is defined by persistently poor blood glucose responses. Up to 70% of prediabetics go on to acquire TIIDM. Additionally, it has been connected to additional symptoms that together make up the metabolic syndrome, such as hypertension, obesity, hypertriglyceridemia, non-alcoholic fatty liver disease, and cardiovascular disease. Therefore, it is thought that keeping blood glucose levels regular is essential for both avoiding and managing the metabolic syndrome (Zeevi *et al.*, 2015). In order to anticipate individual reactions to dietary treatments, bioinformatics technologies are used to create prediction models that incorporate multi-omics data, dietary information, and health outcomes. By enabling individualized dietary recommendations based on a person's genetic composition, metabolic profile, and health objectives, these models enhance the effectiveness of dietary treatments and disease prevention techniques.

## Emerging Trends in Bioinformatics

### Big Data Analytics

Leveraging advanced computational techniques to handle and analyze vast amounts of biological data. The development the unparalleled data in bioinformatics throughout time is a significant issue for administration and storage. It takes effective management of such vast amounts of data to spread knowledge. Information technology advances in computation offer practical analytical approaches to handle this kind of data. (Nagaraj *et al.,* 2018)

### Artificial Intelligence

Artificial intelligence is an attempt to tackle a variety of biological problems by substituting machine intelligence for human intellect. In the field of bioinformatics, recent advancements in artificial intelligence (AI) are expected to be crucial. The scientific community are very excited about machine learning and deep learning, two new areas in biological science that have great promise for the biomedical and health-informatics industries (Raj *et a*l., 2021). Using AI and machine learning to predict disease outcomes, identify potential drug targets, and automate data analysis.

### Cloud Computing

A vast amount of extra multivariate data, typically highly redundant and devoid of any obvious structure, makes up bioinformatics. Experts in the industry are gradually shifting toward machine learning and artificial intelligence approaches in conjunction with cloud computing because it is challenging to examine this data using standard compute techniques. Because cloud computing efficiently addresses the issue of a shortage of processing power and storage space, it has emerged as the most practical choice for bioinformatics data analysis. This chapter examines how cloud computing and the emergence of the intelligent cloud can improve bioinformatics (Sharma *et al.,* 2020). Utilizing cloud platforms to store, share, and process large datasets, facilitating collaboration across research institutions.

### Integrative Omics

Integrating information from transcriptomics, metabolomics, proteomics, and genomes to develop a comprehensive knowledge of biological systems. With its paradigm-shifting impact on a range of fields including molecular medicine, comparative genomics, molecular evolution, drug discovery, and biotechnology, bioinformatics has established itself as a highly influential scientific discipline. Despite the acknowledged potential for a substantial contribution to this discipline, its usefulness in the food science realm is less appreciated. (Holton et al., 2013)

## CONCLUSION

In conclusion, there are still a lot of significant problems and limitations that need to be overcome despite bioinformatics' immense potential to revolutionize the fields of food and nutritional research through better data analysis and insights. Biological complexity, computing requirements, data quality and standardization, integrating diverse data sources, and ethical issues are a few of these issues. Real-world limitations such as the need for interdisciplinary collaboration, budgetary constraints, and availability of advanced tools exacerbate the dilemma.

Coordinated efforts are required to improve computational approaches, establish standardized procedures, and foster interdisciplinary collaboration in order to overcome these challenges. Furthermore, the ethical use of bioinformatics in this field depends on controlling the legal landscape and resolving privacy concerns.

## REFERENCES


Caccia, D., Dugo, M., Callari, M., & Bongarzone, I. (2013). Bioinformatics tools for secretome analysis. *Biochimica et Biophysica Acta (BBA)-. Biochimica et Biophysica Acta. Proteins and Proteomics*, 1834(11), 2442–2453. DOI: 10.1016/j.bbapap.2013.01.039

Capozzi, F., & Bordoni, A. (2013). Foodomics: A new comprehensive approach to food and nutrition. *Genes & Nutrition*, 8(1), 1–4. DOI: 10.1007/s12263-012-0310-x PMID: 22933238

Chelliah, R., Banan-MwineDaliri, E., Khan, I., Wei, S., Elahi, F., Yeon, S.-J., Selvakumar, V., Ofosu, F. K., Rubab, M., Ju, H. H., Rallabandi, H. R., Madar, I. H., Sultan, G., & Oh, D. H. (2022). A review on the application of bioinformatics tools in food microbiome studies. *Briefings in Bioinformatics*, 23(2), bbac007. DOI: 10.1093/bib/bbac007 PMID: 35189636

Desiere, F., German, B., Watzke, H., Pfeifer, A., & Saguy, S. (2001). Bioinformatics and data knowledge: The new frontiers for nutrition and foods. *Trends in Food Science & Technology*, 12(7), 215–229. DOI: 10.1016/S0924-2244(01)00089-9

Duardo-Sanchez, A., & Gonzalez-Diaz, H. (2013). Legal issues for chem-bioinformatics models. *Frontiers in Bioscience*, 5(1), 361–374. DOI: 10.2741/E621 PMID: 23276995

Gill, S. K., Christopher, A. F., Gupta, V., & Bansal, P. (2016). Emerging role of bioinformatics tools and software in evolution of clinical research. *Perspectives in Clinical Research*, 7(3), 115–122. DOI: 10.4103/2229-3485.184782 PMID: 27453827

Gligorijević, V., & Pržulj, N. (2015). Methods for biological data integration: Perspectives and challenges. *Journal of the Royal Society, Interface*, 12(112), 20150571. DOI: 10.1098/rsif.2015.0571 PMID: 26490630

Goujon, M., McWilliam, H., Li, W., Valentin, F., Squizzato, S., Paern, J., & Lopez, R. (2010). A new bioinformatics analysis tools framework at EMBL–EBI. *Nucleic Acids Research*, 38(suppl_2), W695–W699. DOI: 10.1093/nar/gkq313 PMID: 20439314

Healy, M. J., Tong, W., Ostroff, S., Eichler, H. G., Patak, A., Neuspiel, M., Deluyker, H., & Slikker, W.Jr. (2016). Regulatory bioinformatics for food and drug safety. *Regulatory Toxicology and Pharmacology*, 80, 342–347. DOI: 10.1016/j.yrtph.2016.05.021 PMID: 27208439

Holton, T. A., Vijayakumar, V., & Khaldi, N. (2013). Bioinformatics: Current perspectives and future directions for food and nutritional research facilitated by a Food-Wiki database. *Trends in Food Science & Technology*, 34(1), 5–17. DOI: 10.1016/j.tifs.2013.08.009

Jain, K. K. (2004). Role of pharmacoproteomics in the development of personalized medicine. *Pharmacogenomics*, 5(3), 331–336. DOI: 10.1517/phgs.5.3.331.29830 PMID: 15102547

Jan, I., Sharma, P., & Bansal, A. (2023). Microbial Bioinformatics Approach in Food Science. In *Food Microbial Sustainability: Integration of Food Production and Food Safety* (pp. 267–288). Springer Nature Singapore. DOI: 10.1007/978-981-99-4784-3_13

Jeevanandam, J., Agyei, D., Danquah, M. K., & Udenigwe, C. (2022). Food quality monitoring through bioinformatics and big data. In *Future Foods* (pp. 733–744). Academic Press. DOI: 10.1016/B978-0-323-91001-9.00036-0

Jones, D. P., Park, Y., & Ziegler, T. R. (2012). Nutritional metabolomics: Progress in addressing complexity in diet and health. *Annual Review of Nutrition*, 32(1), 183–202. DOI: 10.1146/annurev-nutr-072610-145159 PMID: 22540256

López-Pedrouso, M., Lorenzo, J. M., Alché, J. D. D., Moreira, R., & Franco, D. (2023). Advanced proteomic and bioinformatic tools for predictive analysis of allergens in novel foods. *Biology (Basel)*, 12(5), 714. DOI: 10.3390/biology12050714 PMID: 37237526

Mehmood, M. A., Sehar, U., & Ahmad, N. (2014). Use of bioinformatics tools in different spheres of life sciences. *Journal of Data Mining in Genomics & Proteomics*, 5(2), 1.

Min, S., Lee, B., & Yoon, S. (2017). Deep learning in bioinformatics. *Briefings in Bioinformatics*, 18(5), 851–869. PMID: 27473064

Nagaraj, K., Sharvani, G. S., & Sridhar, A. (2018). Emerging trend of big data analytics in bioinformatics: A literature review. *International Journal of Bioinformatics Research and Applications*, 14(1-2), 144–205. DOI: 10.1504/IJBRA.2018.089175

Ogbe, R. J., Ochalefu, D. O., & Olaniru, O. B. (2016). Bioinformatics advances in genomics-A review. *International Journal of Current Research and Review*, 8(10), 5.

Okoye, C. O., Jiang, H., Nazar, M., Tan, X. L., & Jiang, J. (2024). Redefining modern food analysis: Significance of omics analytical techniques integration, chemometrics and bioinformatics. *Trends in Analytical Chemistry*, 175, 117706. DOI: 10.1016/j.trac.2024.117706

Paul, P., & Saavedra, M. (2020). Agro informatics vis-à-vis Internet of Things (IoT) integration & potentialities—An analysis. *Agro Economist-An International Journal*, 7(1), 13–20. DOI: 10.30954/2394-8159.01.2020.2

Raj, V. S., Priyadarshini, A., Yadav, M. K., Pandey, R. P., Gupta, A., & Vibhuti, A. (2021). Artificial intelligence in bioinformatics. *Biomedical Data Mining for Information Retrieval: Methodologies, Techniques and Applications*, 21-51.

Ram, P. T., Mendelsohn, J., & Mills, G. B. (2012). Bioinformatics and systems biology. *Molecular Oncology*, 6(2), 147–154. DOI: 10.1016/j.molonc.2012.01.008 PMID: 22377422

Rhee, S. Y. (2005). Bioinformatics. Current limitations and insights for the future. *Plant Physiology*, 138(2), 569–570. DOI: 10.1104/pp.104.900153 PMID: 15955919

Saif, R., & Iftikhar, A. (2022). Bioinformatics: Multiomics Applications in Food Science.

Sharma, D. K., Upadhaya, P., & Bhardwaj, S. (2020). Intelligent cloud computing and bioinformatics data analysis. In *Applications of Cloud Computing* (pp. 25–42). Chapman and Hall/CRC. DOI: 10.1201/9781003025696-2

Stephen, S. (2019). Overview of the state of the art in agroinformatics in Hungary and in international contexts. *Animal Breeding and Feeding,68*(3), 185-194.

Sujatha, P. L., Manikkavasagan, I., Kumarasamy, P., & Arvindprabhu, T. (2022). Chapter-6 Bioinformatics Approaches in Food Industry. in *Food Technology and Nutrition*, 83.

Tyagi, E., Sachan, A., & Kumari, P. (2024). Role of Bioinformatics Tools in the Food Processing Industry. *Nonthermal Food Processing, Safety, and Preservation*, 479-507.

Waidha, K. M., Jabalia, N., Singh, D., Jha, A., & Kaur, R. (2015). Bioinformatics approaches in food industry: an overview. In *National Conference on Recent Trends in Biomedical Engineering, Cancer Biology, Bioinformatics and Applied Biotechnology (BECBAB-2015) vol* (Vol. 1, pp. 1-4).

Xu, W. (2018). Application of bioinformatics technology in food processing and food safety assessment.

Zeevi, D., Korem, T., Zmora, N., Israeli, D., Rothschild, D., Weinberger, A., Ben-Yacov, O., Lador, D., Avnit-Sagi, T., Lotan-Pompan, M., Suez, J., Mahdi, J. A., Matot, E., Malka, G., Kosower, N., Rein, M., Zilberman-Schapira, G., Dohnalová, L., Pevsner-Fischer, M., & Segal, E. (2015). Personalized nutrition by prediction of glycemic responses. *Cell*, 163(5), 1079–1094. DOI: 10.1016/j.cell.2015.11.001 PMID: 26590418

# Chapter 9
# The Modern Rise of Bioinformatics Through Ecoinformatics

**Fernando García Cánovas**
*Catholic University of Murcia, Spain*

## ABSTRACT

*Bioinformatics focuses in the use of computer technology to process biological information, particularly genetic data. However, bioinformatics has revolutionized the way scientists analyze such data by addressing the challenges of biodiversity conservation and ecological research. As a result, ecoinformatics emerges from the intersection between ecological research and big data science, enabling more effective analysis, monitoring, and management of biodiversity. Adaptive management and dynamic conservation planning is now relying on the use of real-time data, citizen science and public engagement in conservation biology. The future of biodiversity conservation through ecoinformatics holds enormous potential as advanced computational tools, big data, and innovative technologies continue to evolve, providing more integrated data-management solutions and effective analysis of natural phenomena.*

## INTRODUCTION

As a PhD student in molecular ecology of pollinators in the dawn of the 21th century, my professors brought up to me a trend to a shift in the field of landscape ecology from relying on morphological differences among organisms to genetic ones. At that time, bioinformatics sounded to me as a tool for exploration of organism's genetics, rising a question to my mind about what overlap there should be between

DOI: 10.4018/979-8-3693-3192-7.ch009

bioinformatics and ecology, conservation biology. Years after, a delicious manuscript written by Stéphanie Manel, Michel Schwartz, Gordon Luikart and Pierre Tarberlet, brought to me a new emergent field of land-seascape genetics (Manel et al., 2013). My perspective changed and redirected me to the paradigm shift towards data mining. I know now that the immediate answer is a wide overlap in a rapidly growing field that has many applications by studying interactions between organisms and their environment.

Looking elsewhere for a definition, there is a common agreement about classifying bioinformatics as a scientific discipline that involves the use of computer technology to process biological information (Michener & Jones, 2012). However, that definition is also usually link to genetics as a main focus of study, accepting genetics as the ultimate source of biological information to further integrate it with biological processes at the metabolic level. Despite the fact that bioinformatics has revolutionized the way scientist analyses molecular biological data, the inherent complexity of environmental data poses the most significant challenge. Ecological data documents items of interest such as abundance of individuals, and nucleotide sequences, incorporating scattered climate and environmental descriptors throughout spatial databases from a number of sources.

Bioinformatics is rapidly growing with many known applications in ecology (LaDeau et al., 2017). The diversity of tools and algorithms allows for an optimal analysis of genetic data to identify patters and trend in the distribution of the diversity and variation among species and populations. Researchers benefit from a better understanding about the evolutionary history and identifying those factors that shape the genetic makeup of the species and populations. Results and methods are often used to model the spread of species, providing a better understand of those factors that influence both distribution and abundance. In terms, computational tools model the effects of human activities that ultimately shape the distribution and abundance of species and populations, identifying potential impacts and helping to develop suitable strategies to mitigate and protect biodiversity.

Since the article "Mathematical and computational challenges in population biology and ecosystems science" (Levin et al., 1997), a lot of exciting developments and challenges for studies of complex ecological systems have been achieved, regarding analytical and simulation approaches. From the traditional statistical approaches often distinguished between ANOVA/MANOVA, ecologists as well as geneticist have been exploiting the advances in multivariate analyses. Moreover, increasing computing capabilities have enabled data analysts to explore the comprehensive covariance structures present in complex systems more thoroughly. The enhanced capabilities for covariance analysis, along with improvements in structural equation modeling, have been facilitated more generalized and less constrained analytical methods.

## New Methods, Old Concepts

Spatial genetic structure of species and populations can be described through the combined results of seven main interacting traits and forces (Selkoe et al., 2016): demography, life history of species, spatial factors on migration rates, persistent signals of history, influences of local ecology, evolutionary noise, and up to some degree, study design (Figure 1). Ecoinformatics focuses on finding evidence of the effects of these forces by the study of the genetic structure of species and populations.

*Figure 1. Combination of interacting forces and traits that shape the spatial genetic structure of species and populations. Examples and explanations are further explained (Adapted from Selkoe et al., 2016)*

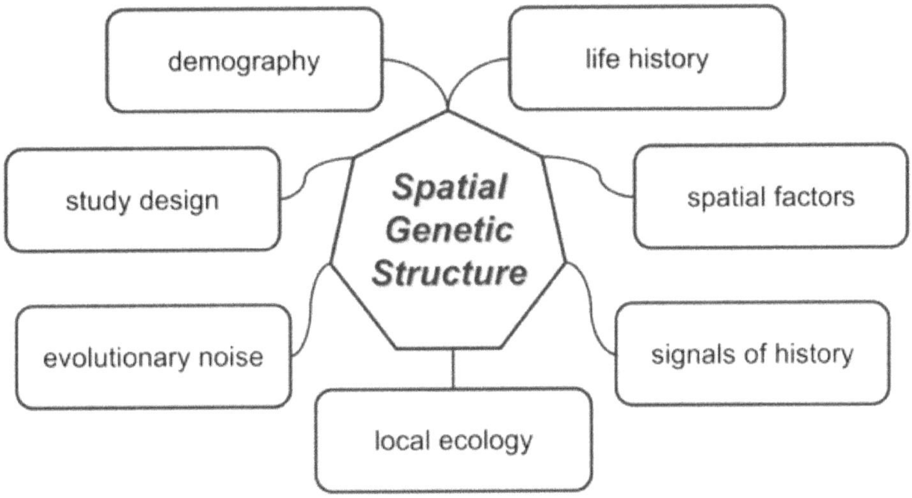


### Demography

Factors such as population size, structure, and dynamics, affect the genetic makeup of species and populations over time (Amorim, 2013). Population size plays a critical role in genetic drift. Smaller the population size, bigger the susceptibility to random fluctuations in allele frequencies, potentially leading to the loss of rare alleles and increased genetic diversity. Effective population size (Ne) is vital concept, as it often differs from the actual population size, making genetic drift more pronounced in populations with smaller Ne.

Migration patterns have deep implications across species and populations (Amorim, 2013). Birth and death rates are crucial to determine population growth, where reproductive success directly influences allele transmission. Populations experiencing expansions or bottlenecks undergo notable changes in allele frequencies. At the extreme of such changes, founder effects arise when a limited number of individuals are capable to establish a new population, leading to a gene pool that may not accurately reflect the genetic diversity of the original population. This phenomenon can result in disproportionately prevalent alleles within the new population.

## Life History Traits

Age at reproduction, lifespan, number of offspring, or parental investment shape the genetic structure and evolution of populations. Understanding life history is essential for the study of the impact of reproductive strategies, longevity, and the timing of critical life events (Selkoe et al., 2016).

The intersection of life history traits and population genetics is particularly evident in reproductive strategies. R-selected species, such as many insects, tend to produce a high volume of offspring with minimal parental investment, resulting in rapid population increases but also elevated offspring mortality rates. In contrast, K-selected species, including most mammals, focus on producing fewer offspring but invest substantial resources in their care, enhancing their survival rates.

Timing in reproduction cycles plays a critical role in gene transmission across generations. Species that reproduce at an earlier age can propagate genetic variations more swiftly, while generation time significantly impacts the rate of mutations. For instance, organisms with shorter generation times, such as bacteria, may experience rapid evolutionary changes, whereas long-lived species, like trees or mammals, tend to evolve at a slower pace. Life history theory addresses the inherent trade-offs between reproductive effort, defined as the energy expended in producing offspring, and other contemporary factors such as longevity.

The evolution of life history traits is significantly influenced by natural selection, which drives alterations in allele frequencies within populations over time. Genetic variants that improve reproductive success in specific environments are favored, leading to the adaptation of life history traits in response to various environmental factors such as predation, resource availability, and competition. Such evolutionary dynamics can result in local adaptation, which will be treated later on this chapter.

## Spatial Factors

Geographic distance is a critical factor. Migration rates generally decrease with increasing distance, a phenomenon that is classically referred to as isolation by distance (IBD). Populations situated closer together are more likely to share migrants, showing genetic similarities; greater distances, showing genetic divergence. Isolated populations contribute to more divergence as distance increases. This relationship can be quantitatively assessed using F-statistics (Holsinger & Weir, 2009), such as $F_{st}$, which indicate the level of genetic differentiation; higher $F_{st}$ values typically reflect greater divergence associated with an increase of geographic distance and reduced migration rates. In metapopulations, a larger, often fragmented population remain interconnected through migration. Migration tends to be unidirectional from high habitat quality areas (sources) to low habitat quality ones (sinks), where reproductive output is constrained, therefore sustaining genetic diversity that might otherwise be at risk of local extinction.

Physical barriers, including mountains, rivers, oceans, and deserts, can significantly prevent exchange of individuals between populations (Manni et al., 2004). Such barriers to gene flow usually lead to population fragmentation and genetic isolation, facilitating allopatric speciation processes even in habitats that appear continuous, but are indeed constrained by other micro-geographic barriers (e.g., steep inclines, bodies of water, inhospitable habitat patches, ...).

The structure of the landscape and its connectivity are both crucial factors. Continuous stretches of suitable habitat facilitate higher rates of migration, promoting gene flow. Natural corridors act as conduits enabling movements and genetic exchange between otherwise isolated populations. In contrast, landscape fragmentation, often resulting from human activities such as urbanization, road construction, and deforestation, can significantly hinder migration by dividing continuous habitats into isolated patches. Additionally, human-made translocations and introductions of species can introduce new genetic material into existing populations, which may decrease genetic differentiation or increase the prevalence of invasive species.

Clinal variations usually describe a gradual replacement of allele frequencies across a geographic area, typically as a response to a continuous environmental gradient (Cánovas et al., 2014). Although migration is not prevented genetic diversity may still exhibit predictable variations in relation to environmental factors due to selection pressures across space. Over time, these gradients can result in genetic clines, where specific alleles become more prevalent in certain regions, probably keeping imprints of local adaptation.

Legacies of historical occurrences (e.g., founder effect, population bottlenecks, population expansions, ...), could remain evident in the genetic compositions of species and populations through long periods of time from centuries to millenia (Amorim, 2013).

## Local Ecology

Local adaptation refers to the evolutionary process whereby populations develop specific characteristics that enhance their survival and reproductive success in particular environments, resulting in the increased frequency of advantageous alleles suited to local conditions (Recknagel and Michener, 2018). The factors influencing this adaptation can encompass a variety of selective pressures, including climatic variations, resource availability, predation risks, interspecies competition, and the prevalence of diseases. Plant populations situated at varying altitudes may adapt by altering their flowering times, while brown algae may develop reproductive strategies to maintain vertical gradients in the shore. Local adaptation can significantly influence the process of speciation, particularly when ecological conditions vary markedly between populations.

The interactions among species within local ecosystems frequently encompass various dynamics, including predator-prey relationships, host-parasite interactions, and mutualistic associations, such as those between plants and their pollinators. These interspecies relationships can catalyze co-evolution, a process in which species undergo reciprocal adaptations over time, resulting in genetic modifications in both parties involved.

## Evolutionary Noise

Stochastic events can significantly influence allele frequencies, genetic diversity, and population structure (Maruyama, 1983). Stochasticity introduces a level of randomness into the evolutionary process through genetic drift, mutation, recombination, and fluctuating selection pressures. These random events can shape the genetic composition of populations by influencing allele frequencies, reducing genetic diversity, and creating genetic differentiation between populations. In small populations, stochastic effects are particularly pronounced and can lead to significant changes in the genetic makeup, often overriding the effects of natural selection.

## Study Design and Bias

The integrity of genetic data is still a paramount (Hurst, 2019). Study design plays a crucial role in any research. Indeed, sampling bias can significantly affect the outcomes of genetic studies. The way a study is structured can significantly influence the findings and interpretations of genetic diversity, population structure, evolutionary history, and adaptive processes. The strategy used in the sampling process and the spatio-temporal distribution of the obtained samples are essential factors that can lead to erroneous conclusions about the existence of population structure or the extent of regional adaptations.

Low-quality samples or missing data can arise from various technical challenges (e.g., suboptimal DNA quality, insufficient sequencing coverage, errors in genotyping, ...), complicating genetic analyses and the reliability of population genetic assessments.

## Innovative Technologies Shaping Molecular Ecology Field

Next-Generation Sequencing (NGS) technologies have changed molecular ecology, by allowing researchers to sequence large amounts of DNA rapidly and affordably (Ismail, 2023). Whole-genome sequencing provides extensive information regarding genetic diversity both within and among species. This technique also facilitates the identification of loci that could confer adaptive advantages and allow the reconstruction of demographic histories of populations. Exome sequencing targets the protein-coding segments of the genome, investigating functionally significant genetic variations and their contributions to organisms' adaptation potentials. Indeed, specific genes can be linked to temperature adaptation by using exome regions, bringing light on potential population responses to the ongoing impacts of climate change.

Restriction site-associated DNA sequencing (RAD-Seq) and Genotyping-by-Sequencing (GBS) focus on sequencing specific portions of the genome previously selected (Ismail, 2023). These approaches offer a less expensive means to recover extensive genome-wide information regarding population structure and genetic variation, avoiding complete genome sequencing. Both techniques have facilitated population genetics and phylogeographic analyses in non-model organisms by producing wide datasets. An illustrative application of RAD-Seq can be found in ecological studies aimed at exploring population connectivity among marine species.

Environmental DNA (eDNA) is a powerful tool for biodiversity monitoring, allowing for the simultaneous detection of multiple species from a single environmental sample without the necessity of direct observation or capture. This technique encompasses the gathering and examination of genetic material released by organisms into their surroundings (e.g., water, soil, and air). The advent of eDNA has significantly transformed molecular ecology by facilitating the identification of rare, elusive, or cryptic species, thereby increasing our understanding about the composition of biotic communities without habitat disturbances.

Genome-editing technologies, particularly clustered regularly interspaced palindromic repeats (CRISPR), changed our comprehension about gene function and adaptation within natural populations by facilitating the precise alteration of specific genes (Gupta & Musunuru, 2014). This application is particularly important in non-model organisms. Researchers can identify targeted gene knockouts, elucidating functional significance of particular genes towards adaptation and ecological dynamics. This technological advancement has profound implications for molecular ecology, by establishing connections between genotype and phenotype and how genetic diversity shapes ecological interactions and evolutionary trajectories. For instance, CRISPR provided insights about the genetic mechanisms underlying pesticide resistance in insect and into the mechanisms of rapid evolutionary changes under anthropogenic selection pressures.

Metagenomics has facilitated the direct sequencing of genetic material from entire communities of microorganisms found in environmental and body samples (Ismail, 2023). The field of microbial ecology has greatly improved our comprehension about ecosystem functions and the complex relationships between microbiomes and their hosts. This is particularly important for biomedical studies where host health is widely affected by the diversity of microbial symbiotic species, including digestive processes and immune responses. The application of metagenomic has been used to investigate co-evolution events and adaptive strategies of hosts and associated microbiomes in response to environmental changes.

## Novel Algorithms and Analytical Techniques

The emergence of landscape and seascape genetics supposed a pivotal evolution in population genetics, emphasizing the importance of spatial ecological predictors (Manel et al., 2003; Riginos & Liggins, 2013). Data typically employed in those disciplines are categorized into two primary groups: first, biological data, which includes genomic sequences, occurrence data for various species or populations, phenotypic information, and in some extend, experimental results; second, environmental data, which encompasses spatio-temporal measurements of biotic and abiotic descriptors of the habitat. These two categories of data can be statistically correlated to uncover potential locally adapted populations, analyze spatial population structures, or discern patterns of connectivity by utilizing a suitable sampling design.

This framework is shifting with the integration of cutting-edge methodologies that require fewer assumptions about the data and promote a more exploratory approach towards biodiversity, such as Bayesian methods and agent-based modeling systems.

New challenges include managing the collective dynamics of diverse individual ensembles and scaling analyses from local areas to broader landscapes throughout time. Central to these challenges is the need to establish connections between phenomena across different spatial and temporal scales, which is a fundamental concern that transcends scientific disciplines and is crucial for the development of algorithms in high-speed computation.

## Species Distribution Modelling

Species Distribution Models (SDMs) are widely applied for decades to predict where species are likely to occur based on environmental descriptors and occurrence of species data (Paradinas et al., 2023). These models have become crucial for conservation planning, biodiversity monitoring, and understanding species' responses to environmental changes (e.g., climate change, habitat loss). A number of techniques have evolved from common multivariate statistical approaches due to increasing computation power, environmental information and combination of techniques.

Maximum entrophy (MaxEnt) is a widely recognized linear-assumption algorithm used for SDM modeling (Phillips eta al., 2017), particularly effective when working with presence-only datasets (descriptors of the occurrence of a species in space). This method estimates the likelihood of a species' occurrence by identifying the most uniform distribution of environmental variables named ecogeographical variables (EGVs). For instance, MaxEnt has been used to project the possible range alterations of species as a response to changing climate conditions. More advanced generalized additive models (GAMs) enhanced generalized linear models (GLMs) as an extension of the multiple linear models by accommodating non-linear relationships between EGVs and the distribution of diversity as occurrences/absences of species and populations (Hirzel & Le Lay, 2008). Relationships between variables are established by smooth functions of higher degrees, becoming extremely flexible. A GAM model should include any kind of non-linear polynomial method, step functions for categorical predictors, and linear models if that seems more appropriate. Because of the additivity, interpretation of the contribution of each predictor can be studied meanwhile other predictors are fixed, outperforming predictions compared to linear modelling.

## Network Analysis

These tools allow ecologists to model and analyze complex species interactions (e.g., food webs, mutualistic networks) and the structure of ecological communities (Tucker, & Duplisea, 2012). Ecological networks are capable of illustrate intricate relationships among species within an ecosystem, encompassing complex interactions such as predation, competition, and mutualism. Various network metrics (e.g., connectance, modularity, and centrality), are commonly estimated to interpret how ecosystems withstand species loss, the significance of keystone species, and the implications of invasive species on the ecological equilibrium. Bayesian networks are probabilistic graphical models that depict ecological relationships through acyclic graphs. These networks find application in ecological risk assessment and environmental management, allowing the evaluation of the effects of diverse stressors on different ecosystem components. A notable example is the use of Bayesian networks to evaluate the risks associated with habitat degradation on populations; the integration of multiple environmental stressors provides a comprehensive risk assessment.

## Agent-Based Modelling

Agent-based models (ABMs) play an important role in wildlife ecology and management (DeAngelis & Diaz, 2019; Soheilypour & Mofrad, 2018). This type of computer model allows to understand how species choose where to live and move around, a truly help in conservation planning. Let's imagine a role play where an autonomous system called agent is capable to 'think' and 'make decisions' on its own like an individual of a species or a grid in a space. These agents can move around a space that can be interpreted as a map. They interact with each other and their surroundings, making decisions based on rules like finding food, avoiding danger, or choosing a place to sleep. All these actions can be tracked to see what happens when animals interact with their environment over time. This is what an agent-based model does in silico to study real-life species and their habitats. ABMs help scientists predict how animals might behave or adapt as a response to an environmental change (e.g., climate change, human activities, ...), becoming an important tool for managers of biodiversity.

There are four main parts into making an ABM for wildlife protection (McLane, et al., 2011); i) the environment represents a map where species live. It can be real (based on actual locations) or made-up. The environment can change over time, like seasons or human development affecting animal habitats; ii) species behavior that means how they act, like choosing where to find food or where to rest, trying to guess how will behave to survive and reproduce; iii) species movement around the map, some might just wander, but others might have a goal like finding food or a mate; iv) learning and memory from past experiences, like remembering good places to find food or avoid danger. By using these models, managers can perform a better planning strategy to protect endangered animals and their habitats. Additionally, hypothetical scenarios can be created to test how species will react to changing conditions, and then use that information to make decisions.

## Machine Learning and AI Techniques

Machine learning and AI techniques are increasingly being applied to ecological datasets to model species distribution, infer spatial and temporal patterns in biodiversity allocation, and predict ecological processes (Wilson, 2024). These two approaches are capable to handle complex, high-dimensional datasets where traditional statistical models may fail.

Ensemble learning methods, such as random forests (RF), are commonly used for both classification and regression tasks (Goldstein et al., 2011; Chen & Ischwaran, 2012). Classification of ecological communities and evaluation of the significance of environmental factors are the main targets for its application. RF provides biodiversity assessments for populations through suitable SDMs that allow to identify critical environmental drivers of species diversity variations in space.

Support vector machines (SVMs) are a tool for classification, particularly in the context of non-linear relationships among variables and complex ecological predictions, such as differentiation of habitats (Hussain & CosioBorda, 2023). SVMs find wide applications in remote sensing, by classifying land cover types or discern ecological patterns from spectral data obtained by satellite imagery analysis. Another illustrative example includes the use of SVMs to detect forest loss or vegetation categorization, thereby enhancing land-use changes that have implications for administrative purposes.

Convolutional neural networks (CNNs) can be classified as deep learning techniques with increasing use for the analysis of image-based data, such as remote sensing and camera trap images, as well as for recognizing patterns in extensive ecological datasets (Borowiec et al., 2022). CNNs are excellent techniques to identify fine-scale features within images and audio recordings (bioacoustics), making them particularly valuable for automated species identification in biodiversity monitoring.

Gradient boosting machines, including XGBoost, represent a class of powerful machine learning algorithms used for both regression analysis and classification tasks (Borowiec et al., 2022; Manley et al., 2023). These algorithms build models sequentially, demonstrating a high level of effectiveness in managing unbalanced datasets.

## In the Realm of Biodiversity and Conservation

Ecoinformatics' applications span all domains of biodiversity analysis and conservation management, including agriculture, biodiversity conservation, and climate change, demonstrating its versatility and importance in contemporary ecological research.

Since 1990, the introduction of biological control and the gradually abandon of chemical control, changed of plague control strategies in greenhouse pepper in the Southeastern of Spain. Agricultural entomology enables researchers and practitioners to improve pest management and incorporate more sustainable agricultural practices such as biological control. Issues in controlling western flower thrips, *Frankliniella occidentalis*, by using chemicals, caused a transition towards biological pest control methods (Cánovas et al., 2006). In fact, biological pest control is now applied in 90% of protected pepper crops in Southeast Spain. *Orius laevigatus* Fieber is a key predator, released for thrips control. Decision-making based on periodic samplings creates a large amount of information, which can be stored and studied. The analysis of complex ecological interactions such as predation, enhances decision-making processes in agriculture. Mathematical modeling is currently employed to estimate pest population dynamics based on presence/absence data collected by agricultural consultants. This predictive capability is essential for proactive pest management, allowing farmers to implement control measures before pest populations reach damaging levels. In that case, spatio-temporal analysis of population dynamics exhibited an inconsistent spatial abundance distribution. The high *F. occidentalis* occurrence (% of occupied flowers) corresponded to areas of late *O. laevigatus* establishment. *O. laevigatus* was reintroduced during April, May, and June in places where it had previously been effectively established. Such a GIS-based release method was developed to optimize *O. laevigatus* releases in accordance to thrips population trends.

One significant application of ecoinformatics in honeybee conservation is the use of molecular tools to monitor genetic diversity and introgression in honeybee populations (Cánovas et al., 2011; Cánovas et al., 2014). The implementation of mitochondrial and nuclear DNA markers to establish and monitor conservation centers for the European populations of domestic honeybee species *Apis mellifera* L., has become essential for preserving local genetic diversity. In terms, local adaptations can enhance the resilience of honeybee populations against diseases and environmental changes. This domestic honeybee is distributed in a wide area with many subspecies, and a number of biogeographical and phylogeographical implications related to natural and anthropological factors. Honeybee populations of the Iberian Peninsula are of particular interest because of the hypothesized hybrid status of the local subspecies *Apis mellifera iberiensis* Engel, 1999, between the European *A. m. mellifera* and the African *A. m. intermissa*. In agreement with this hypothesis, a large analysis of the mitochondrial DNA showed that west European haplotypes gradually decrease from northeast to southwest and the opposite happened with African haplotypes (Cánovas et al., 2014). The gradient is smooth on the east of the Iberian Peninsula and shows the steepest transition on the northwest. Through an extensive analysis of environmental conditions from the last glacial maximum (LGM) until today, by using an extensive range of landscape analysis tools, provided key elements to establish a natural origin for this gradient that probably reflects the colonizing movements of both *A. m. mellifera* and *A. m. intermissa populations.* ABMs allowed to test these temporal and spatial hypotheses about the extension of such gradient and the influence of the environment to shape this geographical cline between species (Bioco et al., 2022). However, nuclear DNA showed a completely different imprint of anthropological influences (Cánovas et al., 2011). There was no clear geographic pattern for Iberian honeybee populations and the expected clinal pattern observed with mitochondrial data had most likely been lost due to the widespread practice of mobile beekeeping and increasing colony trade-off. Genetic homogenization was projected to rise in the coming years. Though this may have a good influence on honey production, putative ecotypes in Iberia are likely to disappear.

Ecoinformatics also plays a significant role in climate change research. By employing ecological niche modeling, researchers can assess species distribution in relation to environmental variables, which is essential for predicting how species may respond to changing climates and estimate the consequences on several aspects of human interactions with the environment. The ability to integrate diverse ecological datasets allows for a more comprehensive understanding of ecological dynamics under various climate scenarios. Moreover, this methodology has been utilized to evaluate and predict climate change impacts on marine fisheries as a consequence of the changes in important environmental systems with critical impact in the economy of local human populations. Wind is a marine coastal factor that is little understood but has a strong interaction with biological productivity as it has been described as one of the major drivers in upwelling systems. An upwelling system can be defined as an oceanographic process where deep, cold, and nutrient-rich waters rise to the surface. These systems play a crucial role in supporting highly productive marine environments. They are often associated with high levels of primary productivity and abundant marine life, including fish populations and other marine organisms.

Long-term (ICOADS: 1960-2010) and short-term (Satellite: 1989-2010) data have provided an opportunity for researchers to analyse wind trends within the Portuguese coast by using temporal and spatial analysis of large series of data. This study revealed that the northerly wind yearly data is non-stationary and highly variable between years (Leitão et al., 2019). The northerly wind intensity increased throughout the 1960s at a linear maximum rate of 0.24 m/sec per decade. That rate seems to be even higher in recent decades (1988-2009), with the wind intensity increasing by a maximum of 0.4 m/sec per decade. This long-term data analysis also allowed to unravel a major increase in northerly winds in winter (January and February), which is the recruitment season for many small and medium-sized pelagic fish. The increase in the intensity of the northerly winds over the past two decades and the past half-century occurred at a higher rate than was estimated by the IPCC for the next century.

A consequence of these results is the impact of climate change that could have severe consequences on the upwelling systems such as the northern branch of the Canary Islands, one of the most important in the world that represents a highly productive ecosystem and commercially important fisheries along the Iberian Portuguese coast (North Atlantic). Fisheries and socioeconomic impacts of climate change were analyzed by examining large collections of commercial species records that are expected to be impacted by changes in oceanographic factors projected through various alternative future climatic scenarios (Leitão & Cánovas, In Press). On one side, the combination of SDMs and spatial analysis approaches allowed the estimation of habitat vulnerability and sensitivity indexes to integrate a species' environmental appropriateness with its anticipated sensitivity to climate change. On the other side, large scale ecological and economic estimates were used to forecast the consequences of climate change on the fishing industry. Results showed that shifts in environmental variables, expected in the future, will not alter general distribution patterns for the studied species. Extensive data on species allocations and detailed environmental data at a large spatial scale, showed that species could find refuges at higher depths without losing distribution range. Predicted changes in habitat suitability values across all species varied between a decrease of 11% and an increase of 7%, with species shifts frequently recorded around ±4% regardless of the species. However, species catches by the middle of this century did not exceed the inter-annual present environmental variability effect. In general, the ecological effects of climate change will not greatly impact the economic revenue of fisheries in this area, although the overall revenue may decrease slightly, up to a maximum of 2.3%. The impact of climate change on diversity availability may not be as severe as previously thought, showing that fisheries revenue may not be at risk.

## The Future of Biodiversity Conservation Through Ecoinformatics

With increasing impacts from human populations, a better description and understanding of natural processes have become critical by incorporating more diverse kind of data. As a response, new, imaginative, integrated data-management solutions should cover all biology knowledge about the natural phenomena from genes to the biosphere (Jones et al., 2006). The future of biodiversity conservation through ecoinformatics holds enormous potential as advanced computational tools, big data, and innovative technologies continue to evolve. Ecoinformatics merges ecological research with data science, enabling more effective analysis, monitoring, and management of biodiversity (Krapivin et al., 2015). Nevertheless, there are key areas where ecoinformatics is shaping the future of biodiversity conservation.

## Improvement in Species Monitoring and Detection

Remote sensing and next generation sequencing techniques have transformed biodiversity monitoring by providing real-time, automated data collection. Combined with machine learning algorithms, these tools can automatically identify species, track population changes, and detect new species or invasive organisms in ecosystems. The integration of SDMs with climate projections enables predictive scenarios for species distribution shift in response to climate change, habitat loss, and other environmental pressures. Biologists can anticipate areas of biodiversity loss or the emergence of new suitable habitats. Remote sensing, combined with real-time data analytic, help in the detection of early signs of ecosystem stress impacts such as deforestation, habitat degradation, or shifts in vegetation health. Machine learning algorithms are providing researchers and managers a tool for the processing of complex ecological datasets to identify key biodiversity hot-spots (e.g., areas of high endemism), and vulnerable areas to human impacts. The Global Biodiversity Information Facility (GBIF) uses big data to identify conservation strategies by highlighting very diverse regions with undocumented species and areas where conservation is urgently needed.

## Adaptive Management and Dynamic Conservation Planning

Real-time conservation decision-making is a chimera in conservation biology and biodiversity management. Dynamic modeling allows simulations across management scenarios in order to predict alternative outcomes. Conservation strategies can then be continuously adjusted based on new data and changing environmental conditions, providing an adaptive management framework for establishing planning strategies. Marine protected areas (MPAs) use dynamic management models that incorporate real-time oceanographic data to adjust boundaries in order to protect species migrations or spawning grounds for larval settlement.

The rise of conservation genomics allows the monitoring of genetic diversity across populations to protect species resilient to environmental changes and diseases. Genomic data have been widely used to identify evolutionary significant units (ESUs), guiding also breeding and reintroduction programs, as well as habitat restoration initiatives.

### Citizen Science and Public Engagement in Biodiversity Conservation

Advances in mobile technologies have empowered public participation in active, real-time biodiversity monitoring through citizen science projects. In fact, citizen scientists are contributing millions of biodiversity records globally, which are used to track species migrations, monitor the spread of invasive species, and document rare species sightings. A list of ongoing citizen projects can be found in EU Citizen Science platform (https://eu-citizen.science). An increasing number of collaborative initiatives are promising a societal transformation that will be inevitably linked to scientific advancements that is also profoundly changing the policy domain in a future democratization of knowledge production.

## Brief Conclusion

Innovative and effective computational strategies are crucial for addressing the intricate complexities inherent in biological systems. These strategies should encompass the management and access of extensive datasets, alongside with the creation of simulation techniques that engage with these data frameworks. In fact, open collaboration projects, such as ecoinformatics.org, will certainly contribute to generate applications and services that will be much beneficial to the environmental and ecological sciences.

## REFERENCES


Amorim, A. (2013). Population Genetics. Brenner's Encyclopedia of Genetics, 407-411. DOI: 10.1016/B978-0-12-374984-0.01195-5

Bioco, J., Cánovas, F., Sousa, P. P., & Fazendeiro, P. (2022). SDSim: A generalized user friendly web ABM system to simulate spatiotemporal distribution of species under environmental scenarios. *Environmental Modelling & Software*, 147, 105234–105245. DOI: 10.1016/j.envsoft.2021.105234

Borowiec, M. L., Dikow, R. B., Frandsen, P. B., McKeeken, A., Valentini, G., & White, A. E. (2022). Deep learning as a tool for ecology and evolution. *Methods in Ecology and Evolution*, 13(8), 1640–1660. DOI: 10.1111/2041-210X.13901

Cánovas, F., De la Rúa, P., Serrano, J., & Galián, J. (2011). Microsatellite variability reveals beekeeping influence on Iberian honeybee populations. *Apidologie*, 42(3), 235–251. DOI: 10.1007/s13592-011-0020-1

Cánovas, F., De la Rúa, P., Serrano, J., & Galián, J. (2014). Analysing and describing a contact area of two distinct evolutionary units. An ecological perspective. *Journal of Insect Conservation*, 18(5), 927–937. DOI: 10.1007/s10841-014-9701-1

Cánovas, F., Lacasa, A., & Sánchez, J. A. (2006). Implementing a Geographical Information System (GIS) for pepper greenhouse natural enemy and pest management in Southeast Spain. *IOBC/WPRS Bulletin*, 29(4), 25–30.

Chen, X., & Ishwaran, H. (2012). Random forests for genomic data analysis. *Genomics*, 99(6), 323–329. DOI: 10.1016/j.ygeno.2012.04.003 PMID: 22546560

DeAngelis, D. L., & Diaz, S. G. (2019). Decision-Making in Agent-Based Modeling: A Current Review and Future Prospectus. *Frontiers in Ecology and Evolution*, 6, 237. DOI: 10.3389/fevo.2018.00237

Goldstein, B. A., Polley, E. C., & Briggs, F. B. (2011). Random forests for genetic association studies. *Statistical Applications in Genetics and Molecular Biology*, 10(1), 32. DOI: 10.2202/1544-6115.1691 PMID: 22889876

Gupta, R. M., & Musunuru, K. (2014). Expanding the genetic editing tool kit: ZFNs, TALENs, & CRISPR-Cas9. *The Journal of Clinical Investigation*, 124(10), 4154–4161. DOI: 10.1172/JCI72992 PMID: 25271723

Hirzel, A. H., & Le Lay, G. (2008). Habitat suitability modelling and niche theory. *Journal of Applied Ecology*, 45(5), 1372–1381. DOI: 10.1111/j.1365-2664.2008.01524.x

Holsinger, K. E., & Weir, B. S. (2009). Genetics in geographically structured populations: Defining, estimating and interpreting F(ST). *Nature Reviews. Genetics*, 10(9), 639–650. DOI: 10.1038/nrg2611 PMID: 19687804

Hurst, L. D. (2019). A century of bias in genetics and evolution. *Heredity*, 123(1), 33–43. DOI: 10.1038/s41437-019-0194-2 PMID: 31189901

Hussain, M. A. (2023). Support vector machine and neural network for enhanced classification algorithm in ecological data. Measurement. *Measurement. Sensors*, 27, 100780. DOI: 10.1016/j.measen.2023.100780

Ismail, H. D. (2023). *Bioinformatics: A Practical Guide to Next Generation Sequencing Data Analysis* (1st ed.). Chapman and Hall/CRC. DOI: 10.1201/9781003355205

Jones, M. B., Schildhauer, M. P., Reichman, O. J., & Bowers, S. (2006). The New Bioinformatics: Integrating Ecological Data from the Gene to the Biosphere. *Annual Review of Ecology, Evolution, and Systematics*, 37(1), 519–544. DOI: 10.1146/annurev.ecolsys.37.091305.110031

Krapivin, V., Varotsos, C., & Soldatov, V. (2015). *New Ecoinformatics Tools in Environmental Science*. Springer Cham. DOI: 10.1007/978-3-319-13978-4

LaDeau, S. L., Han, B. A., Rosi-Marshall, E. J., & Weathers, K. C. (2017). The Next Decade of Big Data in Ecosystem Science. *Ecosystems (New York, N.Y.)*, 20(2), 274–283. DOI: 10.1007/s10021-016-0075-y

Leitão, F., & Cánovas, F. (in press). Predicting climate change impacts on marine fisheries and biodiversity in the Iberian north branch of the canary upwelling system. *Ecological Economics*.

Leitão, F., Relvas, P., Cánovas, F., Baptista, V., & Teodósio, A. (2019). Northerly wind trends along the Portuguese marine coast since 1950. *Theoretical and Applied Climatology*, 137(1-2), 1–19. DOI: 10.1007/s00704-018-2466-9

Manel, S., Schwartz, M., Luikart, G., & Tarberlet, P. (2003). Landscape genetics: Combining landscape ecology and population genetics. *Trends in Ecology & Evolution*, 18(4), 189–197. DOI: 10.1016/S0169-5347(03)00008-9

Manley, W., Tran, T., Prusinski, M., & Brisson, D. (2023). Modeling Tick Populations: An Ecological Test Case for Gradient Boosted Trees. bioRxiv: the preprint server for biology. DOI: 10.1101/2023.03.13.532443

Manni, F., Guérard, E., & Heyer, E. (2004). Geographic patterns of (genetic, morphologic, linguistic) variation: How barriers can be detected by using Monmonier's algorithm. *Human Biology*, 76(2), 173–190. DOI: 10.1353/hub.2004.0034 PMID: 15359530

Maruyama, T. (1983). Stochastic theory of population genetics. *Bulletin of Mathematical Biology*, 45(4), 521–554. DOI: 10.1007/BF02459586

McLane, A. J., Semeniuk, C., McDermid, G. J., & Marceau, D. J. (2011). The role of agent-based models in wildlife ecology and management. *Ecological Modelling*, 222(8), 1544–1556. DOI: 10.1016/j.ecolmodel.2011.01.020

Michener, W., & Jones, M. (2012). Ecoinformatics: Supporting ecology as a data-intensive science. *Trends in Ecology & Evolution*, 27(2), 85–93. DOI: 10.1016/j.tree.2011.11.016 PMID: 22240191

Paradinas, I., Illian, J., & Smout, S. (2023). Understanding spatial effects in species distribution models. *PLoS One*, 18(5), e0285463. DOI: 10.1371/journal.pone.0285463 PMID: 37253039

Phillips, S. J., Anderson, R. P., Dudík, M., Schapire, R. E., & Blair, M. E. (2017). Opening the black box: An open-source release of Maxent. *Ecography*, 40(7), 887–893. DOI: 10.1111/ecog.03049

Recknagel, F., & Michener, W. K. (2018). *Ecological Informatics*. Springer Cham. DOI: 10.1007/978-3-319-59928-1

Riginos, C., & Liggins, L. (2013). Seascape Genetics: Populations, Individuals, and Genes Marooned and Adrift. *Geography Compass*, 7(3), 197–216. DOI: 10.1111/gec3.12032

Selkoe, K., D'Aloia, C., Crandall, E., Iacchei, M., Liggins, L., Puritz, J., Von Der Heyden, S., & Toonen, R. (2016). A decade of seascape genetics: Contributions to basic and applied marine connectivity. *Marine Ecology Progress Series*, 554, 1–19. DOI: 10.3354/meps11792

Soheilypour, M., & Mofrad, M. R. K. (2018). Agent-Based Modeling in Molecular Systems Biology. *BioEssays*, 40(7), 1800020. DOI: 10.1002/bies.201800020 PMID: 29882969

Tucker, A., & Duplisea, D. (2012). Bioinformatics tools in predictive ecology: Applications to fisheries. *Philosophical Transactions of the Royal Society of London. Series B, Biological Sciences*, 367(1586), 279–290. DOI: 10.1098/rstb.2011.0184 PMID: 22144390

Wilson, N. (2024). Artificial intelligence helps drive new frontiers in ecology. *Bioscience*, 74(5), 306–311. DOI: 10.1093/biosci/biae016 PMID: 38854635

# Compilation of References


Abdul Aziz, M., Brini, F., Rouached, H., & Masmoudi, K. (2022). Genetically engineered crops for sustainably enhanced food production systems. DOI: 10.3389/fpls.2022.1027828

Abraham, M. J., Murtola, T., Schulz, R., Páll, S., Smith, J. C., Hess, B., & Lindah, E. (2015). GROMACS: High performance molecular simulations through multi-level parallelism from laptops to supercomputers. *SoftwareX*, 1–2, 19–25. DOI: 10.1016/j.softx.2015.06.001

Adam, Y., Samtal, C., Brandenburg, J. T., Falola, O., & Adebiyi, E. (2021). Performing post-genome-wide association study analysis: Overview, challenges and recommendations. *F1000 Research*, 10, 1002. DOI: 10.12688/f1000research.53962.1 PMID: 35222990

Adelusi, T. I., Oyedele, A.-Q. K., Boyenle, I. D., Ogunlana, A. T., Adeyemi, R. O., Ukachi, C. D., Idris, M. O., Olaoba, O. T., Adedotun, I. O., Kolawole, O. E., Xiaoxing, Y., & Abdul-Hammed, M. (2022). Molecular modeling in drug discovery. *Informatics in Medicine Unlocked*, 29, 100880. DOI: 10.1016/j.imu.2022.100880

Agrawal, P., Singh, H., Srivastava, H. K., Singh, S., Kishore, G., & Raghava, G. P. S. (2019a, February 4). Benchmarking of different molecular docking methods for protein-peptide docking. *BMC Bioinformatics*, 19(13), 426. DOI: 10.1186/s12859-018-2449-y PMID: 30717654

Aiman, S., Ahmad, A., Khan, A., Ali, Y., Malik, A., Alkholief, M., & Jalil, F. (2023). Vaccinomics-aided next-generation novel multi-epitope-based vaccine engineering against multidrug resistant Shigella Sonnei: Immunoinformatics and chemoinformatics approaches. *PLoS One*, 18(11), e0289773. DOI: 10.1371/journal.pone.0289773 PMID: 37992050

Akaçin, İ., Ersoy, Ş., Doluca, O., & Güngörmüşler, M. (2022). Comparing the significance of the utilization of next generation and third generation sequencing technologies in microbial metagenomics. *Microbiological Research*, 264, 127154. DOI: 10.1016/j.micres.2022.127154 PMID: 35961096

Akash, A., Bencurova, E., & Dandekar, T. (2024). How to make DNA data storage more applicable. *Trends in Biotechnology*, 42(1), 17–30. DOI: 10.1016/j.tibtech.2023.07.006 PMID: 37591721

Alamin, M., Sultana, M. H., Xu, H., & Mollah, M. N. H. (2020). Robustification of Linear Regression and Its Application in Genome-Wide Association Studies. *Frontiers in Genetics*, 11, 549. DOI: 10.3389/fgene.2020.00549 PMID: 32582288

Al-Amrani, S., Al-Jabri, Z., Al-Zaabi, A., Alshekaili, J., & Al-Khabori, M. (2021). Proteomics: Concepts and applications in human medicine. *World Journal of Biological Chemistry*, 12(5), 57–69. DOI: 10.4331/wjbc.v12.i5.57 PMID: 34630910

Alberts. (2017). *Biologia molecular da célula*. Artmed Editora.

Alberts, B., Bray, D., Hopkin, K., Johnson, A., Lewis, J., Raff, M., Roberts, K., Walter, P., Andrade, A. E. B., & Renard, G. (2017). *Fundamentos da Biologia Celular* (4ª edição). *The Art of Medication.*

AlChalabi, R., Al-Rahim, A., Omer, D., & Suleiman, A. A. J. (2023). Immunoinformatics design of multi-epitope peptide-based vaccine against Haemophilus influenzae strain using cell division protein. *Network Modeling and Analysis in Health Informatics and Bioinformatics*, 12(1), 1. DOI: 10.1007/s13721-022-00395-x PMID: 36465492

Alex, A., Brundha, M., & Prathap, L. (2020). Sanger sequencing and its recent advances-a review. *PalArch's Journal of Archaeology of Egypt/Egyptology, 17*(7), 698-705.

Alföldi, J., & Lindblad-Toh, K. (2013). Comparative genomics as a tool to understand evolution and disease. *Genome Research*, 23(7), 1063–1068. DOI: 10.1101/gr.157503.113 PMID: 23817047

Ali, M., Pandey, R. K., Khatoon, N., Narula, A., Mishra, A., & Prajapati, V. K. (2017). Exploring dengue genome to construct a multi-epitope based subunit vaccine by utilizing immunoinformatics approach to battle against dengue infection. *Scientific Reports, 7*(1), 1–13. DOI: 10.1038/s41598-017-09199-w

Al-Mekhlal, M., & Ali Khwaja, A. (2019). A synthesis of big data definition and characteristics. *Proceedings - 22nd IEEE International Conference on Computational Science and Engineering and 17th IEEE International Conference on Embedded and Ubiquitous Computing, CSE/EUC 2019*, 314–322. DOI: 10.1109/CSE/EUC.2019.00067

Alquraishi, M. (2019, November 15). AlphaFold at CASP13. *Bioinformatics (Oxford, England)*, 35(22), 4862–4865. DOI: 10.1093/bioinformatics/btz422 PMID: 31116374

Altschul. (1990). Basic local alignment search tool. Journal of Molecular Biology, 215(3), 403-410.

Altschul, S. F., Gish, W., Miller, W., Myers, E. W., & Lipman, D. J. (1990). Basic local alignment search tool. *Journal of Molecular Biology*, 215(3), 403–410. DOI: 10.1016/S0022-2836(05)80360-2 PMID: 2231712

Amorim, A. (2013). Population Genetics. Brenner's Encyclopedia of Genetics, 407-411. DOI: 10.1016/B978-0-12-374984-0.01195-5

Amos, E. O., Araoyinbo, O. S., Akinleye, E. O., Alakanse, S. O., Bamikole, A. O., & Kolawole, O. M. (2023). Immunoinformatics design of multi-epitope vaccine using surface cell antigen OmpB and heat shock protein GroEL against rickettsioses. *Informatics in Medicine Unlocked*, 43, 101411. DOI: 10.1016/j.imu.2023.101411

Anderson, E. (1936). The species problem in Iris. *Annals of the Missouri Botanical Garden*, 23(3), 457–509. DOI: 10.2307/2394164

Anfinsen, C. B. (1973, July 20). Principles that Govern the Folding of Protein Chains. *Science*, 181(4096), 223–230. DOI: 10.1126/science.181.4096.223 PMID: 4124164

Anishchenko, I., Pellock, S. J., Chidyausiku, T. M., Ramelot, T. A., Ovchinnikov, S., Hao, J., Bafna, K., Norn, C., Kang, A., Bera, A. K., DiMaio, F., Carter, L., Chow, C. M., Montelione, G. T., & Baker, D. (2021, December). De novo protein design by deep network hallucination. *Nature*, 600(7889), 547–552. DOI: 10.1038/s41586-021-04184-w PMID: 34853475

Anstine, D. M., & Isayev, O. (2023). Generative models as an emerging paradigm in the chemical sciences. *Journal of the American Chemical Society*, 145(16), 8736–8750. DOI: 10.1021/jacs.2c13467 PMID: 37052978

Arnold, K., Bordoli, L., Kopp, J., & Schwede, T. (2006). The SWISS-MODEL workspace: A web-based environment for protein structure homology modelling. *Bioinformatics (Oxford, England)*, 22(2), 195–201. DOI: 10.1093/bioinformatics/bti770 PMID: 16301204

Athanasopoulou, K., Boti, M. A., Adamopoulos, P. G., Skourou, P. C., & Scorilas, A. (2021). Third-Generation Sequencing: The Spearhead towards the Radical Transformation of Modern Genomics. *Life (Basel, Switzerland)*, 12(1), 30. Advance online publication. DOI: 10.3390/life12010030 PMID: 35054423

Atilgan, E., & Hu, J. (2011, January 1). Improving Protein Docking Using Sustainable Genetic Algorithms. *International Journal of Computer Information Systems and Industrial Management Applications*, 3, 8–8.

Auslander, N., Gussow, A. B., & Koonin, E. V. (2021). Incorporating machine learning into established bioinformatics frameworks. *International Journal of Molecular Sciences*, 22(6), 2903. DOI: 10.3390/ijms22062903 PMID: 33809353

Ayyagari, V. S., Venkateswarulu, T. C., Abraham Peele, K., & Srirama, K. (2022). Design of a multi-epitope-based vaccine targeting M-protein of SARS-CoV2: An immunoinformatics approach. *Journal of Biomolecular Structure & Dynamics*, 40(7), 2963–2977. DOI: 10.1080/07391102.2020.1850357 PMID: 33252008

Bagnoli, F., Baudner, B., Mishra, R. P. N., Bartolini, E., Fiaschi, L., Mariotti, P., Nardi-Dei, V., Boucher, P., & Rappuoli, R. (2011). Designing the Next Generation of Vaccines for Global Public Health. DOI: 10.1089/omi.2010.0127

Bahrami, A. A., Payandeh, Z., Khalili, S., Zakeri, A., & Bandehpour, M. (2019). Immunoinformatics: In Silico Approaches and Computational Design of a Multi-epitope, Immunogenic Protein. *International Reviews of Immunology*, 38(6), 307–322. DOI: 10.1080/08830185.2019.1657426 PMID: 31478759

Bajorath, J. (2002, November). Integration of virtual and high-throughput screening. *Nature Reviews. Drug Discovery*, 1(11), 882–894. DOI: 10.1038/nrd941 PMID: 12415248

Bali, A., & Bali, N. (2022). Role of artificial intelligence in fast-track drug discovery and vaccine development for COVID-19. *Novel AI and Data Science Advancements for Sustainability in the Era of COVID*, 19, 201–229. DOI: 10.1016/B978-0-323-90054-6.00006-4

Baljon, J. J., & Wilson, J. T. (2022). Bioinspired vaccines to enhance MHC class-I antigen cross-presentation. *Current Opinion in Immunology*, 77, 102215. DOI: 10.1016/j.coi.2022.102215 PMID: 35667222

Banaee, M., Zeidi, A., & Faggio, C. (2024). Application of Artificial Intelligence and Machine Learning in Computational Toxicology in Aquatic Toxicology. *Pollution*, 10(1), 210–235.

Bankevich, A., Nurk, S., Antipov, D., Gurevich, A. A., Dvorkin, M., Kulikov, A. S., Lesin, V. M., Nikolenko, S. I., Pham, S., Prjibelski, A. D., Pyshkin, A. V., Sirotkin, A. V., Vyahhi, N., Tesler, G., Alekseyev, M. A., & Pevzner, P. A. (2012, May). SPAdes: A New Genome Assembly Algorithm and Its Applications to Single-Cell Sequencing. *Journal of Computational Biology*, 19(5), 455–477. DOI: 10.1089/cmb.2012.0021 PMID: 22506599

Bao, R., Huang, L., Andrade, J., Tan, W., Kibbe, W. A., Jiang, H., & Feng, G. (2014). Review of current methods, applications, and data management for the bioinformatics analysis of whole exome sequencing. *Cancer Informatics*, 13(Suppl 2), 67–82. DOI: 10.4137/CIN.S13779 PMID: 25288881

Barroso, L. A., Holzle, U., & Ranganathan, P. (2019), The Datacenter as a computer. Designing warehouse-scale machines. https://link.springer.com/book/10.1007/978-3-031-01761-2

Baxevanis, A. D., Bader, G. D., & Wishart, D. S. (2020). *Bioinformatics*. John Wiley & Sons.

Bayat, A. (2002). Bioinformatics. *BMJ (Clinical Research Ed.)*, 324(7344), 1018–1022. DOI: 10.1136/bmj.324.7344.1018 PMID: 11976246

Ben Khedher, M., Ghedira, K., Rolain, J. M., Ruimy, R., & Croce, O. (2022). Application and Challenge of 3rd Generation Sequencing for Clinical Bacterial Studies. *International Journal of Molecular Sciences*, 23(3), 1395. Advance online publication. DOI: 10.3390/ijms23031395 PMID: 35163319

Benson, D. A., Cavanaugh, M., Clark, K., Karsch-Mizrachi, I., Lipman, D. J., Ostell, J., & Sayers, E. W. (2012). GenBank. *Nucleic Acids Research*, 41(D1, n. D1), D36–D42. DOI: 10.1093/nar/gks1195 PMID: 23193287

Berman, H. M. (2000, January 1). The Protein Data Bank. *Nucleic Acids Research*, 28(1), 235–242. DOI: 10.1093/nar/28.1.235 PMID: 10592235

Bhagat, R. T., Butle, S. R., Khobragade, D. S., Wankhede, S. B., Prasad, C. C., Mahure, D. S., & Armarkar, A. V. (2021). Molecular Docking in Drug Discovery. *Journal of Pharmaceutical Research International*, 46–58. DOI: 10.9734/jpri/2021/v33i30B31639

Bhattacharjee, B., & Hallan, V. (2022). Geminivirus-derived vectors as tools for functional genomics. *Frontiers in Microbiology*, 13, 799345. DOI: 10.3389/fmicb.2022.799345 PMID: 35432267

Bhukya, R., & Ashok, A. (2020). Gene expression prediction using deep neural networks. *The International Arab Journal of Information Technology*, 17(3), 422–431. DOI: 10.34028/iajit/17/3/16

Bianconi, I., Aschbacher, R., & Pagani, E. (2023). Current Uses and Future Perspectives of Genomic Technologies in Clinical Microbiology. *Antibiotics (Basel, Switzerland)*, 12(11), 1580. Advance online publication. DOI: 10.3390/antibiotics12111580 PMID: 37998782

Binding, D. B. (n.d.). Available at: https://www.bindingdb.org/bind/

Bioco, J., Cánovas, F., Sousa, P. P., & Fazendeiro, P. (2022). SDSim: A generalized user friendly web ABM system to simulate spatiotemporal distribution of species under environmental scenarios. *Environmental Modelling & Software*, 147, 105234–105245. DOI: 10.1016/j.envsoft.2021.105234

Blanco-González, A., Cabezón, A., Seco-González, A., Conde-Torres, D., Antelo-Riveiro, P., Piñeiro, Á., & Garcia-Fandino, R. (2023). The Role of AI in Drug Discovery: Challenges, Opportunities, and Strategies. *Pharmaceuticals (Basel, Switzerland)*, 16(6), 891. DOI: 10.3390/ph16060891 PMID: 37375838

Boja, E. S. (2014). Integration of omics sciences to advance biology and medicine.

Bonnal, R. J. P., Yates, A., Goto, N., Gautier, L., Willis, S., Fields, C., Katayama, T., & Prins, P. (2019). Sharing programming resources between bio* projects. *Methods in Molecular Biology (Clifton, N.J.)*, 1910, 747–766. DOI: 10.1007/978-1-4939-9074-0_25 PMID: 31278684

Born, J., & Manica, M. (2021). Trends in deep learning for property-driven drug design. *Current Medicinal Chemistry*, 28(38), 7862–7886. DOI: 10.2174/0929867328666210729115728 PMID: 34325627

Bornstein, K., Gryan, G., Chang, E. S., Marchler-Bauer, A., & Schneider, V. A. (2023). The NIH Comparative Genomics Resource: Addressing the promises and challenges of comparative genomics on human health. *BMC Genomics*, 24(1), 575. DOI: 10.1186/s12864-023-09643-4 PMID: 37759191

Borowiec, M. L., Dikow, R. B., Frandsen, P. B., McKeeken, A., Valentini, G., & White, A. E. (2022). Deep learning as a tool for ecology and evolution. *Methods in Ecology and Evolution*, 13(8), 1640–1660. DOI: 10.1111/2041-210X.13901

Borry, P., Bentzen, H. B., Budin-Ljøsne, I., Cornel, M. C., Howard, H. C., Feeney, O., Jackson, L., Mascalzoni, D., Mendes, Á., Peterlin, B., Riso, B., Shabani, M., Skirton, H., Sterckx, S., Vears, D., Wjst, M., & Felzmann, H. (2018). The challenges of the expanded availability of genomic information: An agenda-setting paper. *Journal of Community Genetics*, 9(2), 103–116. DOI: 10.1007/s12687-017-0331-7 PMID: 28952070

Bouazzaoui, A., Abdellatif, A. A. H., Al-Allaf, F. A., Bogari, N. M., Al-Dehlawi, S., & Qari, S. H. (2021). Strategies for Vaccination: Conventional Vaccine Approaches Versus New-Generation Strategies in Combination with Adjuvants. *Pharmaceutics*, 13(2), 1–20. DOI: 10.3390/pharmaceutics13020140 PMID: 33499096

Bravi, B. (2024). Development and use of machine learning algorithms in vaccine target selection. *Npj Vaccines, 9*(1), 1–14. DOI: 10.1038/s41541-023-00795-8

Brittain, H. K., Scott, R., & Thomas, E. (2017). The rise of the genome and personalised medicine. *Clinical Medicine (London, England)*, 17(6), 545–551. DOI: 10.7861/clinmedicine.17-6-545 PMID: 29196356

Brown, T. B., Mann, B., Ryder, N., Subbiah, M., Kaplan, J., Dhariwal, P., Neelakantan, A., Shyam, P., Sastry, G., Askell, A., Agarwal, S., Herbert-Voss, A., Krueger, G., Henighan, T., Child, R., Ramesh, A., Ziegler, D. M., Wu, J., Winter, C., . . . Amodei, D. (2020). Language models are few-shot learners, *NeurIPS 2020,* 33, 1877-1901. https://doi.org//arXiv.2005.14165.DOI: 10.48550

Burley. (2017). *Protein Data Bank* (PDB): the single global macromolecular structure archive. Protein crystallography: methods and protocols, 627-641.

Burton, D. R. (2017). What Are the Most Powerful Immunogen Design Vaccine Strategies? Reverse Vaccinology 2.0 Shows Great Promise. *Cold Spring Harbor Perspectives in Biology*, 9(11), a030262. Advance online publication. DOI: 10.1101/cshperspect.a030262 PMID: 28159875

Burton, H., Adams, M., Bunton, R., & Schröder-Bäck, P. (2008). Developing stakeholder involvement for introducing public health genomics into public policy. *Public Health Genomics*, 12(1), 11–19. DOI: 10.1159/000153426 PMID: 19023186

Cabral, H., Kinoh, H., & Kataoka, K. (2020, December 15). Tumor-Targeted Nanomedicine for Immunotherapy. *Accounts of Chemical Research*, 53(12), 2765–2776. DOI: 10.1021/acs.accounts.0c00518 PMID: 33161717

Caccia, D., Dugo, M., Callari, M., & Bongarzone, I. (2013). Bioinformatics tools for secretome analysis. *Biochimica et Biophysica Acta (BBA)-. Biochimica et Biophysica Acta. Proteins and Proteomics*, 1834(11), 2442–2453. DOI: 10.1016/j.bbapap.2013.01.039

Caffe. (n.d.). Available at: https://caffe.berkeleyvision.org/

Callaway, E. (2020, November 30). 'It will change everything': DeepMind's AI makes gigantic leap in solving protein structures. *Nature*, 588(7837), 203–204. DOI: 10.1038/d41586-020-03348-4 PMID: 33257889

Callaway, E. (2023). AI tools are designing entirely new proteins that could transform medicine. *Nature*, 619(7969), 236–238. DOI: 10.1038/d41586-023-02227-y PMID: 37433935

Callaway, E. (2023a). After AlphaFold: Protein-folding contest seeks next big breakthrough. *Nature*, 613(7942), 13–14. DOI: 10.1038/d41586-022-04438-1 PMID: 36513827

Canatalay, P. J., & Osman, N. U. (2022). A bidirectional LSTM-RNN and GRU method to exon prediction using splice-site mapping. *Applied Sciences (Basel, Switzerland)*, 12(9), 4390. DOI: 10.3390/app12094390

Cano-Gamez, E., & Trynka, G. (2020). From GWAS to function: Using functional genomics to identify the mechanisms underlying complex diseases. *Frontiers in Genetics*, 11, 505357. DOI: 10.3389/fgene.2020.00424 PMID: 32477401

Canouï, E., & Launay, O. (2019). Histoire et principes de la vaccination. *Revue des Maladies Respiratoires*, 36(1), 74–81. DOI: 10.1016/j.rmr.2018.02.015 PMID: 30579659

Cánovas, F., De la Rúa, P., Serrano, J., & Galián, J. (2011). Microsatellite variability reveals beekeeping influence on Iberian honeybee populations. *Apidologie*, 42(3), 235–251. DOI: 10.1007/s13592-011-0020-1

Cánovas, F., De la Rúa, P., Serrano, J., & Galián, J. (2014). Analysing and describing a contact area of two distinct evolutionary units. An ecological perspective. *Journal of Insect Conservation*, 18(5), 927–937. DOI: 10.1007/s10841-014-9701-1

Cánovas, F., Lacasa, A., & Sánchez, J. A. (2006). Implementing a Geographical Information System (GIS) for pepper greenhouse natural enemy and pest management in Southeast Spain. *IOBC/WPRS Bulletin*, 29(4), 25–30.

Capozzi, F., & Bordoni, A. (2013). Foodomics: A new comprehensive approach to food and nutrition. *Genes & Nutrition*, 8(1), 1–4. DOI: 10.1007/s12263-012-0310-x PMID: 22933238

Carugo, O. (2007). Oliviero. Recent progress in measuring structural similarity between proteins. *Current Protein & Peptide Science*, 8(3), 219–241. DOI: 10.2174/138920307780831839 PMID: 17584118

Casas, L., & Saborido-Rey, F. (2023). A review of genomics methods and bioinformatics tools for the analysis of close-kin mark-recapture.

Case, D. A., Cheatham, T. E.III, Darden, T., Gohlke, H., Luo, R., Merz, K. M.Jr, Onufriev, A., Simmerling, C., Wang, B., & Woods, R. J. (2005a). The Amber biomolecular simulation programs. *Journal of Computational Chemistry*, 26(16), 1668–1688. DOI: 10.1002/jcc.20290 PMID: 16200636

Chakraborty, C., Bhattacharya, M., & Dhama, K. (2023). SARS-CoV-2 Vaccines, Vaccine Development Technologies, and Significant Efforts in Vaccine Development during the Pandemic: The Lessons Learned Might Help to Fight against the Next Pandemic. *Vaccines, 11*(3), 682. DOI: 10.3390/vaccines11030682

Chakraborty, C., Sharma, A. R., Bhattacharya, M., & Lee, S. S. (2021). Lessons Learned from Cutting-Edge Immunoinformatics on Next-Generation COVID-19 Vaccine Research. *International Journal of Peptide Research and Therapeutics*, 27(4), 2303–2311. DOI: 10.1007/s10989-021-10254-4 PMID: 34276266

Chan, E. Y. (2005). Advances in sequencing technology. *Mutation Research. Fundamental and Molecular Mechanisms of Mutagenesis*, 573(1-2), 13–40. DOI: 10.1016/j.mrfmmm.2005.01.004 PMID: 15829235

Charifson, P. S., & Walters, W. P. (2002). Filtering databases and chemical libraries. *Molecular Diversity*, 5(4), 185–197. DOI: 10.1023/A:1021316023707 PMID: 12549671

Chatanaka, M. K., Ulndreaj, A., Sohaei, D., & Prassas, I. (2022). Immunoinformatics: Pushing the boundaries of immunology research and medicine. *Immunoinformatics (Amsterdam, Netherlands)*, 5, 100007. DOI: 10.1016/j.immuno.2021.100007

Chauhan, V., Rungta, T., Goyal, K., & Singh, M. P. (2019). Designing a multi-epitope based vaccine to combat Kaposi Sarcoma utilizing immunoinformatics approach. *Scientific Reports, 9*(1), 1–15. DOI: 10.1038/s41598-019-39299-8

Chelliah, R., Banan-MwineDaliri, E., Khan, I., Wei, S., Elahi, F., Yeon, S.-J., Selvakumar, V., Ofosu, F. K., Rubab, M., Ju, H. H., Rallabandi, H. R., Madar, I. H., Sultan, G., & Oh, D. H. (2022). A review on the application of bioinformatics tools in food microbiome studies. *Briefings in Bioinformatics*, 23(2), bbac007. DOI: 10.1093/bib/bbac007 PMID: 35189636

Chen, C., Wang, J., Pan, D., Wang, X., Xu, Y., Yan, J., Wang, L., Yang, X., Yang, M., & Liu, G. P. (2023). Applications of multi-omics analysis in human diseases. *MedComm*, 4(4), e315. DOI: 10.1002/mco2.315 PMID: 37533767

Cheng, C., Fei, Z., & Xiao, P. (2023). Methods to improve the accuracy of next-generation sequencing. *Frontiers in Bioengineering and Biotechnology*, 11, 982111. DOI: 10.3389/fbioe.2023.982111 PMID: 36741756

Cheng, T., Li, Q., Zhou, Z., Wang, Y., & Bryant, S. H. (2012a). Structure-Based Virtual Screening for Drug Discovery: A Problem-Centric Review. *The AAPS Journal*, 14(1), 133–141. DOI: 10.1208/s12248-012-9322-0 PMID: 22281989

Chen, H., Engkvist, O., Wang, Y., Olivecrona, M., & Blaschke, T. (2018). The rise of deep learning in drug discovery. *Drug Discovery Today*, 23(6), 1241–1250. DOI: 10.1016/j.drudis.2018.01.039 PMID: 29366762

Chen, X., & Ishwaran, H. (2012). Random forests for genomic data analysis. *Genomics*, 99(6), 323–329. DOI: 10.1016/j.ygeno.2012.04.003 PMID: 22546560

Chen, X., Zaro, J. L., & Shen, W. C. (2013). Fusion protein linkers: Property, design and functionality. *Advanced Drug Delivery Reviews*, 65(10), 1357–1369. DOI: 10.1016/j.addr.2012.09.039 PMID: 23026637

Chen, Y., Song, J., Ruan, Q., Zeng, X., Wu, L., Cai, L., Wang, X., & Yang, C. (2021). Single-cell sequencing methodologies: From transcriptome to multi-dimensional measurement. *Small Methods*, 5(6), 2100111. DOI: 10.1002/smtd.202100111 PMID: 34927917

Chen, Z., Ain, N., Zhao, Q., & Zhang, X. (2024). From tradition to innovation: Conventional and deep learning frameworks in genome annotation. *Briefings in Bioinformatics*, 25(3), bbae138. DOI: 10.1093/bib/bbae138 PMID: 38581418

Chiu, C. Y., & Miller, S. A. (2019). Clinical metagenomics. *Nature Reviews. Genetics*, 20(6), 341–355. DOI: 10.1038/s41576-019-0113-7 PMID: 30918369

Cho, H. K., Park, S. E., Kim, Y. J., Jo, D. S., Kim, Y. K., Eun, B. W., Lee, T. J., Lee, J., Lee, H., Kim, K. H., Cho, E. Y., Ahn, J. G., & Choi, E. H. (2021). Recommendation for use of diphtheria and tetanus toxoids and acellular pertussis, inactivated poliovirus, Haemophilus influenzae type b conjugate, and hepatitis B vaccine in infants. *Clinical and Experimental Pediatrics*, 64(12), 602–607. DOI: 10.3345/cep.2021.00507 PMID: 34098627

Chothia, C., & Lesk, A. M. (1986). Cyrus; LESK, Arthur M. The relation between the divergence of sequence and structure in proteins. *The EMBO Journal*, 5(4), 823–826. DOI: 10.1002/j.1460-2075.1986.tb04288.x PMID: 3709526

Clough, E., & Barrett, T. (2016). The *gene expression omnibus* database. *Statistical Genomics: Methods and Protocols*, 93–110.

Collins, F. S., & Fink, L. (1995). The Human Genome Project. *Alcohol Health and Research World*, 19(3), 190–195. PMID: 31798046

Congrains, C., Dupuis, J. R., Rodriguez, E. J., Norrbom, A. L., Steck, G., Sutton, B., Nolazco, N., de Brito, R. A., & Geib, S. M. (2023). Phylogenomic analysis provides diagnostic tools for the identification of Anastrepha fraterculus (Diptera: Tephritidae) species complex. *Evolutionary Applications*, 16(9), 1598–1618. DOI: 10.1111/eva.13589 PMID: 37752958

Congressional Budget Office (CBO). (2021). Research and development in the pharmaceutical industry. https://www.cbo.gov/publication/57025

Consortium. (2015, January 28). UniProt: A hub for protein information. *Nucleic Acids Research*, 43(D1, n. D1), D204–D212. DOI: 10.1093/nar/gku989 PMID: 25348405

Cooksey, R. W. (2020). Descriptive Statistics for Summarising Data. In *Illustrating Statistical Procedures: Finding Meaning in Quantitative Data* (pp. 61–139). Springer Singapore. DOI: 10.1007/978-981-15-2537-7_5

Crick, F. (1970). Central dogma of molecular biology. *Nature*, 227(5258), 561–563. DOI: 10.1038/227561a0 PMID: 4913914

Crick, F. H. (1958). On protein synthesis. *Symposia of the Society for Experimental Biology*, 12, 138–163. PMID: 13580867

Crossley, B. M., Bai, J., Glaser, A., Maes, R., Porter, E., Killian, M. L., Clement, T., & Toohey-Kurth, K. (2020). Guidelines for Sanger sequencing and molecular assay monitoring. *Journal of Veterinary Diagnostic Investigation*, 32(6), 767–775. DOI: 10.1177/1040638720905833 PMID: 32070230

CTD. (n.d.). Available at: https://ctdbase.org/downloads/

Cunningham, P., Cord, M., & Delany, S. J. (2008). Supervised learning. In *Machine learning techniques for multimedia: case studies on organization and retrieval* (pp. 21–49). Springer. DOI: 10.1007/978-3-540-75171-7_2

D'Amelio, E., Salemi, S., & D'Amelio, R. (2016). Anti-Infectious Human Vaccination in Historical Perspective. *International Reviews of Immunology*, 35(3), 260–290. DOI: 10.3109/08830185.2015.1082177 PMID: 26606466

Da, C., & Kireev, D. (2014, September 22). Structural Protein–Ligand Interaction Fingerprints (SPLIF) for Structure-Based Virtual Screening: Method and Benchmark Study. *Journal of Chemical Information and Modeling*, 54(9), 2555–2561. DOI: 10.1021/ci500319f PMID: 25116840

Das, A. T., Jeeninga, R. E., & Berkhout, B. (2010). Possible applications for replicating HIV-1 vectors. DOI: 10.2217/hiv.10.20

David, L., Thakkar, A., Mercado, R., & Engkvist, O. (2020). Molecular representations in AI-driven drug discovery: A review and practical guide. *Journal of Cheminformatics*, 12(1), 56. DOI: 10.1186/s13321-020-00460-5 PMID: 33431035

Dayhoff, M. O. (1972). *Atlas of protein sequence and structure*. National Biomedical Research Foundation.

De Groot, A. S., Moise, L., Terry, F., Gutierrez, A. H., Hindocha, P., Richard, G., Hoft, D. F., Ross, T. M., Noe, A. R., Takahashi, Y., Kotraiah, V., Silk, S. E., Nielsen, C. M., Minassian, A. M., Ashfield, R., Ardito, M., Draper, S. J., & Martin, W. D. (2020). Better epitope discovery, precision immune engineering, and accelerated vaccine design using Immunoinformatics tools. *Frontiers in Immunology*, 11, 527882. DOI: 10.3389/fimmu.2020.00442 PMID: 32318055

De Vries, J., Bull, S. J., Doumbo, O., Ibrahim, M., Mercereau-Puijalon, O., Kwiatkowski, D., & Parker, M. (2011). Ethical issues in human genomics research in developing countries. *BMC Medical Ethics*, 12(1), 1–10. DOI: 10.1186/1472-6939-12-5 PMID: 21418562

de Vries, S. J., Rey, J., Schindler, C. E. M., Zacharias, M., & Tuffery, P. (2017). The pepATTRACT web server for blind, large-scale peptide–protein docking. *Nucleic Acids Research*, 45(W1), W361–W364. DOI: 10.1093/nar/gkx335 PMID: 28460116

de Vries, S. J., van Dijk, M., & Bonvin, A. M. J. J. (2010). The HADDOCK web server for data-driven biomolecular docking. *Nature Protocols*, 5(5), 883–897. DOI: 10.1038/nprot.2010.32 PMID: 20431534

De, A., Correia, S., Colombini, E. L., De, A., & Colombini, E. L. (2022). Attention, please! A survey of neural attention models in deep learning. *Artificial Intelligence Review, 55*(8), 6037–6124. DOI: 10.1007/s10462-022-10148-x

Dean, J., Corrado, G. S., Monga, R., Chen, K., Devin, M., Le, Q. V., & Mao, M. Z. (2012). Large-scale distributed deep networks. *NeurIPS 2012*, 25, 1223-1231. https://dl.acm.org/doi/10.5555/2999134.2999271

DeAngelis, D. L., & Diaz, S. G. (2019). Decision-Making in Agent-Based Modeling: A Current Review and Future Prospectus. *Frontiers in Ecology and Evolution*, 6, 237. DOI: 10.3389/fevo.2018.00237

Defresne, M., Barbe, S., & Schiex, T. (2021, January). Protein Design with Deep Learning. *International Journal of Molecular Sciences*, 22(21), 11741. DOI: 10.3390/ijms222111741 PMID: 34769173

Demšar, J. (2004). Orange: From experimental machine learning to interactive data mining. In European Conference On Principles Of Data Mining And Knowledge Discovery. Springer.

Dennett, D. (2004). The age of intelligent machines: Can machines think.

Depristo, M. A., Banks, E., Poplin, R., Garimella, K. V., Maguire, J. R., Hartl, C., Philippakis, A. A., del Angel, G., Rivas, M. A., Hanna, M., McKenna, A., Fennell, T. J., Kernytsky, A. M., Sivachenko, A. Y., Cibulskis, K., Gabriel, S. B., Altshuler, D., & Daly, M. J. (2011, May). A framework for variation discovery and genotyping using next-generation DNA sequencing data. *Nature Genetics*, 43(5), 491–498. DOI: 10.1038/ng.806 PMID: 21478889

Derkach, S. R., Voron'ko, N. G., & Kuchina, Y. A. (2022). Intermolecular Interactions in the Formation of Polysaccharide-Gelatin Complexes: A Spectroscopic Study. *Polymers*, 14(14), 2777. DOI: 10.3390/polym14142777 PMID: 35890554

Desiere, F., German, B., Watzke, H., Pfeifer, A., & Saguy, S. (2001). Bioinformatics and data knowledge: The new frontiers for nutrition and foods. *Trends in Food Science & Technology*, 12(7), 215–229. DOI: 10.1016/S0924-2244(01)00089-9

Dhillon, A., & Verma, G. K. (2020). Convolutional neural network: A review of models, methodologies and applications to object detection. *Progress in Artificial Intelligence*, 9(2), 85–112. DOI: 10.1007/s13748-019-00203-0

Dimitrov, I., Bangov, I., Flower, D. R., & Doytchinova, I. (2014). AllerTOP v.2—A server for in silico prediction of allergens. *Journal of Molecular Modeling*, 20(6), 2278. Advance online publication. DOI: 10.1007/s00894-014-2278-5 PMID: 24878803

Dnyandev, K. M., Babasaheb, G. V., Chandrashekhar, K. V., Chandrakant, M. A., & Vasant, O. K. (2021, May). A Review on Molecular Docking. *International Research Journal of Pure and Applied Chemistry*, 20, 60–68. DOI: 10.9734/irjpac/2021/v22i330396

Dos Santos, L. M., Mariano, D., & De Melo-Minardi, R. C. (2024). O impacto da Inteligência artificial nas ciências da vida através da Bioinformática. *Revista da Universidade Federal de Minas Gerais, 30*, e47996.

Dos Santos, L. M., Mariano, D., & De Melo-Minardi, R. C. (2024). O impacto da Inteligência artificial nas ciências da vida através da Bioinformática. *Revista da Universidade Federal de Minas Gerais, 30*.

Doytchinova, I. A., & Flower, D. R. (2007). VaxiJen: A server for prediction of protective antigens, tumour antigens and subunit vaccines. *BMC Bioinformatics*, 8(1), 1–7. DOI: 10.1186/1471-2105-8-4 PMID: 17207271

Doytchinova, I. A., & Flower, D. R. (2008). Bioinformatic Approach for Identifying Parasite and Fungal Candidate Subunit Vaccines. *The Open Vaccine Journal*, 1(1), 22–26. https://www.jenner.ac.uk/VaxiJen. DOI: 10.2174/1875035400801010022

DrugBank. (n.d.). Available at: https://go.drugbank.com/releases/latest

Duardo-Sanchez, A., & Gonzalez-Diaz, H. (2013). Legal issues for chem-bioinformatics models. *Frontiers in Bioscience*, 5(1), 361–374. DOI: 10.2741/E621 PMID: 23276995

Duda, R. O., Hart, P. E., & Stork, D. G. (2012). *Pattern Classification*. Wiley.

Du, L., Liu, Q., Fan, Z., Tang, J., Zhang, X., Price, M., Yue, B., & Zhao, K. (2021). Pyfastx: A robust Python package for fast random access to sequences from plain and gzipped FASTA/Q files. *Briefings in Bioinformatics*, 22(4), bbaa368. DOI: 10.1093/bib/bbaa368 PMID: 33341884

Durland, J., & Ahmadian-Moghadam, H. (2024). Genetics, Mutagenesis. In *StatPearls*. StatPearls Publishing.

Durmaz, V., Köchl, K., Krassnigg, A., Parigger, L., Hetmann, M., Singh, A., Nutz, D., Korsunsky, A., Kahler, U., König, C., Chang, L., Krebs, M., Bassetto, R., Pavkov-Keller, T., Resch, V., Gruber, K., Steinkellner, G., & Gruber, C. C. (2022). Structural bioinformatics analysis of SARS-CoV-2 variants reveals higher hACE2 receptor binding affinity for Omicron B.1.1.529 spike RBD compared to wild type reference. *Scientific Reports*, 12(1), 14534. DOI: 10.1038/s41598-022-18507-y PMID: 36008461

Düwell, M. (2012). Bioethics: Methods, theories, domains. *Bioethics: Methods, Theories, Domains*, 1–308. DOI: 10.4324/9780203084205

Eberhardt, J., Santos-Martins, D., Tillack, A. F., & Forli, S. (2021). AutoDock Vina 1.2.0: New Docking Methods, Expanded Force Field, and Python Bindings. *Journal of Chemical Information and Modeling*, 61(8), 3891–3898. DOI: 10.1021/acs.jcim.1c00203 PMID: 34278794

Eckhardt, M., Hultquist, J. F., Kaake, R. M., Hüttenhain, R., & Krogan, N. J. (2020). A systems approach to infectious disease. *Nature Reviews. Genetics*, 21(6), 339–354. DOI: 10.1038/s41576-020-0212-5 PMID: 32060427

Eid, J., Fehr, A., Gray, J., Luong, K., Lyle, J., Otto, G., Peluso, P., Rank, D., Baybayan, P., Bettman, B., Bibillo, A., Bjornson, K., Chaudhuri, B., Christians, F., Cicero, R., Clark, S., Dalal, R., deWinter, A., Dixon, J., & Turner, S. (2009, January 2). Real-time DNA sequencing from single polymerase molecules. *Science*, 323(5910), 133–138. DOI: 10.1126/science.1162986 PMID: 19023044

Ejigu, G. F., & Jung, J. (2020). Review on the Computational Genome Annotation of Sequences Obtained by Next-Generation Sequencing. *Biology (Basel)*, 9(9), 295. Advance online publication. DOI: 10.3390/biology9090295 PMID: 32962098

EM, M., & DN, L. (2010). Treatment of visceral leishmaniasis. *Journal of Global Infectious Diseases, 2*(2), 280. DOI: 10.4103/0974-777X.62883

Emmert-Streib, F., & Dehmer, M. (2019). Understanding Statistical Hypothesis Testing: The Logic of Statistical Inference. Retrieved from https://www.mdpi.com/2504-4990/1/3/54

Erdmann, A., Rehmann-Sutter, C., & Bozzaro, C. (2021). Patients' and professionals' views related to ethical issues in precision medicine: A mixed research synthesis. *BMC Medical Ethics*, 22(1), 1–18. DOI: 10.1186/s12910-021-00682-8 PMID: 34465328

Erickson, J., & Kempf, D. (1994). Structure-based design of symmetric inhibitors of HIV-1 protease. *Archives of Virology. Supplementum*, 9, 19–29. DOI: 10.1007/978-3-7091-9326-6_3 PMID: 8032250

Esplin, E. D., Oei, L., & Snyder, M. P. (2014). Personalized sequencing and the future of medicine: Discovery, diagnosis and defeat of disease. *Pharmacogenomics*, 15(14), 1771–1790. DOI: 10.2217/pgs.14.117 PMID: 25493570

Excler, J.-L., Saville, M., Berkley, S., & Kim, J. H. (2021). Vaccine development for emerging infectious diseases. *Nature Medicine*, 27(4), 591–600. DOI: 10.1038/s41591-021-01301-0 PMID: 33846611

Fang, S., Zhang, L., Guo, J., Niu, Y., Wu, Y., Li, H., & Sun, X. (2018). NONCODEV5: A comprehensive annotation database for long non-coding RNAs. *Nucleic Acids Research*, 46(D1), D308–D314. DOI: 10.1093/nar/gkx1107 PMID: 29140524

Fan, J., Fu, A., & Zhang, L. (2019a, June 1). Progress in molecular docking. *Quantitative Biology*, 7(2), 83–89. DOI: 10.1007/s40484-019-0172-y

Fan, Y., Xiong, H., & Sun, G. (2023). DeepASDPred: A CNN-LSTM-based deep learning method for autism spectrum disorders risk RNA identification. *BMC Bioinformatics*, 24(1), 261. DOI: 10.1186/s12859-023-05378-x PMID: 37349705

Fassio, A. V., Shub, L., Ponzoni, L., McKinley, J., O'Meara, M. J., Ferreira, R. S., Keiser, M. J., & Minardi, R. C. M. (2022). Prioritizing virtual screening with interpretable interaction fingerprints. *Journal of Chemical Information and Modeling*, 62(18), 4300–4318. DOI: 10.1021/acs.jcim.2c00695 PMID: 36102784

Feltes. (2015). *Bioinformática: Da biologia à flexibilidade molecular*. SBBq.

Feng, L. B., Grosse, S. D., Green, R. F., Fink, A. K., & Sawicki, G. S. (2018). Precision Medicine In Action: The Impact Of Ivacaftor On Cystic Fibrosis-Related Hospitalizations. *Health Affairs (Project Hope)*, 37(5), 773–779. DOI: 10.1377/hlthaff.2017.1554 PMID: 29733727

Ferrari, D. G., & De Castro Silva, L. N. (2017). *Introdução a mineração de dados*. Saraiva Educação S.A.

Field, M., Hardcastle, N., Jameson, M., Aherne, N., & Holloway, L. (2021, July). Machine learning applications in radiation oncology. *Physics and Imaging in Radiation Oncology*, 19, 13–24. DOI: 10.1016/j.phro.2021.05.007 PMID: 34307915

Flynn, E. D., & Lappalainen, T. (2022). Functional characterization of genetic variant effects on expression. *Annual Review of Biomedical Data Science*, 5(1), 119–139. DOI: 10.1146/annurev-biodatasci-122120-010010 PMID: 35483347

Fukunaga, K. (2013). *Introduction to statistical pattern recognition*. Elsevier.

Galanis, K. A., Nastou, K. C., Papandreou, N. C., Petichakis, G. N., Pigis, D. G., & Iconomidou, V. A. (2021). Linear B-Cell Epitope Prediction for In Silico Vaccine Design: A Performance Review of Methods Available via Command-Line Interface. *International Journal of Molecular Sciences*, 22(6), 1–19. DOI: 10.3390/ijms22063210 PMID: 33809918

Galasso, I. (2024). Precision medicine for whom? public health outputs from "Genomics England" and "All of Us" to make up for upstream and downstream exclusion. *The American Journal of Bioethics*, 24(3), 71–85. DOI: 10.1080/15265161.2023.2180108 PMID: 36876959

Gao, C. (2021). Genome engineering for crop improvement and future agriculture. *Cell*, 184(6), 1621–1635. DOI: 10.1016/j.cell.2021.01.005 PMID: 33581057

Gao, W., Mahajan, S. P., Sulam, J., & Gray, J. J. (2020). Deep learning in protein structural modeling and design. *Patterns (New York, N.Y.)*, 1(9), 100142. DOI: 10.1016/j.patter.2020.100142 PMID: 33336200

Gardy, J. L., & Loman, N. J. (2018). Towards a genomics-informed, real-time, global pathogen surveillance system. *Nature Reviews. Genetics*, 19(1), 9–20. DOI: 10.1038/nrg.2017.88 PMID: 29129921

Gauthier, J. (2019). A brief history of bioinformatics. *Briefings in Bioinformatics, 20*(6).

Geman, D., Ochs, M., Price, N. D., Tomasetti, C., & Younes, L. (2015). An argument for mechanism-based statistical inference in cancer. *Human Genetics*, 134(5), 479–495. DOI: 10.1007/s00439-014-1501-x PMID: 25381197

Gentile, F., Yaacoub, J. C., Gleave, J., Fernandez, M., Ton, A.-T., Ban, F., Stern, A., & Cherkasov, A. (2022, March). Artificial intelligence–enabled virtual screening of ultra-large chemical libraries with deep docking. *Nature Protocols*, 17(3), 672–697. DOI: 10.1038/s41596-021-00659-2 PMID: 35121854

Ghosh, A., Larrondo-Petrie, M. M., & Pavlovic, M. (2023). Revolutionizing Vaccine Development for COVID-19: A Review of AI-Based Approaches. *Information, 14*(12), 665. DOI: 10.3390/info14120665

Ghurye, J., & Pop, M. (2019). Modern technologies and algorithms for scaffolding assembled genomes. *PLoS Computational Biology*, 15(6), e1006994. DOI: 10.1371/journal.pcbi.1006994 PMID: 31166948

Giani, A. M., Gallo, G. R., Gianfranceschi, L., & Formenti, G. (2020). Long walk to genomics: History and current approaches to genome sequencing and assembly. *Computational and Structural Biotechnology Journal*, 18, 9–19. DOI: 10.1016/j.csbj.2019.11.002 PMID: 31890139

Gill, S. K., Christopher, A. F., Gupta, V., & Bansal, P. (2016). Emerging role of bioinformatics tools and software in evolution of clinical research. *Perspectives in Clinical Research*, 7(3), 115–122. DOI: 10.4103/2229-3485.184782 PMID: 27453827

Giordano, D., Biancaniello, C., Argenio, M. A., & Facchiano, A. (2022, May). Drug Design by Pharmacophore and Virtual Screening Approach. *Pharmaceuticals (Basel, Switzerland)*, 15(5), 646. DOI: 10.3390/ph15050646 PMID: 35631472

Gligorijević, V., & Pržulj, N. (2015). Methods for biological data integration: Perspectives and challenges. *Journal of the Royal Society, Interface*, 12(112), 20150571. DOI: 10.1098/rsif.2015.0571 PMID: 26490630

Goldstein, B. A., Polley, E. C., & Briggs, F. B. (2011). Random forests for genetic association studies. *Statistical Applications in Genetics and Molecular Biology*, 10(1), 32. DOI: 10.2202/1544-6115.1691 PMID: 22889876

Gomes, L. G. R., Dutra, J. C. F., Profeta, R., Dias, M. V., García, G. J. Y., Rodrigues, D. L. N., Neto, A. G., Aburjaile, F. F., Tiwari, S., Soares, S. C., Azevedo, V., & Jaiswal, A. K. (2023a). Systematic review of reverse vaccinology and immunoinformatics data for non-viral sexually transmitted infections. *Anais da Academia Brasileira de Ciências*, 95(suppl 2), e20230617. DOI: 10.1590/0001-3765202320230617 PMID: 38055447

Gomes, L. G. R., Rodrigues, T. C. V., Jaiswal, A. K., Santos, R. G., Kato, R. B., Barh, D., Alzahrani, K. J., Banjer, H. J., Soares, S. de C., Azevedo, V., & Tiwari, S. (2022). In Silico Designed Multi-Epitope Immunogen "Tpme-VAC/LGCM-2022" May Induce Both Cellular and Humoral Immunity against Treponema pallidum Infection. *Vaccines*, 10(7), 1019. Advance online publication. DOI: 10.3390/vaccines10071019 PMID: 35891183

Goodfellow, I., Bengio, Y., & Courville, A. (2016). Deep learning. MIT Press. http://www.deeplearningbook.org

Goodman, K. W. (2020). Ethics in Health Informatics. *Yearbook of Medical Informatics*, 29(1), 26–31. DOI: 10.1055/s-0040-1701966 PMID: 32303095

Google Deepmind. (n.d.). *AlphaFold: a solution to a 50-year-old grand challenge in biology*. https://deepmind.google/discover/blog/alphafold-a-solution-to-a-50-year-old-grand-challenge-in-biology/

Goujon, M., McWilliam, H., Li, W., Valentin, F., Squizzato, S., Paern, J., & Lopez, R. (2010). A new bioinformatics analysis tools framework at EMBL–EBI. *Nucleic Acids Research*, 38(suppl_2), W695–W699. DOI: 10.1093/nar/gkq313 PMID: 20439314

Graff, E., Shakhnovich, E. I., & Coley, C. W. (2021). Accelerating high-throughput virtual screening through molecular pool-based active learning. *Chemical Science (Cambridge)*, 12(22), 7866–7881. DOI: 10.1039/D0SC06805E PMID: 34168840

Graw, S., Chappell, K., Washam, C. L., Gies, A., Bird, J., Robeson, M. S., & Byrum, S. D. (2021). Multi-omics data integration considerations and study design for biological systems and disease. *Molecular Omics*, 17(2), 170–185. DOI: 10.1039/D0MO00041H PMID: 33347526

Greenbaum, J., Sidney, J., Chung, J., Brander, C., Peters, B., & Sette, A. (2011). Functional classification of class II human leukocyte antigen (HLA) molecules reveals seven different supertypes and a surprising degree of repertoire sharing across supertypes. *Immunogenetics*, 63(6), 325–335. DOI: 10.1007/s00251-011-0513-0 PMID: 21305276

Green, E. D., Gunter, C., Biesecker, L. G., Di Francesco, V., Easter, C. L., Feingold, E. A., Felsenfeld, A. L., Kaufman, D. J., Ostrander, E. A., Pavan, W. J., Phillippy, A. M., Wise, A. L., Dayal, J. G., Kish, B. J., Mandich, A., Wellington, C. R., Wetterstrand, K. A., Bates, S. A., Leja, D., & Manolio, T. A. (2020). Strategic vision for improving human health at The Forefront of Genomics. *Nature*, 586(7831), 683–692. DOI: 10.1038/s41586-020-2817-4 PMID: 33116284

Green, S., Prainsack, B., & Sabatello, M. (2023). Precision medicine and the problem of structural injustice. *Medicine, Health Care, and Philosophy*, 26(3), 433–450. DOI: 10.1007/s11019-023-10158-8 PMID: 37231234

Groom. (2016). The Cambridge structural database. Acta Crystallographica Section B: Structural Science. Crystal Engineering and Materials, 72(2), 171-179.

Grosdidier, A., Zoete, V., & Michielin, O. (2011). SwissDock, a protein-small molecule docking web service based on EADock DSS. *Nucleic Acids Research*, 39(suppl_2), W270–W277. DOI: 10.1093/nar/gkr366 PMID: 21624888

Guarra, F., & Colombo, G. (2023). Computational Methods in Immunology and Vaccinology: Design and Development of Antibodies and Immunogens. *Journal of Chemical Theory and Computation*, 19(16), 5315–5333. DOI: 10.1021/acs.jctc.3c00513 PMID: 37527403

Guedes, I. A., Barreto, A. M. S., Marinho, D., Krempser, E., Kuenemann, M. A., Sperandio, O., Dardenne, L. E., & Miteva, M. A. (2021, February 4). New machine learning and physics-based scoring functions for drug discovery. *Scientific Reports*, 11(1), 3198. DOI: 10.1038/s41598-021-82410-1 PMID: 33542326

Gu, J., & Bourne, P. E. (2011). *Structural Bioinformatics*. John Wiley & Sons.

Guo, G. (2003). KNN model-based approach in classification. In *On the move to meaningful internet systems 2003: coopis, doa, and odbase: otm confederated international conferences, coopis, doa, and odbase 2003, catania, sicily, italy, november 3-7, 2003. Proceedings*. Springer. DOI: 10.1007/978-3-540-39964-3_62

Gupta, N., & Verma, V. K. (2019). Next-Generation Sequencing and Its Application: Empowering in Public Health Beyond Reality. In Arora, P. K. (Ed.), *Microbial Technology for the Welfare of Society* (pp. 313–341). Springer Singapore. DOI: 10.1007/978-981-13-8844-6_15

Gupta, P. D. (2015). Pharmacogenetics, pharmacogenomics and ayurgenomics for personalized medicine: A paradigm shift. *Indian Journal of Pharmaceutical Sciences*, 77(2), 135–141. DOI: 10.4103/0250-474X.156543 PMID: 26009644

Gupta, R. M., & Musunuru, K. (2014). Expanding the genetic editing tool kit: ZFNs, TALENs, & CRISPR-Cas9. *The Journal of Clinical Investigation*, 124(10), 4154–4161. DOI: 10.1172/JCI72992 PMID: 25271723

Gupta, S., Kapoor, P., Chaudhary, K., Gautam, A., Kumar, R., & Raghava, G. P. S. (2013). In Silico Approach for Predicting Toxicity of Peptides and Proteins. *PLoS One*, 8(9), e73957. DOI: 10.1371/journal.pone.0073957 PMID: 24058508

Gu, W., Meng, F., Haag, R., & Zhong, Z. (2021, January). Actively targeted nanomedicines for precision cancer therapy: Concept, construction, challenges and clinical translation. *Journal of Controlled Release*, 329, 676–695. DOI: 10.1016/j.jconrel.2020.10.003 PMID: 33022328

Gwinn, M., Maccannell, D. R., & Khabbaz, R. F. (2017, February 22). Integrating Advanced Molecular Technologies into Public Health. *Journal of Clinical Microbiology*, 55(3), 703–714. DOI: 10.1128/JCM.01967-16 PMID: 28031438

Hagen, J. B. (2000). The origins of bioinformatics. *Nature Reviews. Genetics*, 1(3), 231–236. DOI: 10.1038/35042090 PMID: 11252753

Hajj Hussein, I., Chams, N., Chams, S., El Sayegh, S., Badran, R., Raad, M., Gerges-Geagea, A., Leone, A., & Jurjus, A. (2015). Vaccines Through Centuries: Major Cornerstones of Global Health. *Frontiers in Public Health*, 3, 167945. DOI: 10.3389/fpubh.2015.00269 PMID: 26636066

Handel, A., La Gruta, N. L., & Thomas, P. G. (2019). Simulation modelling for immunologists. *Nature Reviews Immunology, 20*(3), 186–195. DOI: 10.1038/s41577-019-0235-3

Harvey, W. T., Carabelli, A. M., Jackson, B., Gupta, R. K., Thomson, E. C., Harrison, E. M., & Peacock, S. J. (2021). SARS-CoV-2 variants, spike mutations and immune escape. *Nature Reviews. Microbiology*, 19(7), 409–424. DOI: 10.1038/s41579-021-00573-0 PMID: 34075212

Hasegawa, H., & Holm, L. (2009). Advances and pitfalls of protein structural alignment. *Current Opinion in Structural Biology*, 19(3), 341–348. DOI: 10.1016/j.sbi.2009.04.003 PMID: 19481444

Hasin, Y., Seldin, M., & Lusis, A. (2017, May 5). Multi-omics approaches to disease. *Genome Biology*, 18(1), 83. DOI: 10.1186/s13059-017-1215-1 PMID: 28476144

Hassan, M., Awan, F. M., Naz, A., deAndrés-Galiana, E. J., Alvarez, O., Cernea, A., Fernández-Brillet, L., Fernández-Martínez, J. L., & Kloczkowski, A. (2022). Innovations in Genomics and Big Data Analytics for Personalized Medicine and Health Care: A Review. *International Journal of Molecular Sciences*, 23(9), 4645. Advance online publication. DOI: 10.3390/ijms23094645 PMID: 35563034

Hawkins, R. D., Hon, G. C., & Ren, B. (2010). Next-generation genomics: An integrative approach. *Nature Reviews. Genetics*, 11(7), 476–486. DOI: 10.1038/nrg2795 PMID: 20531367

Healy, M. J., Tong, W., Ostroff, S., Eichler, H. G., Patak, A., Neuspiel, M., Deluyker, H., & Slikker, W.Jr. (2016). Regulatory bioinformatics for food and drug safety. *Regulatory Toxicology and Pharmacology*, 80, 342–347. DOI: 10.1016/j.yrtph.2016.05.021 PMID: 27208439

Heather, J. M., & Chain, B. (2016). The sequence of sequencers: The history of sequencing DNA. *Genomics*, 107(1), 1–8. DOI: 10.1016/j.ygeno.2015.11.003 PMID: 26554401

Hershberg, E. A., Stevens, G., Diesh, C., Xie, P., De Jesus Martinez, T., Buels, R., Stein, L., & Holmes, I. (2021). JBrowseR: An R interface to the JBrowse 2 genome browser. *Bioinformatics (Oxford, England)*, 37(21), 3914–3915. DOI: 10.1093/bioinformatics/btab459 PMID: 34196689

Hilleman, M. R. (2000). Vaccines in historic evolution and perspective: A narrative of vaccine discoveries. *Vaccine*, 18(15), 1436–1447. DOI: 10.1016/S0264-410X(99)00434-X PMID: 10618541

Hirzel, A. H., & Le Lay, G. (2008). Habitat suitability modelling and niche theory. *Journal of Applied Ecology*, 45(5), 1372–1381. DOI: 10.1111/j.1365-2664.2008.01524.x

Holliday, J., Sani, N., & Willett, P. (2018, June). Ligand-based virtual screening using a genetic algorithm with data fusion. *Match (Mülheim an der Ruhr, Germany)*, 80(3, 6).

Hollingsworth, S. A., & Dror, R. O. (2018, September 19). Molecular dynamics simulation for all. *Neuron*, 99(6), 1129–1143. DOI: 10.1016/j.neuron.2018.08.011 PMID: 30236283

Holmes, C., McDonald, F., Jones, M., Ozdemir, V., & Graham, J. E. (2010). Standardization and Omics Science: Technical and Social Dimensions Are Inseparable and Demand Symmetrical Study. DOI: 10.1089/omi.2010.0022

Holm, L., & Sander, C. (1993). Structural alignment of globins, phycocyanins and colicin A. *FEBS Letters*, 315(3), 301–306. DOI: 10.1016/0014-5793(93)81183-Z PMID: 8422921

Holsinger, K. E., & Weir, B. S. (2009). Genetics in geographically structured populations: Defining, estimating and interpreting F(ST). *Nature Reviews. Genetics*, 10(9), 639–650. DOI: 10.1038/nrg2611 PMID: 19687804

Holton, T. A., Vijayakumar, V., & Khaldi, N. (2013). Bioinformatics: Current perspectives and future directions for food and nutritional research facilitated by a Food-Wiki database. *Trends in Food Science & Technology*, 34(1), 5–17. DOI: 10.1016/j.tifs.2013.08.009

Hood, L., & Rowen, L. (2013, September 13). The Human Genome Project: Big science transforms biology and medicine. *Genome Medicine*, 5(9), 79. DOI: 10.1186/gm483 PMID: 24040834

Hoogstrate, Y., Jenster, G. W., & Werken, H. J. (2021). FASTAFS: File system virtualisation of random access compressed FASTA files. *BMC Bioinformatics*, 22(1), 1–12. DOI: 10.1186/s12859-021-04455-3 PMID: 34724897

Huang, B., Kong, L., Wang, C., Ju, F., Zhang, Q., Zhu, J., Gong, T., Zhang, H., Yu, C., Zheng, W.-M., & Bu, D. (2023). Protein Structure Prediction: Challenges, Advances, and the Shift of Research Paradigms. *Genomics, Proteomics & Bioinformatics*, 21(5), 913–925. DOI: 10.1016/j.gpb.2022.11.014 PMID: 37001856

Huang, P.-S., Boyken, S. E., & Baker, D. (2016). The coming of age of de novo protein design. *Nature*, 537(7620), 320–327. DOI: 10.1038/nature19946 PMID: 27629638

Huber, W., Carey, V. J., Gentleman, R., Anders, S., Carlson, M., Carvalho, B. S., Bravo, H. C., Davis, S., Gatto, L., Girke, T., Gottardo, R., Hahne, F., Hansen, K. D., Irizarry, R. A., Lawrence, M., Love, M. I., MacDonald, J., Obenchain, V., Oleś, A. K., & Morgan, M. (2015). Orchestrating high-throughput genomic analysis with Bioconductor. *Nature Methods*, 12(2), 115–121. DOI: 10.1038/nmeth.3252 PMID: 25633503

Hunt, M., Newbold, C., Berriman, M., & Otto, T. D. (2014). A comprehensive evaluation of assembly scaffolding tools. *Genome Biology*, 15(3), R42. DOI: 10.1186/gb-2014-15-3-r42 PMID: 24581555

Hurst, L. D. (2019). A century of bias in genetics and evolution. *Heredity*, 123(1), 33–43. DOI: 10.1038/s41437-019-0194-2 PMID: 31189901

Hussain, M. A. (2023). Support vector machine and neural network for enhanced classification algorithm in ecological data. Measurement. *Measurement. Sensors*, 27, 100780. DOI: 10.1016/j.measen.2023.100780

Hutchins, J. R. (2014). What's that gene (or protein)? Online resources for exploring functions of genes, transcripts, and proteins. *Molecular Biology of the Cell*, 25(8), 1187–1201. DOI: 10.1091/mbc.e13-10-0602 PMID: 24723265

Iacoangeli, A., Marcatili, P., Deane, C., Lesk, A. M., Pastore, A., & Teichmann, S. A. (2022, April). Editorial: A Journey Through 50 Years of Structural Bioinformatics in Memoriam of Cyrus Chothia. *Frontiers in Molecular Biosciences*, 9, 28. DOI: 10.3389/fmolb.2022.885318

Inácio, M. M., Moreira, A. L. E., Cruz-Leite, V. R. M., Mattos, K., Silva, L. O. S., Venturini, J., Ruiz, O. H., Ribeiro-Dias, F., Weber, S. S., Soares, C. M. de A., & Borges, C. L. (2023). Fungal Vaccine Development: State of the Art and Perspectives Using Immunoinformatics. *Journal of Fungi, 9*(6), 633. DOI: 10.3390/jof9060633

Ip, C. L. C., Loose, M., Tyson, J. R., de Cesare, M., Brown, B. L., Jain, M., Leggett, R. M., Eccles, D. A., Zalunin, V., Urban, J. M., Piazza, P., Bowden, R. J., Paten, B., Mwaigwisya, S., Batty, E. M., Simpson, J. T., Snutch, T. P., Birney, E., Buck, D., & Olsen, H. E. (2015, October 15). MinION Analysis and Reference Consortium: Phase 1 data release and analysis. *F1000 Research*, 4, 1075. DOI: 10.12688/f1000research.7201.1 PMID: 26834992

Iqbal, N., & Kumar, P. (2023). From data science to bioscience: Emerging era of bioinformatics applications, tools and challenges. *Procedia Computer Science*, 218, 1516–1528. DOI: 10.1016/j.procs.2023.01.130

Irvine, D. J., & Dane, E. L. (2020, May). Enhancing cancer immunotherapy with nanomedicine. *Nature Reviews. Immunology*, 20(5), 321–334. DOI: 10.1038/s41577-019-0269-6 PMID: 32005979

Ishack, S., & Lipner, S. R. (2021). Bioinformatics and immunoinformatics to support COVID-19 vaccine development. *Journal of Medical Virology*, 93(9), 5209–5211. DOI: 10.1002/jmv.27017 PMID: 33851735

Ismail, H. D. (2023). *Bioinformatics: A Practical Guide to Next Generation Sequencing Data Analysis* (1st ed.). Chapman and Hall/CRC. DOI: 10.1201/9781003355205

Iupac, C., Org, N., & Comm, I.-I. (1975). Nomenclature of α-amino acids. *Biochemistry*, 14(2), 449–462. DOI: 10.1021/bi00673a036

Jabłoński, M. (2021). Intramolecular Hydrogen Bonding 2021. *Molecules (Basel, Switzerland)*, 26(20), 6319. DOI: 10.3390/molecules26206319 PMID: 34684899

Jain, K. K. (2004). Role of pharmacoproteomics in the development of personalized medicine. *Pharmacogenomics*, 5(3), 331–336. DOI: 10.1517/phgs.5.3.331.29830 PMID: 15102547

Jaiswal, A. K., Tiwari, S., Jamal, S. B., Barh, D., Azevedo, V., & Soares, S. C. (2017a). An In Silico Identification of Common Putative Vaccine Candidates against Treponema pallidum: A Reverse Vaccinology and Subtractive Genomics Based Approach. *International Journal of Molecular Sciences, 18*(2), 402. DOI: 10.3390/ijms18020402

Jäkel, S., & Williams, A. (2020). What have advances in transcriptomic technologies taught us about human white matter pathologies? *Frontiers in Cellular Neuroscience*, 14, 238. DOI: 10.3389/fncel.2020.00238 PMID: 32848627

Jan, I., Sharma, P., & Bansal, A. (2023). Microbial Bioinformatics Approach in Food Science. In *Food Microbial Sustainability: Integration of Food Production and Food Safety* (pp. 267–288). Springer Nature Singapore. DOI: 10.1007/978-981-99-4784-3_13

Jarvis, M. C. (2023). Hydrogen bonding and other non-covalent interactions at the surfaces of cellulose microfibrils. *Cellulose (London, England)*, 30(2), 667–687. DOI: 10.1007/s10570-022-04954-3

Jeevanandam, J., Agyei, D., Danquah, M. K., & Udenigwe, C. (2022). Food quality monitoring through bioinformatics and big data. In *Future Foods* (pp. 733–744). Academic Press. DOI: 10.1016/B978-0-323-91001-9.00036-0

Jensen, K. K., Andreatta, M., Marcatili, P., Buus, S., Greenbaum, J. A., Yan, Z., Sette, A., Peters, B., & Nielsen, M. (2018). Improved methods for predicting peptide binding affinity to MHC class II molecules. *Immunology*, 154(3), 394–406. DOI: 10.1111/imm.12889 PMID: 29315598

Jensen, M. A., Ferretti, V., Grossman, R. L., & Staudt, L. M. (2017). The NCI Genomic Data Commons as an engine for precision medicine. *Blood*, 130(4), 453–459. DOI: 10.1182/blood-2017-03-735654 PMID: 28600341

Jeon, J., Nim, S., Teyra, J., Datti, A., Wrana, J. L., Sidhu, S. S., Moffat, J., & Kim, P. M. (2014, July 30). A systematic approach to identify novel cancer drug targets using machine learning, inhibitor design and high-throughput screening. *Genome Medicine*, 6(7), 57. DOI: 10.1186/s13073-014-0057-7 PMID: 25165489

Jiang, D., Hsieh, C.-Y., Wu, Z., Kang, Y., Wang, J., Wang, E., Liao, B., Shen, C., Xu, L., Wu, J., Cao, D., & Hou, T. (2021). InteractionGraphNet: A novel and efficient deep graph representation learning framework for accurate protein–ligand interaction predictions. *Journal of Medicinal Chemistry*, 64(24), 18209–18232. DOI: 10.1021/acs.jmedchem.1c01830 PMID: 34878785

Jiang, S., Feher, M., Williams, C., Cole, B., & Shaw, D. E. (2020, September 28). AutoPH4: An Automated Method for Generating Pharmacophore Models from Protein Binding Pockets. *Journal of Chemical Information and Modeling*, 60(9), 4326–4338. DOI: 10.1021/acs.jcim.0c00121 PMID: 32639159

Jiang, Z. H. U. (2008). Refining homology models by combining replica-exchange molecular dynamics and statistical potentials. *Proteins*, 72(4), 1171–1188. DOI: 10.1002/prot.22005 PMID: 18338384

John, G., Sahajpal, N. S., Mondal, A. K., Ananth, S., Williams, C., Chaubey, A., Rojiani, A. M., & Kolhe, R. (2021). Next-Generation sequencing (NGS) in COVID-19: A tool for SARS-CoV-2 diagnosis, monitoring new strains and phylodynamic modeling in molecular epidemiology. *Current Issues in Molecular Biology*, 43(2), 845–867. DOI: 10.3390/cimb43020061 PMID: 34449545

Johnson, J. A., & Cavallari, L. H. (2015). Warfarin pharmacogenetics. *Trends in Cardiovascular Medicine*, 25(1), 33–41. DOI: 10.1016/j.tcm.2014.09.001 PMID: 25282448

Jones, D. P., Park, Y., & Ziegler, T. R. (2012). Nutritional metabolomics: Progress in addressing complexity in diet and health. *Annual Review of Nutrition*, 32(1), 183–202. DOI: 10.1146/annurev-nutr-072610-145159 PMID: 22540256

Jones, M. B., Schildhauer, M. P., Reichman, O. J., & Bowers, S. (2006). The New Bioinformatics: Integrating Ecological Data from the Gene to the Biosphere. *Annual Review of Ecology, Evolution, and Systematics*, 37(1), 519–544. DOI: 10.1146/annurev.ecolsys.37.091305.110031

Jovic, D., Liang, X., Zeng, H., Lin, L., Xu, F., & Luo, Y. (2022). Single-cell RNA sequencing technologies and applications: A brief overview. *Clinical and Translational Medicine*, 12(3), e694. DOI: 10.1002/ctm2.694 PMID: 35352511

Jowett, T. (1999). Analysis of protein and gene expression. *Methods in Cell Biology*, 59, 63–85. DOI: 10.1016/S0091-679X(08)61821-X PMID: 9891356

Juengst, E., McGowan, M. L., Fishman, J. R., & Settersten, R. A.Jr. (2016). From "personalized" to "precision" medicine: The ethical and social implications of rhetorical reform in genomic medicine. *The Hastings Center Report*, 46(5), 21–33. DOI: 10.1002/hast.614 PMID: 27649826

Jumper, J. (2021). Highly accurate protein structure prediction with AlphaFold. *Nature, 596*(7873), 583–589.

Jumper, J., Evans, R., Pritzel, A., Green, T., Figurnov, M., Ronneberger, O., Tunyasuvunakool, K., Bates, R., Žídek, A., Potapenko, A., Bridgland, A., Meyer, C., Kohl, S. A. A., Ballard, A. J., Cowie, A., Romera-Paredes, B., Nikolov, S., Jain, R., Adler, J., … Hassabis, D. (2021). Highly accurate protein structure prediction with AlphaFold. *Nature, 596*(7873), 583–589. DOI: 10.1038/s41586-021-03819-2

Jung, H., Ventura, T., Chung, J. S., Kim, W.-J., Nam, B.-H., Kong, H. J., Kim, Y.-O., Jeon, M.-S., & Eyun, S.-i. (2020). Twelve quick steps for genome assembly and annotation in the classroom. *PLoS Computational Biology*, 16(11), e1008325. DOI: 10.1371/journal.pcbi.1008325 PMID: 33180771

Junqueira, L. C., & Carneiro, J. (2023). *Biologia Celular e Molecular* (10ª edição). Guanabara Koogan.

Kanehisa, M., & Bork, P. (2003). Bioinformatics in the post-sequence era. *Nature Genetics*, 33(3), 305–310. DOI: 10.1038/ng1109 PMID: 12610540

Karczewski, K. J., Weisburd, B., Thomas, B., Solomonson, M., Ruderfer, D. M., Kavanagh, D., Hamamsy, T., Lek, M., Samocha, K. E., Cummings, B. B., Birnbaum, D., Daly, M. J., & MacArthur, D. G. (2017). The ExAC browser: Displaying reference data information from over 60 000 exomes. *Nucleic Acids Research*, 45(D1), D840–d845. DOI: 10.1093/nar/gkw971 PMID: 27899611

Karim, M. R., Islam, T., Shajalal, M., Beyan, O., Lange, C., Cochez, M., Rebholz-Schuhmann, D., & Decker, S. (2023). Explainable AI for Bioinformatics: Methods, Tools and Applications. *Briefings in Bioinformatics*, 24(5), bbad236. Advance online publication. DOI: 10.1093/bib/bbad236 PMID: 37478371

Karolchik, D., Hinrichs, A. S., & Kent, W. J. (2011). The UCSC Genome Browser. *Curr Protoc Hum Genet*. DOI: 10.1002/0471142905.hg1806s71

Kathryn, T. (2021). Highly accurate protein structure prediction for the human proteome. *Nature*, 596(7873), 590–596. DOI: 10.1038/s41586-021-03828-1 PMID: 34293799

Kaur, T., Madgulkar, A., Bhalekar, M., & Asgaonkar, K. (2019, March 1). Molecular Docking in Formulation and Development. *Current Drug Discovery Technologies*, 16(1), 30–39. DOI: 10.2174/1570163815666180219112421 PMID: 29468973

Kaushik, R., Kant, R., & Christodoulides, M. (2023). Artificial intelligence in accelerating vaccine development - current and future perspectives. *Frontiers in Bacteriology*, 2, 1258159. DOI: 10.3389/fbrio.2023.1258159

Kelley, L. A., Mezulis, S., Yates, C. M., Wass, M. N., & Sternberg, M. J. E. (2015). The Phyre2 web portal for protein modeling, prediction and analysis. *Nature Protocols, 10*(6), 845–858. DOI: 10.1038/nprot.2015.053

Keras. (n.d.). Available at: https://keras.io/

Kessler, C. (2018). Genomics and Precision Medicine: Implications for Critical Care. *AACN Advanced Critical Care*, 29(1), 28–35. DOI: 10.4037/aacnacc2018521 PMID: 29496711

Khalid, K., & Poh, C. L. (2023). The Promising Potential of Reverse Vaccinology-Based Next-Generation Vaccine Development over Conventional Vaccines against Antibiotic-Resistant Bacteria. *Vaccines, 11*(7), 1264. DOI: 10.3390/vaccines11071264

Khan, S. U., Ahemad, N., Chuah, L.-H., Naidu, R., & Htar, T. T. (2022). G protein-coupled estrogen receptor-1: Homology modeling approaches and application in screening new GPER-1 modulators. *Journal of Biomolecular Structure & Dynamics*, 40(7), 3325–3335. DOI: 10.1080/07391102.2020.1844059 PMID: 33164654

Khoshnood, S., Ghanavati, R., Shirani, M., Ghahramanpour, H., Sholeh, M., Shariati, A., Sadeghifard, N., & Heidary, M. (2022). Viral vector and nucleic acid vaccines against COVID-19: A narrative review. *Frontiers in Microbiology*, 13, 984536. Advance online publication. DOI: 10.3389/fmicb.2022.984536 PMID: 36118203

Khoury. (2014). Protein folding and de novo protein design for biotechnological applications. Trends in Biotechnology, 32(2), 99-109.

Kimber, T. B., Chen, Y., & Volkamer, A. (2021). Deep Learning in Virtual Screening: Recent Applications and Developments. *International Journal of Molecular Sciences, 22*(9). *International Journal of Molecular Sciences*, 9(9), 4435. Advance online publication. DOI: 10.3390/ijms22094435 PMID: 33922714

Kim, J., Park, S., Min, D., & Kim, W. (2021). A comprehensive survey of recent drug Discovery using deep learning. *International Journal of Molecular Sciences*, 22(18), 9983. DOI: 10.3390/ijms22189983 PMID: 34576146

Kinganda-Lusamaki, E., Black, A., Mukadi, D. B., Hadfield, J., Mbala-Kingebeni, P., Pratt, C. B., & Bisento, N. (2021). Integration of genomic sequencing into the response to the Ebola virus outbreak in Nord Kivu, Democratic Republic of the Congo. *Nature Medicine*, 27(4), 710–716. DOI: 10.1038/s41591-021-01302-z PMID: 33846610

Kitchen, D. B., Decornez, H., Furr, J. R., & Bajorath, J. (2004). Docking and scoring in virtual screening for drug discovery: Methods and applications. *Nature Reviews. Drug Discovery*, 3(11), 935–949. DOI: 10.1038/nrd1549 PMID: 15520816

Kloczkowski, A., Jernigan, R., Wu, Z., Song, G., Yang, L., Kolinski, A., & Pokarowski, P. (2009). Distance matrix-based approach to protein structure prediction. *Journal of Structural and Functional Genomics*, 10(1), 67–81. DOI: 10.1007/s10969-009-9062-2 PMID: 19224393

Koboldt, D. C., Steinberg, K. M., Larson, D. E., Wilson, R. K., & Mardis, E. R. (2013). The next-generation sequencing revolution and its impact on genomics. *Cell*, 155(1), 27–38. DOI: 10.1016/j.cell.2013.09.006 PMID: 24074859

Koes, D. R., & Camacho, C. J. (2014, September 30). Shape-Based Virtual Screening with Volumetric Aligned Molecular Shapes. *Journal of Computational Chemistry*, 35(25), 1824–1834. DOI: 10.1002/jcc.23690 PMID: 25049193

Kolluri, S., Lin, J., Liu, R., Zhang, Y., & Zhang, W. (2022, January). Machine Learning and Artificial Intelligence in Pharmaceutical Research and Development: A Review. *The AAPS Journal*, 24(1, 4), 19. DOI: 10.1208/s12248-021-00644-3 PMID: 34984579

Kondratyuk, N., Nikolskiy, V., Pavlov, D., & Stegailov, V. (2021). GPU-accelerated molecular dynamics: State-of-art software performance and porting from Nvidia CUDA to AMD HIP. *International Journal of High Performance Computing Applications*, 35(4), 312–324. DOI: 10.1177/10943420211008288

Korendovych, I. V., & DeGrado, W. F. (2020). De novo protein design, a retrospective. *Quarterly Reviews of Biophysics*, 53, e3. DOI: 10.1017/S0033583519000131

Kotha, K., & Clancy, J. P. (2013). Ivacaftor treatment of cystic fibrosis patients with the G551D mutation: A review of the evidence. *Therapeutic Advances in Respiratory Disease*, 7(5), 288–296. DOI: 10.1177/1753465813502115 PMID: 24004658

Koutroumbas, K., & Theodoridis, S. (2008). *Pattern recognition*. Academic Press.

Kozakov, D., Hall, D. R., Xia, B., Porter, K. A., Padhorny, D., Yueh, C., Beglov, D., & Vajda, S. (2017). The ClusPro web server for protein–protein docking. *Nature Protocols*, 12(2), 255–278. DOI: 10.1038/nprot.2016.169 PMID: 28079879

Krapivin, V., Varotsos, C., & Soldatov, V. (2015). *New Ecoinformatics Tools in Environmental Science*. Springer Cham. DOI: 10.1007/978-3-319-13978-4

Krieger. (2003). Homology modeling. Structural Bioinformatics, 44, 509-523.

Krishna, C., Kumar, D., & Kushwaha, D. S. (2023). A Comprehensive Survey on Pandemic Patient Monitoring System: Enabling Technologies, Opportunities, and Research Challenges. *Wireless Personal Communications*, 131(3), 2125–2172. DOI: 10.1007/s11277-023-10535-9 PMID: 37360140

Krishna, R., Wang, J., Ahern, W., Sturmfels, P., Venkatesh, P., Kalvet, I., Lee, G. R., Morey-Burrows, F. S., Anishchenko, I., Humphreys, I. R., McHugh, R., Vafeados, D., Li, X., Sutherland, G. A., Hitchcock, A., Hunter, C. N., Kang, A., Brackenbrough, E., Bera, A. K., & Baker, D. (2024). Generalized biomolecular modeling and design with RoseTTAFold All-Atom. *Science*, 384(6693), eadl2528. Advance online publication. DOI: 10.1126/science.adl2528 PMID: 38452047

Kröger, K., Seidel, M., & Gauglitz, G. (2003). Optical Sensing Methods in High-Throughput Screening. In Potyrailo, R. A., & Amis, E. J. (Eds.), *High-Throughput Analysis: A Tool for Combinatorial Materials Science* (pp. 261–288). Springer US. DOI: 10.1007/978-1-4419-8989-5_13

Kryukov, K., Ueda, M. T., Nakagawa, S., & Imanishi, T. (2020). Sequence Compression Benchmark (SCB) database—A comprehensive evaluation of reference-free compressors for FASTA-formatted sequences. *GigaScience*, 9(7), giaa072. Advance online publication. DOI: 10.1093/gigascience/giaa072 PMID: 32627830

Kumuthini, J., Chimenti, M., Nahnsen, S., Peltzer, A., Meraba, R., McFadyen, R., Wells, G., Taylor, D., Maienschein-Cline, M., Li, J. L., Thimmapuram, J., Murthy-Karuturi, R., & Zass, L. (2020). Ten simple rules for providing effective bioinformatics research support. *PLoS Computational Biology*, 16(3), e1007531. DOI: 10.1371/journal.pcbi.1007531 PMID: 32214318

Kwoji, I. D. (2023). 'Multi-omics' data integration: applications in probiotics studies. *NPJ Science of Food, 7*(1), 25.

Kwok, A. J., Mentzer, A., & Knight, J. C. (2021). Host genetics and infectious disease: New tools, insights and translational opportunities. *Nature Reviews. Genetics*, 22(3), 137–153. DOI: 10.1038/s41576-020-00297-6 PMID: 33277640

LaDeau, S. L., Han, B. A., Rosi-Marshall, E. J., & Weathers, K. C. (2017). The Next Decade of Big Data in Ecosystem Science. *Ecosystems (New York, N.Y.)*, 20(2), 274–283. DOI: 10.1007/s10021-016-0075-y

Lamond, A. I. (1993). The spliceosome. *BioEssays*, 15(9), 595–603. DOI: 10.1002/bies.950150905 PMID: 8240312

Lancaster, E. M., Jablons, D., & Kratz, J. R. (2020). Applications of Next-Generation Sequencing in Neoantigen Prediction and Cancer Vaccine Development. *Genetic Testing and Molecular Biomarkers*, 24(2), 59–66. DOI: 10.1089/gtmb.2018.0211 PMID: 30907630

Lander, E. S., Linton, L. M., Birren, B., Nusbaum, C., Zody, M. C., Baldwin, J., Devon, K., Dewar, K., Doyle, M., FitzHugh, W., Funke, R., Gage, D., Harris, K., Heaford, A., Howland, J., Kann, L., Lehoczky, J., LeVine, R., McEwan, P., & Morgan, M. J. (2001, February). Initial sequencing and analysis of the human genome. *Nature*, 409(6822), 860–921. DOI: 10.1038/35057062 PMID: 11237011

Lari, A., Lari, N., & Biabangard, A. (2022). Immunoinformatics Approach to Design a Novel Subunit Vaccine Against Visceral Leishmaniasis. *International Journal of Peptide Research and Therapeutics*, 28(1), 34. DOI: 10.1007/s10989-021-10344-3 PMID: 34931120

Larsen, M. V., Lundegaard, C., Lamberth, K., Buus, S., Lund, O., & Nielsen, M. (2007). Large-scale validation of methods for cytotoxic T-lymphocyte epitope prediction. *BMC Bioinformatics*, 8(1), 1–12. DOI: 10.1186/1471-2105-8-424 PMID: 17973982

LeCun, Y., Bengio, Y., & Hinton, G. (2015). Deep learning. *Nature*, 521(7553), 436–444. DOI: 10.1038/nature14539 PMID: 26017442

Lecun, Y., Boser, B., Denker, J. S., Henderson, D., Howard, R. E., Hubbard, W., & Jackel, L. D. (1989, December 1). Backpropagation Applied to Handwritten Zip Code Recognition. *Neural Computation*, 1(4), 541–551. DOI: 10.1162/neco.1989.1.4.541

Lee, S. S. J. (2021). Obligations of the "Gift": Reciprocity and Responsibility in Precision Medicine. *The American Journal of Bioethics*, 21(4), 57–66. DOI: 10.1080/15265161.2020.1851813 PMID: 33325811

Lee, T. I., & Young, R. A. (2000). Transcription of eukaryotic protein-coding genes. *Annual Review of Genetics*, 34(1), 77–137. DOI: 10.1146/annurev.genet.34.1.77 PMID: 11092823

Leinonen, R., Akhtar, R., Birney, E., Bower, L., Cerdeno-Tárraga, A., Cheng, Y., Cleland, I., Faruque, N., Goodgame, N., Gibson, R., Hoad, G., Jang, M., Pakseresht, N., Plaister, S., Radhakrishnan, R., Reddy, K., Sobhany, S., Ten Hoopen, P., Vaughan, R., & Cochrane, G. (2011). The European Nucleotide Archive. *Nucleic Acids Research*, 39(Database issue), D28–D31. DOI: 10.1093/nar/gkq967 PMID: 20972220

Leitão, F., & Cánovas, F. (in press). Predicting climate change impacts on marine fisheries and biodiversity in the Iberian north branch of the canary upwelling system. *Ecological Economics*.

Leitão, F., Relvas, P., Cánovas, F., Baptista, V., & Teodósio, A. (2019). Northerly wind trends along the Portuguese marine coast since 1950. *Theoretical and Applied Climatology*, 137(1-2), 1–19. DOI: 10.1007/s00704-018-2466-9

Lei, Y., Meng, Y., Guo, X., Ning, K., Bian, Y., Li, L., & Dong, Y. (2022). Overview of structural variation calling: Simulation, identification, and visualization. *Computers in Biology and Medicine*, 145, 105534. DOI: 10.1016/j.compbiomed.2022.105534 PMID: 35585730

Lei, Y., Tang, R., Xu, J., Wang, W., Zhang, B., Liu, J., Yu, X., & Shi, S. (2021). Applications of single-cell sequencing in cancer research: Progress and perspectives. *Journal of Hematology & Oncology*, 14(1), 91. DOI: 10.1186/s13045-021-01105-2 PMID: 34108022

Leman, J. K., Weitzner, B. D., Lewis, S. M., Adolf-Bryfogle, J., Alam, N., Alford, R. F., Aprahamian, M., Baker, D., Barlow, K. A., Barth, P., Basanta, B., Bender, B. J., Blacklock, K., Bonet, J., Boyken, S. E., Bradley, P., Bystroff, C., Conway, P., Cooper, S., & Bonneau, R. (2020). Macromolecular modeling and design in Rosetta: Recent methods and frameworks. *Nature Methods*, 17(7), 665–680. DOI: 10.1038/s41592-020-0848-2 PMID: 32483333

Lemmen, C., & Lengauer, T. (2000). Computational methods for the structural alignment of molecules. *Journal of Computer-Aided Molecular Design*, 14(3), 215–232. DOI: 10.1023/A:1008194019144 PMID: 10756477

Lesk. (2019). *Introduction to bioinformatics*. Oxford University Press.

Lesk, A. M. (2017). *Introduction to genomics*. Oxford University Press. DOI: 10.1093/hesc/9780198754831.001.0001

Levinthal, C. (1969). How to fold graciously. In Mossbauer Spectroscopy in Biological Systems: Proceedings of a Meeting Held at Allerton House, Monticello, Illinois (pp. 22-24). Retrieved from https://web.archive.org/web/20110523080407/http://www-miller.ch.cam.ac.uk/levinthal/levinthal.html

Levinthal, C. (1968). Are there pathways for protein folding? *Journal de Chimie Physique*, 65, 44–45. DOI: 10.1051/jcp/1968650044

Liao, Y., Liu, Z., Zhang, Y., Lu, P., Wen, L., & Tang, F. (2023). High-throughput and high-sensitivity full-length single-cell RNA-seq analysis on third-generation sequencing platform. *Cell Discovery*, 9(1), 5. DOI: 10.1038/s41421-022-00500-4 PMID: 36631434

Li, J., & Liu, C. (2019). Coding or Noncoding, the Converging Concepts of RNAs. *Frontiers in Genetics*, 10, 496. DOI: 10.3389/fgene.2019.00496 PMID: 31178900

Lim, C. P., Kok, B. H., Lim, H. T., Chuah, C., Abdul Rahman, B., Abdul Majeed, A. B., Wykes, M., Leow, C. H., & Leow, C. Y. (2023). Recent trends in next generation immunoinformatics harnessed for universal coronavirus vaccine design. *Pathogens and Global Health*, 117(2), 134–151. DOI: 10.1080/20477724.2022.2072456 PMID: 35550001

Lin, H. H., Zhang, G. L., Tongchusak, S., Reinherz, E. L., & Brusic, V. (2008). Evaluation of MHC-II peptide binding prediction servers: applications for vaccine research. *BMC Bioinformatics, 9*(12), 1–10. DOI: 10.1186/1471-2105-9-S12-S22

Lin, M. J., Svensson-Arvelund, J., Lubitz, G. S., Marabelle, A., Melero, I., Brown, B. D., & Brody, J. D. (2022). Cancer vaccines: the next immunotherapy frontier. *Nature Cancer, 3*(8), 911–926. DOI: 10.1038/s43018-022-00418-6

Lin, D., Cho, Y., Fan, H., & Zhang, Y. (2024). From explainable to interactive AI: A literature review on current trends and future directions. *Environmental Science and Ecotechnology*, 13, 100248. DOI: 10.1016/j.dsp.2024.01.001

Lindahl, E. R. (2008). Molecular Dynamics Simulations. In A. Kukol (Org.), *Molecular Modeling of Proteins* (p. 3–23). Humana Press. DOI: 10.1007/978-1-59745-177-2_1

Lindahl, E., Hess, B., & van der Spoel, D. (2001). GROMACS 3.0: A package for molecular simulation and trajectory analysis. *Journal of Molecular Modeling*, 7(8), 306–317. DOI: 10.1007/s008940100045

Lin, J., & Ngiam, K. Y. (2023). How data science and AI-based technologies impact genomics. *Singapore Medical Journal*, 64(1), 59–66. DOI: 10.4103/singaporemedj.SMJ-2021-438 PMID: 36722518

Lin, X., Li, X., & Lin, X. (2020, January). A Review on Applications of Computational Methods in Drug Screening and Design. *Molecules (Basel, Switzerland)*, 25(6), 1375. DOI: 10.3390/molecules25061375 PMID: 32197324

Lin, Z., Akin, H., Rao, R., Hie, B., Zhu, Z., Lu, W., Smetanin, N., Verkuil, R., Kabeli, O., Shmueli, Y., dos Santos Costa, A., Fazel-Zarandi, M., Sercu, T., Candido, S., & Rives, A. (2023, March 17). Evolutionary-scale prediction of atomic-level protein structure with a language model. *Science*, 379(6637), 1123–1130. DOI: 10.1126/science.ade2574 PMID: 36927031

Lipinski, C. A., Lombardo, F., Dominy, B. W., & Feeney, P. J. (2001, March 1). Experimental and computational approaches to estimate solubility and permeability in drug discovery and development settings. *Advanced Drug Delivery Reviews*, 46(1–3), 3–26. DOI: 10.1016/S0169-409X(00)00129-0 PMID: 11259830

Lipton, Z. C. (2016). The mythos of model interpretability. (*WHI 2016*) *arXiv*, 1606.03490 DOI: 10.48550/arXiv.1606.03490

Liu, Y., & Ye, Q. (2022). Nucleic Acid Vaccines against SARS-CoV-2. *Vaccines, 10*(11), 1849. DOI: 10.3390/vaccines10111849

Liu, J., Guo, Z., Wu, T., Roy, R. S., Chen, C., & Cheng, J. (2023). Improving AlphaFold2-based protein tertiary structure prediction with MULTICOM in CASP15. *Communications Chemistry*, 6(1), 188. DOI: 10.1038/s42004-023-00991-6 PMID: 37679431

Liu, L. (2012, July 5). Comparison of Next-Generation Sequencing Systems. *BioMed Research International*, 2012, e251364. PMID: 22829749

Liu, L., Yu, W., Cai, K., Ma, S., Wang, Y., Ma, Y., & Zhao, H. (2023). Identification of vaccine candidates against rhodococcus equi by combining pangenome analysis with a reverse vaccinology approach. *Heliyon*, 9(8), e18623. Advance online publication. DOI: 10.1016/j.heliyon.2023.e18623 PMID: 37576287

Liu, M., Li, C., Chen, R., Cao, D., & Zeng, X. (2024). Geometric deep learning for drug discovery. *Expert Systems with Applications*, 240, 122498. DOI: 10.1016/j.eswa.2023.122498

Liu, Z., Du, J., Fang, J., Yin, Y., Xu, G., & Xie, L. (2019). Deepscreening: A deep learning-based screening web server for accelerating drug discovery. *Database (Oxford)*, 2019, baz104. Advance online publication. DOI: 10.1093/database/baz104 PMID: 31608949

Logsdon, G. A., Vollger, M. R., & Eichler, E. E. (2020). Long-read human genome sequencing and its applications. *Nature Reviews. Genetics*, 21(10), 597–614. DOI: 10.1038/s41576-020-0236-x PMID: 32504078

López-Pedrouso, M., Lorenzo, J. M., Alché, J. D. D., Moreira, R., & Franco, D. (2023). Advanced proteomic and bioinformatic tools for predictive analysis of allergens in novel foods. *Biology (Basel)*, 12(5), 714. DOI: 10.3390/biology12050714 PMID: 37237526

Lowe, R., Shirley, N., Bleackley, M., Dolan, S., & Shafee, T. (2017). Transcriptomics technologies. *PLoS Computational Biology*, 13(5), e1005457. DOI: 10.1371/journal.pcbi.1005457 PMID: 28545146

Lo, Y.-C., Rensi, S. E., Torng, W., & Altman, R. B. (2018). Machine learning in chemoinformatics and drug discovery. *Drug Discovery Today*, 23(8), 1538–1546. DOI: 10.1016/j.drudis.2018.05.010 PMID: 29750902

Lu, D., Lu, T., & Cao, S. (2012). Individualized cancer chemotherapy by detecting cancer biomarkers. *Metabolomics*, 2(5), e121.

Lu, W., Wu, Q., Zhang, J., Rao, J., Li, C., & Zheng, S. (2022). TANKBind: trigonometry-aware neural networKs for drug-protein binding structure prediction. *bioRxiv,* 2022.06.06.495043 DOI: 10.1101/2022.06.06.495043

Lyne, P. D. (2002). Structure-based virtual screening: An overview. *Drug Discovery Today*, 7(20), 1047–1055. DOI: 10.1016/S1359-6446(02)02483-2 PMID: 12546894

Ma, D.-L., Shiu-Hin Chan, D., & Leung, C.-H. (2011a). Molecular docking for virtual screening of natural product databases. *Chemical Science (Cambridge)*, 2(9), 1656–1665. DOI: 10.1039/C1SC00152C

Madhavan, A., Arun, K. B., Sindhu, R., Binod, P., Kim, S. H., & Pandey, A. (2019). Tailoring of microbes for the production of high value plant-derived compounds: From pathway engineering to fermentative production. *Biochimica et Biophysica Acta (BBA)-. Biochimica et Biophysica Acta. Proteins and Proteomics*, 1867(11), 140262. DOI: 10.1016/j.bbapap.2019.140262 PMID: 31404685

Madhavan, A., Arun, K., Sindhu, R., Krishnamoorthy, J., Reshmy, R., Sirohi, R., Pugazhendi, A., Awasthi, M. K., Szakacs, G., & Binod, P. (2021). Customized yeast cell factories for biopharmaceuticals: From cell engineering to process scale up. *Microbial Cell Factories*, 20(1), 124. DOI: 10.1186/s12934-021-01617-z PMID: 34193127

Maia, E. H. B., Assis, L. C., de Oliveira, T. A., da Silva, A. M., & Taranto, A. G. (2020, April). Structure-Based Virtual Screening: From Classical to Artificial Intelligence. *Frontiers in Chemistry*, 8, 28. DOI: 10.3389/fchem.2020.00343 PMID: 32411671

Ma, J., & Wang, S. (2014). Algorithms, applications, and challenges of protein structure alignment. *Advances in Protein Chemistry and Structural Biology*, 94, 121–175. DOI: 10.1016/B978-0-12-800168-4.00005-6 PMID: 24629187

Makhoba, X. H. (2020). Potential Impact of the Multi-Target Drug Approach in the Treatment of Some Complex Diseases. Drug Design, Development and Therapy, 14, 3235–3249.

Malaina, I., Martínez, L., Montoya, J. M., Alonso, S., Boyano, M. D., Asumendi, A., Izu, R., Sanchez-Diez, A., Cancho-Galan, G., & de la Fuente, , I. (2023). A Universal Antigen-Ranking Method to Design Personalized Vaccines Targeting Neoantigens against Melanoma. *Life (Chicago, Ill.)*, 13(1), 155. DOI: 10.3390/life13010155 PMID: 36676104

Malvezzi, A. (2008). Modelos de virtual Screening de inibidores da cruzaína: desenvolvimento e validação experimental. Universidade de São Paulo.

Manel, S., Schwartz, M., Luikart, G., & Tarberlet, P. (2003). Landscape genetics: Combining landscape ecology and population genetics. *Trends in Ecology & Evolution*, 18(4), 189–197. DOI: 10.1016/S0169-5347(03)00008-9

Manley, W., Tran, T., Prusinski, M., & Brisson, D. (2023). Modeling Tick Populations: An Ecological Test Case for Gradient Boosted Trees. bioRxiv: the preprint server for biology. DOI: 10.1101/2023.03.13.532443

Manni, F., Guérard, E., & Heyer, E. (2004). Geographic patterns of (genetic, morphologic, linguistic) variation: How barriers can be detected by using Monmonier's algorithm. *Human Biology*, 76(2), 173–190. DOI: 10.1353/hub.2004.0034 PMID: 15359530

Mariano, D. C. B. (2016). SIMBA: a web tool for managing bacterial genome assembly generated by Ion PGM sequencing technology. *BMC Bioinformatics, 17*(18), 456.

Mariano, D. (2020). A brief history of bioinformatics told by data visualization. In *Brazilian Symposium On Bioinformatics*. Springer. DOI: 10.1007/978-3-030-65775-8_22

Mariano, D. (2020). Classificação de dados. In *Data Mining. 1*. Sagah.

Mariano, D. (2021). Métricas de avaliação em machine learning: acurácia, sensibilidade, precisão, especificidade e F-score. In *Revista Bioinfo. 1* (Vol. 1). Alfahelix. DOI: 10.51780/978-6-599-275326-15

Mariano, D. (2022). Revista BIOINFO. In *AlphaFold e a busca pelo Santo Graal da Biologia Molecular. 2*. Alfahelix.

Mariano, D. C., Pereira, F. L., Ghosh, P., Barh, D., Figueiredo, H. C. P., Silva, A., Ramos, R. T. J., & Azevedo, V. A. C. (2015). MapRepeat: An approach for effective assembly of repetitive regions in prokaryotic genomes. *Bioinformation*, 11(6), 276–279. DOI: 10.6026/97320630011276 PMID: 26229287

Marini, F., Ludt, A., Linke, J., & Strauch, K. (2021). GeneTonic: An R/Bioconductor package for streamlining the interpretation of RNA-seq data. *BMC Bioinformatics*, 22(1), 610. DOI: 10.1186/s12859-021-04461-5 PMID: 34949163

Markina, Y. V., Gerasimova, E. V., Markin, A. M., Glanz, V. Y., Wu, W. K., Sobenin, I. A., & Orekhov, A. N. (2020). Sialylated Immunoglobulins for the Treatment of Immuno-Inflammatory Diseases. *International Journal of Molecular Sciences*, 21(15), 1–17. DOI: 10.3390/ijms21155472 PMID: 32751832

Martina, M., De Rosa, V., Magon, G., Acquadro, A., Barchi, L., Barcaccia, G., De Paoli, E., Vannozzi, A., & Portis, E. (2024). Revitalizing agriculture: Next-generation genotyping and -omics technologies enabling molecular prediction of resilient traits in the Solanaceae family. *Frontiers in Plant Science*, 15, 1278760. DOI: 10.3389/fpls.2024.1278760 PMID: 38375087

Martinelli, D. D. (2022). In silico vaccine design: A tutorial in immunoinformatics. *Healthcare Analytics*, 2, 100044. DOI: 10.1016/j.health.2022.100044

Martínez-García, M., & Hernández-Lemus, E. (2021). Data Integration Challenges for Machine Learning in Precision Medicine. *Frontiers in Medicine*, 8, 784455. DOI: 10.3389/fmed.2021.784455 PMID: 35145977

Martinez-Olivares, C. E., Hernández-Pando, R., & Mixcoha, E. (2023). In silico EsxG EsxH rational epitope selection: Candidate epitopes for vaccine design against pulmonary tuberculosis. *PLoS One*, 18(4), e0284264. DOI: 10.1371/journal.pone.0284264 PMID: 37079575

Martini, S., Nielsen, M., Peters, B., & Sette, A. (2020). The Immune Epitope Database and Analysis Resource Program 2003–2018: Reflections and outlook. *Immunogenetics*, 72(1–2), 57–76. DOI: 10.1007/s00251-019-01137-6 PMID: 31761977

Martins. (2021). Propedia: a database for protein–peptide identification based on a hybrid clustering algorithm. BMC Bioinformatics, 22, 1-20.

Marti-Renom. (2009). Structure comparison and alignment. Structural Bioinformatics, 2.

Maruyama, T. (1983). Stochastic theory of population genetics. *Bulletin of Mathematical Biology*, 45(4), 521–554. DOI: 10.1007/BF02459586

Masignani, V., Pizza, M., & Moxon, E. R. (2019). The Development of a Vaccine Against Meningococcus B Using Reverse Vaccinology. *Frontiers in Immunology*, 10(MAR), 751. Advance online publication. DOI: 10.3389/fimmu.2019.00751 PMID: 31040844

Mattick, J. S., Amaral, P. P., Carninci, P., Carpenter, S., Chang, H. Y., Chen, L.-L., & Fitzgerald, K. A. (2023). Long non-coding RNAs: Definitions, functions, challenges and recommendations. *Nature Reviews. Molecular Cell Biology*, 24(6), 430–447. DOI: 10.1038/s41580-022-00566-8 PMID: 36596869

Mbala-Kingebeni, P., Aziza, A., Di Paola, N., Wiley, M. R., Makiala-Mandanda, S., Caviness, K., & Prieto, K. (2019). Medical countermeasures during the 2018 Ebola virus disease outbreak in the North Kivu and Ituri Provinces of the Democratic Republic of the Congo: A rapid genomic assessment. *The Lancet. Infectious Diseases*, 19(6), 648–657. DOI: 10.1016/S1473-3099(19)30118-5 PMID: 31000464

McGrath, S., & Ghersi, D. (2016). Building towards precision medicine: Empowering medical professionals for the next revolution. *BMC Medical Genomics*, 9(1), 1–6. DOI: 10.1186/s12920-016-0183-8 PMID: 27160306

McGuire, A. L., Caulfield, T., & Cho, M. K. (2008). Research ethics and the challenge of whole-genome sequencing. *Nature Reviews. Genetics*, 9(2), 152–156. DOI: 10.1038/nrg2302 PMID: 18087293

McInnes, C. (2007). Virtual screening strategies in drug discovery. Current Opinion in Chemical Biology, Analytical Techniques / Mechanisms, 11(5), 494–502.

McLane, A. J., Semeniuk, C., McDermid, G. J., & Marceau, D. J. (2011). The role of agent-based models in wildlife ecology and management. *Ecological Modelling*, 222(8), 1544–1556. DOI: 10.1016/j.ecolmodel.2011.01.020

McMahon. (2005). The C-type lectin fold as an evolutionary solution for massive sequence variation. Nature Structural & Molecular Biology, 12(10), 886-892.

McNutt, A. T., Francoeur, P., Aggarwal, R., Masuda, T., Meli, R., Ragoza, M., Sunseri, J., & Koes, D. R. (2021). GNINA 1.0: Molecular docking with deep learning. *Journal of Cheminformatics*, 13(1), 43. DOI: 10.1186/s13321-021-00522-2 PMID: 34108002

Meagher, K. M., & Lee, L. M. (2016). Integrating public health and deliberative public bioethics: Lessons from the human genome project ethical, legal, and social implications program. *Public Health Reports*, 131(1), 44–51. DOI: 10.1177/003335491613100110 PMID: 26843669

Mehmood, M. A., Sehar, U., & Ahmad, N. (2014). Use of bioinformatics tools in different spheres of life sciences. *Journal of Data Mining in Genomics & Proteomics*, 5(2), 1.

Melo, A. R. da S., de Macêdo, L. S., Invenção, M. da C. V., de Moura, I. A., da Gama, M. A. T. M., de Melo, C. M. L., Silva, A. J. D., Batista, M. V. de A., & Freitas, A. C. de. (2022). Third-Generation Vaccines: Features of Nucleic Acid Vaccines and Strategies to Improve Their Efficiency. *Genes, 13*(12), 2287. DOI: 10.3390/genes13122287

Meng, L., Striegel, A., & Milenković, T. (2016). Local versus global biological network alignment. *Bioinformatics (Oxford, England)*, 32(20), 3155–3164. DOI: 10.1093/bioinformatics/btw348 PMID: 27357169

Metzker, M. L. (2005, January 12). Emerging technologies in DNA sequencing. *Genome Research*, 15(12), 1767–1776. DOI: 10.1101/gr.3770505 PMID: 16339375

Metzker, M. L. (2010, January). Sequencing technologies — The next generation. *Nature Reviews. Genetics*, 11(1), 31–46. DOI: 10.1038/nrg2626 PMID: 19997069

Michalik, M., Djahanshiri, B., Leo, J. C., & Linke, D. (2016). Reverse Vaccinology: The Pathway from Genomes and Epitope Predictions to Tailored Recombinant Vaccines. *Methods in Molecular Biology (Clifton, N.J.)*, 1403, 87–106. DOI: 10.1007/978-1-4939-3387-7_4 PMID: 27076126

Micheel, J., Safrastyan, A., & Wollny, D. (2021). Advances in non-coding RNA sequencing. *Non-Coding RNA*, 7(4), 70. DOI: 10.3390/ncrna7040070 PMID: 34842804

Michener, W. K. (2015). Ten Simple Rules for Creating a Good Data Management Plan. *PLoS Computational Biology*, 11(10), e1004525. DOI: 10.1371/journal.pcbi.1004525 PMID: 26492633

Michener, W., & Jones, M. (2012). Ecoinformatics: Supporting ecology as a data-intensive science. *Trends in Ecology & Evolution*, 27(2), 85–93. DOI: 10.1016/j.tree.2011.11.016 PMID: 22240191

Miller, J. R., Koren, S., & Sutton, G. (2010, June). Assembly algorithms for next-generation sequencing data. *Genomics*, 95(6), 315–327. DOI: 10.1016/j.ygeno.2010.03.001 PMID: 20211242

Minchin, S., & Lodge, J. (2019). Understanding biochemistry: Structure and function of nucleic acids. *Essays in Biochemistry*, 63(4), 433–456. DOI: 10.1042/EBC20180038 PMID: 31652314

Min, S., Lee, B., & Yoon, S. (2017). Deep learning in bioinformatics. *Briefings in Bioinformatics*, 18(5), 851–869. DOI: 10.1093/bib/bbw068 PMID: 27473064

Min, S., Lee, B., & Yoon, S. (2017, September 1). Deep learning in bioinformatics. *Briefings in Bioinformatics*, 18(5), 851–869. PMID: 27473064

Mirdita, M., Schütze, K., Moriwaki, Y., Heo, L., Ovchinnikov, S., & Steinegger, M. (2022). ColabFold: Making Protein folding accessible to all. *Nature Methods*, 19(6), 679–682. DOI: 10.1038/s41592-022-01488-1 PMID: 35637307

Mitra, D., Mitra, D., Sabri Bensaad, M., Sinha, S., Pant, K., Pant, M., Priyadarshini, A., Singh, P., Dassamiour, S., Hambaba, L., Panneerselvam, P., & Das Mohapatra, P. K. (2022). Evolution of bioinformatics and its impact on modern bio-science in the twenty-first century: Special attention to pharmacology, plant science and drug discovery. *Computational Toxicology (Amsterdam, Netherlands)*, 24, 100248. DOI: 10.1016/j.comtox.2022.100248

*Molecular Dynamics*. (n.d.). Recuperado 5 de abril de 2024, de https://link.springer.com/book/10.1007/978-3-319-16375-8

Montavon, G., Samek, W., & Müller, K. R. (2018). Methods for interpreting and understanding deep neural networks. *Digital Signal Processing*, 73, 1–15. DOI: 10.1016/j.dsp.2017.10.011

Morales-Helguera, A. (2015). Virtual screening tailored ensembles of QSAR models for the discovery of dual $A_{2A}$ Adenosine Receptor Antagonists / Monoamine Oxidase B Inhibitors. Proceedings of MOL2NET, International Conference on Multidisciplinary Sciences. https://sciforum.net/conference/MOL2NET-1/paper/3361

Morange, M. (2006). The protein side of the central dogma: Permanence and change. *History and Philosophy of the Life Sciences*, 513–524. PMID: 18351049

Motamedi, H., Alvandi, A., Fathollahi, M., Ari, M. M., Moradi, S., Moradi, J., & Abiri, R. (2023). In silico designing and immunoinformatics analysis of a novel peptide vaccine against metallo-beta-lactamase (VIM and IMP) variants. *PLoS One*, 18(7), e0275237. DOI: 10.1371/journal.pone.0275237 PMID: 37471423

Motta, S., & Brusic, V. (2004). *Mathematical Modelling of the Immune System*. DOI: 10.1007/978-3-642-18734-6_10

Moult, J., Pedersen, J. T., Judson, R., & Fidelis, K. (1995). A large-scale experiment to assess protein structure prediction methods. *Proteins*, 23(3), ii–iv. DOI: 10.1002/prot.340230303 PMID: 8710822

Mousavi Sagharchi, S., & Mahmoudi Nasab, A. (2023). *Investigation of Rifampicin and Isoniazid Resistance Mutated Genes in Mycobacterium tuberculosis Isolated from Patients.* Paper presented at the ICMAAA 2023: International Conference on Mechanisms of Action of Antimicrobial Agents, Venice, Italy.

Moyle, P. M., & Toth, I. (2013a). Modern subunit vaccines: Development, components, and research opportunities. *ChemMedChem*, 8(3), 360–376. DOI: 10.1002/cmdc.201200487 PMID: 23316023

Mrozek, D., Stępień, K., Grzesik, P., & Małysiak-Mrozek, B. (2021). A Large-Scale and Serverless Computational Approach for Improving Quality of NGS Data Supporting Big Multi-Omics Data Analyses. *Frontiers in Genetics*, 12, 699280. DOI: 10.3389/fgene.2021.699280 PMID: 34326863

Muhammed, M. T., & Aki-Yalcin, E. (2019). Homology modeling in drug discovery: Overview, current applications, and future perspectives. *Chemical Biology & Drug Design*, 93(1), 12–20. DOI: 10.1111/cbdd.13388 PMID: 30187647

Munshi, P., & Guru Row, T. N. (2006). Intra- and intermolecular interactions in small bioactive molecules: Cooperative features from experimental and theoretical charge-density analysis. *Acta Crystallographica. Section B, Structural Science*, 62(4), 612–626. DOI: 10.1107/S0108768106017393 PMID: 16840811

Myszczynska, M. (2020). Applications of machine learning to diagnosis and treatment of neurodegenerative diseases. Nature Reviews Neurology, 16(8), 440–456.

Nagaraj, K., Sharvani, G. S., & Sridhar, A. (2018). Emerging trend of big data analytics in bioinformatics: A literature review. *International Journal of Bioinformatics Research and Applications*, 14(1-2), 144–205. DOI: 10.1504/IJBRA.2018.089175

Nam, N. N., Do, H. D. K., Loan Trinh, K. T., & Lee, N. Y. (2023). Metagenomics: An Effective Approach for Exploring Microbial Diversity and Functions. *Foods*, 12(11), 2140. Advance online publication. DOI: 10.3390/foods12112140 PMID: 37297385

Necsulea, A. (2020). Phylogenomics and genome annotation.

Nelson, D. L. (2014). Princípios de bioquímica de Lehninger. *The Art of Medication*.

Nelson, D. L., Cox, M. M., Termignoni, C., Renard, G., Pereira, M. L. S., & Farias, S. E. (2014). *Lehninger principles of biochemistry* (6ª edição). *The Art of Medication*.

Nelson, D. L., Lehninger, A. L., & Cox, M. M. (2008). *Lehninger Principles of Biochemistry*. W. H. Freeman.

Nelson, M. T., Humphrey, W., Gursoy, A., Dalke, A., Kalé, L. V., Skeel, R. D., & Schulten, K. (1996). NAMD: A Parallel, Object-Oriented Molecular Dynamics Program. *The International Journal of Supercomputer Applications and High Performance Computing*, 10(4), 251–268. DOI: 10.1177/109434209601000401

Neves, B. J., Braga, R. C., Melo-Filho, C. C., Moreira-Filho, J. T., Muratov, E. N., & Andrade, C. H. (2018, November 13). QSAR-Based Virtual Screening: Advances and Applications in Drug Discovery. *Frontiers in Pharmacology*, 9, 1275. DOI: 10.3389/fphar.2018.01275 PMID: 30524275

Nguyen, N. H. (2024). Genetics and Genomics of Infectious Diseases in Key Aquaculture Species. *Biology (Basel)*, 13(1), 29. DOI: 10.3390/biology13010029 PMID: 38248460

Nichols, J. A., Herbert Chan, H. W., & Baker, M. A. B. (2018, September 4). Machine learning: Applications of artificial intelligence to imaging and diagnosis. *Biophysical Reviews*, 11(1), 111–118. DOI: 10.1007/s12551-018-0449-9 PMID: 30182201

Nielsen, M., Andreatta, M., Peters, B., & Buus, S. (2020b). Immunoinformatics: Predicting Peptide-MHC Binding. *Annual Review of Biomedical Data Science, 3*, 191–215. https://doi.org/DOI: 10.1146/ANNUREV-BIODATASCI-021920-100259/CITE/REFWORKS

Nielsen, M., Andreatta, M., Peters, B., & Buus, S. (2020a). Immunoinformatics: Predicting Peptide-MHC Binding. *Annual Review of Biomedical Data Science*, 3(1), 191–215. DOI: 10.1146/annurev-biodatasci-021920-100259 PMID: 37427310

Nielsen, T. K., Forero-Junco, L. M., Kot, W., Moineau, S., Hansen, L. H., & Riber, L. (2023). Detection of nucleotide modifications in bacteria and bacteriophages: Strengths and limitations of current technologies and software. *Molecular Ecology*, 32(6), 1236–1247. DOI: 10.1111/mec.16679 PMID: 36052951

Nusrat, S., Harbig, T., & Gehlenborg, N. (2019). Tasks, Techniques, and Tools for Genomic Data Visualization. *Computer Graphics Forum*, 38(3), 781–805. DOI: 10.1111/cgf.13727 PMID: 31768085

O'Donoghue, S. I. (2021). Grand challenges in bioinformatics data visualization. *Frontiers in Bioinformatics*, 1, 669186. DOI: 10.3389/fbinf.2021.669186 PMID: 36303723

Ogbe, R. J., Ochalefu, D. O., & Olaniru, O. B. (2016). Bioinformatics advances in genomics-A review. *International Journal of Current Research and Review*, 8(10), 5.

Okazaki, R., Okazaki, T., Sakabe, K., Sugimoto, K., & Sugino, A. (1968). Mechanism of DNA chain growth. I. Possible discontinuity and unusual secondary structure of newly synthesized chains. *Proceedings of the National Academy of Sciences of the United States of America*, 59(2), 598–605. DOI: 10.1073/pnas.59.2.598 PMID: 4967086

Okoye, C. O., Jiang, H., Nazar, M., Tan, X. L., & Jiang, J. (2024). Redefining modern food analysis: Significance of omics analytical techniques integration, chemometrics and bioinformatics. *Trends in Analytical Chemistry*, 175, 117706. DOI: 10.1016/j.trac.2024.117706

Oli, A. N., Obialor, W. O., & Ositadimma, M. (2020a). Immunoinformatics and vaccine development: An overview. *ImmunoTargets and Therapy*, 9, 13–30. DOI: 10.2147/ITT.S241064 PMID: 32161726

Olivares, F., Loyola, R., Olmedo, B., Miccono, M. Á., Aguirre, C., Vergara, R., & Mora, R. (2021). CRISPR/Cas9 targeted editing of genes associated with fungal susceptibility in Vitis vinifera L. Cv. Thompson Seedless using Geminivirus-derived replicons. *Frontiers in Plant Science*, 12, 791030. DOI: 10.3389/fpls.2021.791030 PMID: 35003180

Oliveri, S., Ferrari, F., Manfrinati, A., & Pravettoni, G. (2018). A systematic review of the psychological implications of genetic testing: A comparative analysis among cardiovascular, neurodegenerative and cancer diseases. *Frontiers in Genetics*, 9, 322074. DOI: 10.3389/fgene.2018.00624 PMID: 30619456

Opo, F. A. D. M., Rahman, M. M., Ahammad, F., Ahmed, I., Bhuiyan, M. A., & Asiri, A. M. (2021, February 18). Structure based pharmacophore modeling, virtual screening, molecular docking and ADMET approaches for identification of natural anti-cancer agents targeting XIAP protein. *Scientific Reports*, 11(1), 4049. DOI: 10.1038/s41598-021-83626-x PMID: 33603068

Ozden, B., Kryshtafovych, A., & Karaca, E. (2023). The impact of AI-based modeling on the accuracy of protein assembly prediction: Insights from CASP15. *Proteins*, 91(12), 1636–1657. https://doi.org/doi. DOI: 10.1002/prot.26598 PMID: 37861057

Paiva, V. (2022). Protein structural bioinformatics: An overview. *Computers in Biology and Medicine, 147*.

Palumbo, O., Trequattrini, F., Cimini, A., Tsurumaki, A., Navarra, M. A., & Paolone, A. (2021). Inter- and Intramolecular Interactions in Ether-Functionalized Ionic Liquids. *The Journal of Physical Chemistry B*, 125(9), 2380–2388. DOI: 10.1021/acs.jpcb.0c11429 PMID: 33625218

Pan. (2021). Recent advances in de novo protein design: Principles, methods, and applications. Journal of Biological Chemistry, 296.

Paradinas, I., Illian, J., & Smout, S. (2023). Understanding spatial effects in species distribution models. *PLoS One*, 18(5), e0285463. DOI: 10.1371/journal.pone.0285463 PMID: 37253039

Park, J.-Y., & Harris, D. (2003, April 24). Construction and assessment of models of CYP2E1: Predictions of metabolism from docking, molecular dynamics, and density functional theoretical calculations. *Journal of Medicinal Chemistry*, 46(9), 1645–1660. DOI: 10.1021/jm020538a PMID: 12699383

Patel, L., Shukla, T., Huang, X., Ussery, D. W., & Wang, S. (2020). Machine learning methods in drug discovery. *Molecules (Basel, Switzerland)*, 25(22), 5277. DOI: 10.3390/molecules25225277 PMID: 33198233

Pattan, V., Kashyap, R., Bansal, V., Candula, N., Koritala, T., & Surani, S. (2021). Genomics in medicine: A new era in medicine. *World Journal of Methodology*, 11(5), 231–242. DOI: 10.5662/wjm.v11.i5.231 PMID: 34631481

Paul, D., Sanap, G., Shenoy, S., Kalyane, D., Kalia, K., & Tekade, R. K. (2021, January). TEKADE RK. Artificial intelligence in drug discovery and development. *Drug Discovery Today*, 26(1), 80–93. DOI: 10.1016/j.drudis.2020.10.010 PMID: 33099022

Paul, P., & Saavedra, M. (2020). Agro informatics vis-à-vis Internet of Things (IoT) integration & potentialities—An analysis. *Agro Economist-An International Journal*, 7(1), 13–20. DOI: 10.30954/2394-8159.01.2020.2

Pedregosa, F. (2011). Scikit-learn: Machine learning in Python. *The Journal of Machine Learning Research, 12*, 2825–2830.

Pellegrino, E. D. (1992). Intersections of Western Biomedical Ethics and World Culture: Problematic and Possibility. *Cambridge Quarterly of Healthcare Ethics*, 1(3), 191–196. DOI: 10.1017/S0963180100000360 PMID: 11651289

Peng, K., Safonova, Y., Shugay, M., Popejoy, A. B., Rodriguez, O. L., Breden, F., Brodin, P., Burkhardt, A. M., Bustamante, C., Cao-Lormeau, V. M., Corcoran, M. M., Duffy, D., Fuentes-Guajardo, M., Fujita, R., Greiff, V., Jönsson, V. D., Liu, X., Quintana-Murci, L., Rossetti, M., … Mangul, S. (2021). Diversity in immunogenomics: the value and the challenge. *Nature Methods, 18*(6), 588–591. DOI: 10.1038/s41592-021-01169-5

Peng, H., Huang, N., Qi, J., Xie, P., Xu, C., Wang, J., & Yang, C. (2003, November 3). Identification of novel inhibitors of BCR-ABL tyrosine kinase via virtual screening. *Bioorganic & Medicinal Chemistry Letters*, 13(21), 3693–3699. DOI: 10.1016/j.bmcl.2003.08.014 PMID: 14552760

Pereira, F. (2019). Metagenomics: A gateway to drug discovery. In *Advances in Biological Science Research* (pp. 453-468). Elsevier.

Pereira, J., Simpkin, A. J., Hartmann, M. D., Rigden, D. J., Keegan, R. M., & Lupas, A. N. (2021). High-accuracy protein structure prediction in CASP14. *Proteins*, 89(12), 1687–1699. DOI: 10.1002/prot.26171 PMID: 34218458

Pharmaceutical Research and Manufacturers of America. (2020). PhRMA annual membership survey. https://phrma.org/research-and-development/2020-phrma-annual-membership-survey

Pharm, G. K. B. (n.d.). Available at: https://www.pharmgkb.org/downloads

Phillips, J. C., Braun, R., Wang, W., Gumbart, J., Tajkhorshid, E., Villa, E., Chipot, C., Skeel, R. D., Kalé, L., & Schulten, K. (2005). Scalable molecular dynamics with NAMD. *Journal of Computational Chemistry*, 26(16), 1781–1802. DOI: 10.1002/jcc.20289 PMID: 16222654

Phillips, S. J., Anderson, R. P., Dudík, M., Schapire, R. E., & Blair, M. E. (2017). Opening the black box: An open-source release of Maxent. *Ecography*, 40(7), 887–893. DOI: 10.1111/ecog.03049

Pierce, B. G., Wiehe, K., Hwang, H., Kim, B.-H., Vreven, T., & Weng, Z. (2014). ZDOCK server: Interactive docking prediction of protein–protein complexes and symmetric multimers. *Bioinformatics (Oxford, England)*, 30(12), 1771–1773. DOI: 10.1093/bioinformatics/btu097 PMID: 24532726

Pinzi, L., & Rastelli, G. (2019). Molecular Docking: Shifting Paradigms in Drug Discovery. *International Journal of Molecular Sciences*, 20(18), 4331. DOI: 10.3390/ijms20184331 PMID: 31487867

Pires, D. E. (2011). Cutoff Scanning Matrix (CSM): structural classification and function prediction by protein inter-residue distance patterns. *BMC Genomics, 12*(4).

Pires, D. E. V., de Melo-Minardi, R. C., da Silveira, C. H., Campos, F. F., & Meira, W.Jr. (2013, April 1). aCSM: Noise-free graph-based signatures to large-scale receptor-based ligand prediction. *Bioinformatics (Oxford, England)*, 29(7), 855–861. DOI: 10.1093/bioinformatics/btt058 PMID: 23396119

Poland, G. A., Kennedy, R. B., McKinney, B. A., Ovsyannikova, I. G., Lambert, N. D., Jacobson, R. M., & Oberg, A. L. (2013). *Vaccinomics, adversomics, and the immune response network theory: individualized vaccinology in the 21st century.* Paper presented at the Seminars in immunology. DOI: 10.1016/j.smim.2013.04.007

Poland, G. A., Ovsyannikova, I. G., & Kennedy, R. B. (2018). Personalized vaccinology: A review. *Vaccine*, 36(36), 5350–5357. DOI: 10.1016/j.vaccine.2017.07.062 PMID: 28774561

Pollard, A. J., & Bijker, E. M. (2020). A guide to vaccinology: from basic principles to new developments. *Nature Reviews Immunology, 21*(2), 83–100. DOI: 10.1038/s41577-020-00479-7

Pollastri, M. P. (2010, June). Overview on the Rule of Five. *Current Protocols in Pharmacology*, 49(1). DOI: 10.1002/0471141755.ph0912s49 PMID: 22294375

Ponomarenko, J., Bui, H. H., Li, W., Fusseder, N., Bourne, P. E., Sette, A., & Peters, B. (2008). ElliPro: A new structure-based tool for the prediction of antibody epitopes. *BMC Bioinformatics*, 9(1), 1–8. DOI: 10.1186/1471-2105-9-514 PMID: 19055730

Power, G. M., Vaughan, A. M., Qiao, L., Clemente, N. S., Pescarini, J. M., Paixão, E. S., Lobkowicz, L., Raja, A. I., Souza, A. P., Barreto, M. L., & Brickley, E. B. (2022). Socioeconomic risk markers of arthropod-borne virus (arbovirus) infections: A systematic literature review and meta-analysis. *BMJ Global Health*, 7(4), 7735. DOI: 10.1136/bmjgh-2021-007735 PMID: 35428678

Prado, L. C. da S., Giacchetto Felice, A., Rodrigues, T. C. V., Tiwari, S., Andrade, B. S., Kato, R. B., Oliveira, C. J. F., Silva, M. V., Barh, D., Azevedo, V. A. C., Jaiswal, A. K., & Soares, S. C. (2022). New putative therapeutic targets against Serratia marcescens using reverse vaccinology and subtractive genomics. *Journal of Biomolecular Structure & Dynamics*, 40(20), 10106–10121. DOI: 10.1080/07391102.2021.1942211 PMID: 34192477

Privitera, G. F., Alaimo, S., Ferro, A., & Pulvirenti, A. (2022). Virus finding tools: Current solutions and limitations. *Briefings in Bioinformatics*, 23(4), bbac235. DOI: 10.1093/bib/bbac235 PMID: 35753694

PyTorch. Available at: https://pytorch.org/

Qiang, X. (2020, January 17). CPPred-FL: A sequence-based predictor for large-scale identification of cell-penetrating peptides by feature representation learning. *Briefings in Bioinformatics*, 21(1), 11–23. PMID: 30239616

Rácz, A., Bajusz, D., & Héberger, K. (2018, October 4). Life beyond the Tanimoto coefficient: Similarity measures for interaction fingerprints. *Journal of Cheminformatics*, 10(1), 48. DOI: 10.1186/s13321-018-0302-y PMID: 30288626

Radway, R. M., Sethi, K., Chen, W. C., Kwon, J., Liu, S., Wu, T. F., Beigne, E., Shulaker, M. M., Wong, H. S. P., & Mitra, S. (2021). The Future of Hardware Technologies for Computing: N3XT 3D MOSAIC, Illusion Scaleup, Co-Design. *Technical Digest - International Electron Devices Meeting, IEDM,* 25.4.1-25.4.4. DOI: 10.1109/IEDM19574.2021.9720647

Ragoza, M., Hochuli, J., Idrobo, E., Sunseri, J., & Koes, D. R. (2017). Protein-ligand scoring with convolutional neural networks. *Journal of Chemical Information and Modeling*, 57(4), 942–957. DOI: 10.1021/acs.jcim.6b00740 PMID: 28368587

Rahman, M., & Schellhorn, H. E. (2023). Metabolomics of infectious diseases in the era of personalized medicine. *Frontiers in Molecular Biosciences*, 10, 1120376. DOI: 10.3389/fmolb.2023.1120376 PMID: 37275959

Raj, V. S., Priyadarshini, A., Yadav, M. K., Pandey, R. P., Gupta, A., & Vibhuti, A. (2021). Artificial intelligence in bioinformatics. *Biomedical Data Mining for Information Retrieval: Methodologies, Techniques and Applications*, 21-51.

Ram, P. T., Mendelsohn, J., & Mills, G. B. (2012). Bioinformatics and systems biology. *Molecular Oncology*, 6(2), 147–154. DOI: 10.1016/j.molonc.2012.01.008 PMID: 22377422

Rapin, N., Lund, O., Bernaschi, M., & Castiglione, F. (2010). Computational Immunology Meets Bioinformatics: The Use of Prediction Tools for Molecular Binding in the Simulation of the Immune System. *PLoS One*, 5(4), 9862. DOI: 10.1371/journal.pone.0009862 PMID: 20419125

Rappuoli, R. (2000). Reverse vaccinology. *Current Opinion in Microbiology*, 3(5), 445–450. DOI: 10.1016/S1369-5274(00)00119-3 PMID: 11050440

Rawat, S. S., Keshri, A. K., Kaur, R., & Prasad, A. (2023a). Immunoinformatics Approaches for Vaccine Design: A Fast and Secure Strategy for Successful Vaccine Development. *Vaccines, 11*(2), 221. DOI: 10.3390/vaccines11020221

Rebets, Y., Kormanec, J., Luzhetskyy, A., Bernaerts, K., & Anné, J. (2017). Cloning and expression of metagenomic DNA in Streptomyces lividans and subsequent fermentation for optimized production. *Metagenomics: Methods and Protocols*, 99-144.

Recknagel, F., & Michener, W. K. (2018). *Ecological Informatics*. Springer Cham. DOI: 10.1007/978-3-319-59928-1

Reuter, J. A., Spacek, D. V., & Snyder, M. P. (2015, May 21). High-Throughput Sequencing Technologies. *Molecular Cell*, 58(4), 586–597. DOI: 10.1016/j.molcel.2015.05.004 PMID: 26000844

Rhee, S. Y. (2005). Bioinformatics. Current limitations and insights for the future. *Plant Physiology*, 138(2), 569–570. DOI: 10.1104/pp.104.900153 PMID: 15955919

Riginos, C., & Liggins, L. (2013). Seascape Genetics: Populations, Individuals, and Genes Marooned and Adrift. *Geography Compass*, 7(3), 197–216. DOI: 10.1111/gec3.12032

Ritchie, M. E., Phipson, B., Wu, D., Hu, Y., Law, C. W., Shi, W., & Smyth, G. K. (2015). limma powers differential expression analyses for RNA-sequencing and microarray studies. *Nucleic Acids Research*, 43(7), e47. DOI: 10.1093/nar/gkv007 PMID: 25605792

Rodrigues, T. C. V., Jaiswal, A. K., De Sarom, A., Oliveira, L. D. C., Oliveira, C. J. F., Ghosh, P., Tiwari, S., Miranda, F. M., Benevides, L. D. J., Azevedo, V. A. D. C., & Soares, S. D. C. (2019). Reverse vaccinology and subtractive genomics reveal new therapeutic targets against Mycoplasma pneumoniae: A causative agent of pneumonia. *Royal Society Open Science*, 6(7), 190907. Advance online publication. DOI: 10.1098/rsos.190907 PMID: 31417766

Rosario-Cruz, R., Domínguez-García, D. I., López-Silva, S., & Rosario-Domínguez, F. (2023). Immunoinformatics and tick vaccinology. *Exploration of Immunology*, 3(1), 1–16. DOI: 10.37349/ei.2023.00085

Rosati, D., Palmieri, M., Brunelli, G., Morrione, A., Iannelli, F., Frullanti, E., & Giordano, A. (2024). Differential gene expression analysis pipelines and bioinformatic tools for the identification of specific biomarkers: A review. *Computational and Structural Biotechnology Journal*, 23, 1154–1168. DOI: 10.1016/j.csbj.2024.02.018 PMID: 38510977

Rosenquist, R., Cuppen, E., Buettner, R., Caldas, C., Dreau, H., Elemento, O., . . . Jobanputra, V. (2022). *Clinical utility of whole-genome sequencing in precision oncology.* Paper presented at the Seminars in Cancer Biology. DOI: 10.1016/j.semcancer.2021.06.018

Roychowdhury, R., Das, S. P., Gupta, A., Parihar, P., Chandrasekhar, K., Sarker, U., Kumar, A., Ramrao, D. P., & Sudhakar, C. (2023). Multi-Omics Pipeline and Omics-Integration Approach to Decipher Plant's Abiotic Stress Tolerance Responses. *Genes*, 14(6), 1281. Advance online publication. DOI: 10.3390/genes14061281 PMID: 37372461

Rudrapal, M., & Chetia, D. (2020, July 15). Virtual Screening, Molecular Docking and QSAR Studies in Drug Discovery and Development Programme. *Journal of Drug Delivery and Therapeutics*, 10(4), 225–233. DOI: 10.22270/jddt.v10i4.4218

Russell, S. J., & Norvig, P. (2016). *Artificial intelligence: a modern approach*. Pearson.

Sadybekov, A. V., & Katritch, V. (2023, April). Computational approaches streamlining drug discovery. *Nature*, 616(7958), 673–685. DOI: 10.1038/s41586-023-05905-z PMID: 37100941

Sahito, J. H., Zhang, H., Gishkori, Z. G. N., Ma, C., Wang, Z., Ding, D., Zhang, X., & Tang, J. (2024). Advancements and Prospects of Genome-Wide Association Studies (GWAS) in Maize. *International Journal of Molecular Sciences*, 25(3), 1918. Advance online publication. DOI: 10.3390/ijms25031918 PMID: 38339196

Saif, R., & Iftikhar, A. (2022). Bioinformatics: Multiomics Applications in Food Science.

Saleh. (2017). Design and development of some viral protease inhibitors by QSAR and molecular modeling studies. In *Viral proteases and their inhibitors* (pp. 25–58). Academic Press. DOI: 10.1016/B978-0-12-809712-0.00002-2

Saliou, P., & Girard, M. (2005). De Jenner et Pasteur à la vaccinologie. *Therapie*, 60(3), 201–204. DOI: 10.2515/therapie:2005024 PMID: 16128259

Samuel, H., Nweke-Maraizu, U., & Etim, E. (2023). Understanding Intermolecular and Intramolecular Hydrogen Bonds. *Spectroscopic and Computational Approaches.*, 5, 439–465. DOI: 10.48309/JCR.2023.407989.1235

Sanger, F., Nicklen, S., & Coulson, A. R. (1977, December). DNA sequencing with chain-terminating inhibitors. *Proceedings of the National Academy of Sciences of the United States of America*, 74(12), 5463–5467. DOI: 10.1073/pnas.74.12.5463 PMID: 271968

Sanjeev, A., Mattaparthi, V. S. K., & Kaushik, S. (2019). Nucleic-Acid Structure Database. In S. Ranganathan, M. Gribskov, K. Nakai, & C. Schönbach (Orgs.), *Encyclopedia of Bioinformatics and Computational Biology* (p. 567–574). Academic Press. DOI: 10.1016/B978-0-12-809633-8.20285-9

Santana, K., do Nascimento, L. D., Lima e Lima, A., Damasceno, V., Nahum, C., Braga, R. C., & Lameira, J. (2021, April 29). Applications of Virtual Screening in Bioprospecting: Facts, Shifts, and Perspectives to Explore the Chemo-Structural Diversity of Natural Products. *Frontiers in Chemistry*, 9, 662688. DOI: 10.3389/fchem.2021.662688 PMID: 33996755

Santos, L. M. dos, Mariano, D., & Melo-Minardi, R. C. de. (2023). O impacto da inteligência artificial nas ciências da vida através da bioinformática. *Revista da UFMG*, 30. https://doi.org/DOI: 10.35699/2965-6931.2023.47996

Sapoval, N., Aghazadeh, A., Nute, M. G., Antunes, D. A., Balaji, A., Baraniuk, R., Barberan, C. J., Dannenfelser, R., Dun, C., Edrisi, M., Elworth, R. A. L., Kille, B., Kyrillidis, A., Nakhleh, L., Wolfe, C. R., Yan, Z., Yao, V., & Treangen, T. J. (2022). Current progress and open challenges for applying deep learning across the biosciences. *Nature Communications*, 13(1), 1728. DOI: 10.1038/s41467-022-29268-7 PMID: 35365602

Saravanan, K., Panigrahi, M., Kumar, H., Rajawat, D., Nayak, S. S., Bhushan, B., & Dutt, T. (2022). Role of genomics in combating COVID-19 pandemic. *Gene*, 823, 146387. DOI: 10.1016/j.gene.2022.146387 PMID: 35248659

Sasidharan, R., & Chothia, C. (2007). The selection of acceptable protein mutations. *Proceedings of the National Academy of Sciences of the United States of America*, 104(24), 10080–10085. DOI: 10.1073/pnas.0703737104 PMID: 17540730

Satam, H., Joshi, K., Mangrolia, U., Waghoo, S., Zaidi, G., Rawool, S., & Das, G. (2023). Next-generation sequencing technology: Current trends and advancements. *Biology (Basel)*, 12(7), 997. DOI: 10.3390/biology12070997 PMID: 37508427

Schatz, M. C., Philippakis, A. A., Afgan, E., Banks, E., Carey, V. J., Carroll, R. J., & Wuichet, K. (2022). Inverting the model of genomics data sharing with the NHGRI Genomic Data Science Analysis, Visualization, and Informatics Lab-space. *Cell Genomics*, 2(1), 100085. Advance online publication. DOI: 10.1016/j.xgen.2021.100085 PMID: 35199087

Schilbert, H. M., Rempel, A., & Pucker, B. (2020). Comparison of Read Mapping and Variant Calling Tools for the Analysis of Plant NGS Data. *Plants*, 9(4), 439. Advance online publication. DOI: 10.3390/plants9040439 PMID: 32252268

Schneider, G. (2010, April). Virtual screening: An endless staircase? *Nature Reviews. Drug Discovery*, 9(4), 273–276. DOI: 10.1038/nrd3139 PMID: 20357802

Schneider, G., & Böhm, H.-J. (2002). Virtual screening and fast automated docking methods. *Drug Discovery Today*, 7(1), 64–70. DOI: 10.1016/S1359-6446(01)02091-8 PMID: 11790605

Schwarz, U. I., Gulilat, M., & Kim, R. B. (2019). The Role of Next-Generation Sequencing in Pharmacogenetics and Pharmacogenomics. *Cold Spring Harbor Perspectives in Medicine*, 9(2), a033027. Advance online publication. DOI: 10.1101/cshperspect.a033027 PMID: 29844222

Seddon, M. P., Cosgrove, D. A., Packer, M. J., & Gillet, V. J. (2019, January 28). Alignment-Free Molecular Shape Comparison Using Spectral Geometry: The Framework. *Journal of Chemical Information and Modeling*, 59(1), 98–116. DOI: 10.1021/acs.jcim.8b00676 PMID: 30462505

Selkoe, K., D'Aloia, C., Crandall, E., Iacchei, M., Liggins, L., Puritz, J., Von Der Heyden, S., & Toonen, R. (2016). A decade of seascape genetics: Contributions to basic and applied marine connectivity. *Marine Ecology Progress Series*, 554, 1–19. DOI: 10.3354/meps11792

Senior, A. W., Evans, R., Jumper, J., Kirkpatrick, J., Sifre, L., Green, T., Qin, C., Žídek, A., Nelson, A. W. R., Bridgland, A., Penedones, H., Petersen, S., Simonyan, K., Crossan, S., Kohli, P., Jones, D. T., Silver, D., Kavukcuoglu, K., & Hassabis, D. (2020, January). Improved protein structure prediction using potentials from deep learning. *Nature*, 577(7792), 706–710. DOI: 10.1038/s41586-019-1923-7 PMID: 31942072

Sepulveda, J. L. (2020). Using R and Bioconductor in Clinical Genomics and Transcriptomics. *The Journal of Molecular Diagnostics*, 22(1), 3–20. DOI: 10.1016/j.jmoldx.2019.08.006 PMID: 31605800

Sette, A., & Rappuoli, R. (2010). Reverse vaccinology: Developing vaccines in the era of genomics. *Immunity*, 33(4), 530–541. DOI: 10.1016/j.immuni.2010.09.017 PMID: 21029963

Shannon, P., Markiel, A., Ozier, O., Baliga, N. S., Wang, J. T., Ramage, D., Amin, N., Schwikowski, B., & Ideker, T. (2003). Cytoscape: A software environment for integrated models of biomolecular interaction networks. *Genome Research*, 13(11), 2498–2504. DOI: 10.1101/gr.1239303 PMID: 14597658

Sharma, D. K., Upadhaya, P., & Bhardwaj, S. (2020). Intelligent cloud computing and bioinformatics data analysis. In *Applications of Cloud Computing* (pp. 25–42). Chapman and Hall/CRC. DOI: 10.1201/9781003025696-2

Shawan, M. M. A. K., Sharma, A. R., Halder, S. K., Al Arian, T., Shuvo, M. N., Sarker, S. R., & Hasan, M. A. (2023). Advances in Computational and Bioinformatics Tools and Databases for Designing and Developing a Multi-Epitope-Based Peptide Vaccine. *International Journal of Peptide Research and Therapeutics*, 29(4), 60. DOI: 10.1007/s10989-023-10535-0 PMID: 37251529

Shen, C., Ding, J., Wang, Z., Cao, D., Ding, X., & Hou, T. (2019). From machine learning to deep learning: Advances in scoring functions for protein–ligand docking. *Wiley Interdisciplinary Reviews. Computational Molecular Science*, 10(1), e1429. Advance online publication. DOI: 10.1002/wcms.1429

Shen, C., Zhang, X., Hsieh, C.-Y., Deng, Y., Wang, D., Xu, L., Wu, J., Li, D., Kang, Y., Hou, T., & Pan, P. (2023). A generalized protein–ligand scoring framework with balanced scoring, docking, ranking and screening powers. *Chemical Science (Cambridge)*, 14(30), 8129–8146. DOI: 10.1039/D3SC02044D PMID: 37538816

Shen, X., Jiang, C., Wen, Y., Li, C., & Lu, Q. (2022). A brief review on deep learning applications in genomic studies. *Frontiers in Systems Biology*, 2, 877717. DOI: 10.3389/fsysb.2022.877717

Shi, H., Yan, K. K., Ding, L., Qian, C., Chi, H., & Yu, J. (2020). Network Approaches for Dissecting the Immune System. *iScience*, 23(8), 101354. DOI: 10.1016/j.isci.2020.101354 PMID: 32717640

Shi, L., & Wang, Z. (2019). Computational Strategies for Scalable Genomics Analysis. *Genes*, 10(12), 1017. Advance online publication. DOI: 10.3390/genes10121017 PMID: 31817630

Shin, W.-H., Zhu, X., Bures, M., & Kihara, D. (2015, July 16). Three-Dimensional Compound Comparison Methods and Their Application in Drug Discovery. *Molecules (Basel, Switzerland)*, 20(7), 12841–12862. DOI: 10.3390/molecules200712841 PMID: 26193243

Shrestha, A., & Mahmood, A. (2019). Review of deep learning algorithms and architectures. *IEEE Access : Practical Innovations, Open Solutions*, 7, 53040–53065. DOI: 10.1109/ACCESS.2019.2912200

Singh, J., & Banerjee, R. (2019). A study on single and multi-layer perceptron neural network. *3rd International Conference on Computing Methodologies and Communication (ICCMC)*, 35–40. DOI: 10.1109/ICCMC.2019.8819775

Sinha, D., Maurya, A. K., Abdi, G., Majeed, M., Agarwal, R., Mukherjee, R., Ganguly, S., Aziz, R., Bhatia, M., Majgaonkar, A., Seal, S., Das, M., Banerjee, S., Chowdhury, S., Adeyemi, S. B., & Chen, J. T. (2023). Integrated Genomic Selection for Accelerating Breeding Programs of Climate-Smart Cereals. *Genes*, 14(7), 1484. Advance online publication. DOI: 10.3390/genes14071484 PMID: 37510388

Skerrett-Byrne Anthony, D. (2023). Transcriptomics. In Bradshaw, R. A., Hart, G. W., & Stahl, P. D. (Eds.), *Encyclopedia of Cell Biology* (2nd ed., pp. 363–371). Academic Press. DOI: 10.1016/B978-0-12-821618-7.00157-7

Soheilypour, M., & Mofrad, M. R. K. (2018). Agent-Based Modeling in Molecular Systems Biology. *BioEssays*, 40(7), 1800020. DOI: 10.1002/bies.201800020 PMID: 29882969

Sohl-Dickstein, J., Weiss, E., Maheswaranathan, N., & Ganguli, S. (2015). Deep unsupervised learning using nonequilibrium thermodynamics. In F. Bach & D. Blei (Eds.), *Proceedings of the 32nd International Conference on Machine Learning*. Lille, France: PMLR. Available at: https://proceedings.mlr.press/v37/sohl-dickstein15.html

Solís-Moruno, M., Batlle-Masó, L., Bonet, N., Aróstegui, J. I., & Casals, F. (2023). Somatic genetic variation in healthy tissue and non-cancer diseases. *European Journal of Human Genetics*, 31(1), 48–54. DOI: 10.1038/s41431-022-01213-8 PMID: 36289407

Song, L., Hobaugh, M. R., Shustak, C., Cheley, S., Bayley, H., & Gouaux, J. E. (1996). Structure of staphylococcal α-hemolysin, a heptameric transmembrane pore. *Science*, 274(5294), 1859–1865. DOI: 10.1126/science.274.5294.1859 PMID: 8943190

Soni, S., Gill, V. J. S., Anusheel, , Singh, J., Chhabra, J., Gill, G. J. S., & Bakshi, R. (2023). Dengue, Chikungunya, and Zika: The Causes and Threats of Emerging and Re-emerging Arboviral Diseases. *Cureus*, 15(7). Advance online publication. DOI: 10.7759/cureus.41717 PMID: 37575782

Sorkun, M. C. (2020). An artificial intelligence-aided virtual screening recipe for two-dimensional materials discovery. NPJ Computational Materials, 6(1), 106.

Soto, J. A., Gálvez, N. M. S., Andrade, C. A., Ramírez, M. A., Riedel, C. A., Kalergis, A. M., & Bueno, S. M. (2022). BCG vaccination induces cross-protective immunity against pathogenic microorganisms. *Trends in Immunology*, 43(4), 322–335. DOI: 10.1016/j.it.2021.12.006 PMID: 35074254

Soubeyrand, B. (2018). De la fabrication d'un vaccin à sa mise à disposition en pharmacie. *Revue des Maladies Respiratoires*, 35(10), 1005–1019. DOI: 10.1016/j.rmr.2018.07.003 PMID: 30266457

Spyrou, M. A., Bos, K. I., Herbig, A., & Krause, J. (2019). Ancient pathogen genomics as an emerging tool for infectious disease research. *Nature Reviews. Genetics*, 20(6), 323–340. DOI: 10.1038/s41576-019-0119-1 PMID: 30953039

Srivastava, A., Mahmood, A., & Srivastava, R. (2017). A comparative analysis of SVM random forest methods for protein function prediction. *2017 International Conference on Current Trends in Computer, Electrical, Electronics and Communication (CTCEEC)*, 1008-1010. DOI: 10.1109/CTCEEC.2017.8455066

Staudemeyer, R. C., & Morris, E. R. (2019). *Understanding LSTM — a tutorial into Long Short-Term Memory Recurrent Neural Networks*. https://arxiv.org/abs/1909.09586v1

Stephen, S. (2019). Overview of the state of the art in agroinformatics in Hungary and in international contexts. *Animal Breeding and Feeding,68*(3), 185-194.

Strianese, O., Rizzo, F., Ciccarelli, M., Galasso, G., D'Agostino, Y., Salvati, A., Del Giudice, C., Tesorio, P., & Rusciano, M. R. (2020). Precision and personalized medicine: How genomic approach improves the management of cardiovascular and neurodegenerative disease. *Genes*, 11(7), 747. DOI: 10.3390/genes11070747 PMID: 32640513

Subramanian, I., Verma, S., Kumar, S., Jere, A., & Anamika, K. (2020). Multi-omics data integration, interpretation, and its application. *Bioinformatics and Biology Insights*, 14, 1177932219899051. DOI: 10.1177/1177932219899051 PMID: 32076369

Sujatha, P. L., Manikkavasagan, I., Kumarasamy, P., & Arvindprabhu, T. (2022). Chapter-6 Bioinformatics Approaches in Food Industry. in *Food Technology and Nutrition*, 83.

Sunny, S., & Jayaraj, P. B. (2022). Protein–Protein Docking: Past, Present, and Future. *The Protein Journal*, 41(1), 1–26. DOI: 10.1007/s10930-021-10031-8 PMID: 34787783

Sun, T., Zhou, B., Lai, L., & Pei, J. (2017). Sequence-based prediction of protein-protein interaction using a deep-learning algorithm. *BMC Bioinformatics*, 18(1), 277. Advance online publication. DOI: 10.1186/s12859-017-1700-2 PMID: 28545462

Szklarczyk, D., Franceschini, A., Wyder, S., Forslund, K., Heller, D., Huerta-Cepas, J., Simonovic, M., Roth, A., Santos, A., Tsafou, K. P., Kuhn, M., Bork, P., Jensen, L. J., & von Mering, C. (2015). STRING v10: Protein–protein interaction networks, integrated over the tree of life. *Nucleic Acids Research*, 43(D1), D447–D452. DOI: 10.1093/nar/gku1003 PMID: 25352553

Telatin, A., Fariselli, P., & Birolo, G. (2021). SeqFu: A suite of utilities for the robust and reproducible manipulation of sequence files. *Bioengineering (Basel, Switzerland)*, 8(5), 59. DOI: 10.3390/bioengineering8050059 PMID: 34066939

TensorFlow. (n.d.). Available at: https://www.tensorflow.org/

Thak, E. J., Yoo, S. J., Moon, H. Y., & Kang, H. A. (2020). Yeast synthetic biology for designed cell factories producing secretory recombinant proteins. *FEMS Yeast Research*, 20(2), foaa009. DOI: 10.1093/femsyr/foaa009 PMID: 32009173

The Times. (2024). Deepmind finds biology's 'holy grail' with answer to protein problem. Available at https://www.thetimes.co.uk/article/deepmind-finds-biology-s-holy-grail-with-answer-to-protein-problem-htg6s7qlq

Thieffry, D., & Sarkar, S. (1998). Forty years under the central dogma. *Trends in Biochemical Sciences*, 23(8), 312–316. DOI: 10.1016/S0968-0004(98)01244-4 PMID: 9757833

Thornton, J. M., Laskowski, R. A., & Borkakoti, N. (2021, October). AlphaFold heralds a data-driven revolution in biology and medicine. *Nature Medicine*, 27(10), 1666–1669. DOI: 10.1038/s41591-021-01533-0 PMID: 34642488

Tomar, N., & De, R. K. (2010). Immunoinformatics: An integrated scenario. *Immunology*, 131(2), 153–168. DOI: 10.1111/j.1365-2567.2010.03330.x PMID: 20722763

Tomar, N., & De, R. K. (2014). Immunoinformatics: A brief review. *Methods in Molecular Biology (Clifton, N.J.)*, 1184, 23–55. DOI: 10.1007/978-1-4939-1115-8_3 PMID: 25048118

Torchala, M., Moal, I. H., Chaleil, R. A. G., Fernandez-Recio, J., & Bates, P. A. (2013). SwarmDock: A server for flexible protein–protein docking. *Bioinformatics (Oxford, England)*, 29(6), 807–809. DOI: 10.1093/bioinformatics/btt038 PMID: 23343604

Torrisi, M., Pollastri, G., & Le, Q. (2020). Deep learning methods in protein structure prediction. *Computational and Structural Biotechnology Journal*, 18, 1301–1310. DOI: 10.1016/j.csbj.2019.12.011 PMID: 32612753

Tripathi, A., & Bankaitis, V. A. (2017). Molecular Docking: From Lock and Key to Combination Lock. *Journal of Molecular Medicine and Clinical Applications*, 2(1). Advance online publication. DOI: 10.16966/2575-0305.106 PMID: 29333532

Tropsha, A., Isayev, O., Varnek, A., Schneider, G., & Cherkasov, A. (2024). Integrating QSAR modelling and deep learning in drug discovery: The emergence of deep QSAR. *Nature Reviews. Drug Discovery*, 23(2), 141–155. DOI: 10.1038/s41573-023-00832-0 PMID: 38066301

Trott, O., & Olson, A. J. (2010). AutoDock Vina: Improving the speed and accuracy of docking with a new scoring function, efficient optimization, and multithreading. *Journal of Computational Chemistry*, 31(2), 455–461. DOI: 10.1002/jcc.21334 PMID: 19499576

Tucker, A., & Duplisea, D. (2012). Bioinformatics tools in predictive ecology: Applications to fisheries. *Philosophical Transactions of the Royal Society of London. Series B, Biological Sciences*, 367(1586), 279–290. DOI: 10.1098/rstb.2011.0184 PMID: 22144390

Tuena, C., Semonella, M., Fernández-Álvarez, J., Colombo, D., & Cipresso, P. (2020). Predictive precision medicine: towards the computational challenge. *P5 eHealth: An Agenda for the Health Technologies of the Future*, 71-86.

Turon, G., Hlozek, J., Woodland, J. G., Kumar, A., Chibale, K., & Duran-Frigola, M. (2023, September 15). First fully-automated AI/ML virtual screening cascade implemented at a drug discovery centre in Africa. *Nature Communications*, 14(1), 5736. DOI: 10.1038/s41467-023-41512-2 PMID: 37714843

Tuszynska, I., Magnus, M., Jonak, K., Dawson, W., & Bujnicki, J. M. (2015, July 1). NPDock: A web server for protein–nucleic acid docking. *Nucleic Acids Research*, 43(W1, n. W1), W425–W430. DOI: 10.1093/nar/gkv493 PMID: 25977296

Tutte, W. T. (2001). *Graph theory* (Vol. 21). Cambridge University Press.

Tyagi, E., Sachan, A., & Kumari, P. (2024). Role of Bioinformatics Tools in the Food Processing Industry. *Nonthermal Food Processing, Safety, and Preservation*, 479-507.

Uddin, M., Mustafa, F., Rizvi, T. A., Loney, T., Al Suwaidi, H., Al-Marzouqi, A. H. H., & Stefanini, C. (2020). SARS-CoV-2/COVID-19: Viral genomics, epidemiology, vaccines, and therapeutic interventions. *Viruses*, 12(5), 526. DOI: 10.3390/v12050526 PMID: 32397688

Ulschmid, C. M., Patel, J., Pan, A. Y., Liegl, M., & Holland, K. E. (2023). Intralesional measles, mumps, and rubella vaccine after failure of intralesional Candida antigen for the treatment of recalcitrant pediatric warts. *Pediatric Dermatology*, 40(6), 1057–1059. DOI: 10.1111/pde.15415 PMID: 37596908

Uszczynska-Ratajczak, B., Lagarde, J., Frankish, A., Guigó, R., & Johnson, R. (2018). Towards a complete map of the human long non-coding RNA transcriptome. *Nature Reviews. Genetics*, 19(9), 535–548. DOI: 10.1038/s41576-018-0017-y PMID: 29795125

Uversky, V. N., & Finkelstein, A. V. (2019). Life in Phases: Intra- and Inter- Molecular Phase Transitions in Protein Solutions. *Biomolecules*, 9(12), 842. DOI: 10.3390/biom9120842 PMID: 31817975

Van De Sande, B., Lee, J. S., Mutasa-Gottgens, E., Naughton, B., Bacon, W., Manning, J., Wang, Y., Pollard, J., Mendez, M., Hill, J., Kumar, N., Cao, X., Chen, X., Khaladkar, M., Wen, J., Leach, A., & Ferran, E. (2023, June). Applications of single-cell RNA sequencing in drug discovery and development. *Nature Reviews. Drug Discovery*, 22(6), 496–520. DOI: 10.1038/s41573-023-00688-4 PMID: 37117846

Van De Waterbeemd. (1997). Glossary of terms used in computational drug design (IUPAC Recommendations 1997). Pure and Applied Chemistry, 69(5), 1137-1152.

van der Poll, T., Shankar-Hari, M., & Wiersinga, W. J. (2021). The immunology of sepsis. *Immunity*, 54(11), 2450–2464. DOI: 10.1016/j.immuni.2021.10.012 PMID: 34758337

Van Der Spoel, D., Lindahl, E., Hess, B., Groenhof, G., Mark, A. E., & Berendsen, H. J. C. (2005). GROMACS: Fast, flexible, and free. *Journal of Computational Chemistry*, 26(16), 1701–1718. DOI: 10.1002/jcc.20291 PMID: 16211538

Varkey, B. (2021). Principles of Clinical Ethics and Their Application to Practice. *Medical Principles and Practice : International Journal of the Kuwait University. Medical Principles and Practice*, 30(1), 17–28. DOI: 10.1159/000509119 PMID: 32498071

Venter, J. C. (2001). The sequence of the human genome. *Science, 291*(5507), 1304–1351.

Verdonk, M. L., Cole, J. C., Hartshorn, M. J., Murray, C. W., & Taylor, R. D. (2003). Improved protein–ligand docking using GOLD. *Proteins*, 52(4), 609–623. DOI: 10.1002/prot.10465 PMID: 12910460

Vinatzer, B. A., Heath, L. S., Almohri, H. M. J., Stulberg, M. J., Lowe, C., & Li, S. (2019). Cyberbiosecurity Challenges of Pathogen Genome Databases. *Frontiers in Bioengineering and Biotechnology*, 7, 106. DOI: 10.3389/fbioe.2019.00106 PMID: 31157218

Vita, R., Mahajan, S., Overton, J. A., Dhanda, S. K., Martini, S., Cantrell, J. R., Wheeler, D. K., Sette, A., & Peters, B. (2019). The Immune Epitope Database (IEDB): 2018 update. *Nucleic Acids Research*, 47(D1), D339–D343. DOI: 10.1093/nar/gky1006 PMID: 30357391

Vyas, V. K., Ukawala, R. D., Chintha, C., & Ghate, M. (2012). Homology modeling a fast tool for drug discovery: Current perspectives. *Indian Journal of Pharmaceutical Sciences*, 74(1), 1. DOI: 10.4103/0250-474X.102537 PMID: 23204616

Vyas, V., Jain, A., Jain, A., & Gupta, A. (2008). Virtual Screening: A Fast Tool for Drug Design. *Scientia Pharmaceutica, 76*(3). *Scientia Pharmaceutica*, 3(3), 333–360. Advance online publication. DOI: 10.3797/scipharm.0803-03

Wade, C., & Glynn, K. (2020). *Hands-On Gradient Boosting with XGBoost and scikit-learn: Perform accessible machine learning and extreme gradient boosting with Python*. Packt Publishing.

Waidha, K. M., Jabalia, N., Singh, D., Jha, A., & Kaur, R. (2015). Bioinformatics approaches in food industry: an overview. In *National Conference on Recent Trends in Biomedical Engineering, Cancer Biology, Bioinformatics and Applied Biotechnology (BECBAB-2015) vol* (Vol. 1, pp. 1-4).

Walters, W. P., Stahl, M. T., & Murcko, M. A. (1998, April 1). Virtual screening—An overview. *Drug Discovery Today*, 3(4), 160–178. DOI: 10.1016/S1359-6446(97)01163-X

Wang, J., Li, Z., & Zhang, J. (2022). Visualizing the knowledge structure and evolution of bioinformatics. *BMC Bioinformatics, 23*(8), 404.

Wang, D., & Farhana, A. (2024). Biochemistry, RNA Structure. In *StatPearls*. StatPearls Publishing. https://www.ncbi.nlm.nih.gov/books/NBK558999/

Wang, R. C., & Wang, Z. (2023). Precision Medicine: Disease Subtyping and Tailored Treatment. *Cancers (Basel)*, 15(15), 3837. Advance online publication. DOI: 10.3390/cancers15153837 PMID: 37568653

Wang, S., Sun, S.-T., Zhang, X.-Y., Ding, H.-R., Yuan, Y., He, J.-J., Wang, M.-S., Yang, B., & Li, Y.-B. (2023). The evolution of single-cell RNA sequencing technology and application: Progress and perspectives. *International Journal of Molecular Sciences*, 24(3), 2943. DOI: 10.3390/ijms24032943 PMID: 36769267

Watson, J. L., Juergens, D., Bennett, N. R., Trippe, B. L., Yim, J., Eisenach, H. E., Ahern, W., Borst, A. J., Ragotte, R. J., Milles, L. F., Wicky, B. I. M., Hanikel, N., Pellock, S. J., Courbet, A., Sheffler, W., Wang, J., Venkatesh, P., Sappington, I., Torres, S. V., & Baker, D. (2023). De novo design of protein structure and function with RFdiffusion. *Nature*, 620(7976), 1089–1100. DOI: 10.1038/s41586-023-06415-8 PMID: 37433327

Webb, B., & Sali, A. (2017). Protein Structure Modeling with MODELLER. In M. Kaufmann, C. Klinger, & A. Savelsbergh (Orgs.), *Functional Genomics: Methods and Protocols* (p. 39–54). Springer. DOI: 10.1007/978-1-4939-7231-9_4

Webb, A. R., Copsey, K. D., & Cawley, G. (2011). *Statistical pattern recognition* (Vol. 2). Wiley Online Library. DOI: 10.1002/9781119952954

Weininger, D. (1988). SMILES, a chemical language and information system. 1. Introduction to methodology and encoding rules. *Journal of Chemical Information and Computer Sciences*, 28(1), 31–36. DOI: 10.1021/ci00057a005

Welser, J., Pitera, J. W., & Goldberg, C. (2018). Future Computing Hardware for AI. *Technical Digest - International Electron Devices Meeting, IEDM,* 1.3.1-1.3.6. DOI: 10.1109/IEDM.2018.8614482

Wermuth, C. G., Ganellin, C. R., Lindberg, P., & Mitscher, L. A. (1998, January 1). Glossary of terms used in medicinal chemistry (IUPAC Recommendations 1998). *Pure and Applied Chemistry*, 70(5), 1129–1143. DOI: 10.1351/pac199870051129

Wilson, N. (2024). Artificial intelligence helps drive new frontiers in ecology. *Bioscience*, 74(5), 306–311. DOI: 10.1093/biosci/biae016 PMID: 38854635

*World Health Organization (WHO)*. (n.d.). Retrieved March 28, 2024, from https://www.who.int/home

Wu, F., Zhao, S., Yu, B., Chen, Y. M., Wang, W., Song, Z. G., Hu, Y., Tao, Z. W., Tian, J. H., Pei, Y. Y., Yuan, M. L., Zhang, Y. L., Dai, F. H., Liu, Y., Wang, Q. M., Zheng, J. J., Xu, L., Holmes, E. C., & Zhang, Y. Z. (2020). A new coronavirus associated with human respiratory disease in China. *Nature, 579*(7798), 265–269. DOI: 10.1038/s41586-020-2008-3

Xia, X. (2017). Bioinformatics and drugs discovery. *Current Topics in Medicinal Chemistry*, 17(15), 1709–1726. DOI: 10.2174/1568026617666161116143440 PMID: 27848897

Xu, W. (2018). Application of bioinformatics technology in food processing and food safety assessment.

Xu, C., Liu, R., Huang, S., Li, W., Li, Z., & Luo, H.-B. (2023). 3D-SMGE: A pipeline for scaffold-based molecular generation and evaluation. *Briefings in Bioinformatics*, 24(6), 1–11. DOI: 10.1093/bib/bbad327 PMID: 37756591

Xu, D., & Zhang, Y. (2012). Ab initio protein structure assembly using continuous structure fragments and optimized knowledge-based force field. *Proteins*, 80(7), 1715–1735. DOI: 10.1002/prot.24065 PMID: 22411565

Xu, X., Huang, M., & Zou, X. (2018). Docking-based inverse virtual screening: Methods, applications, and challenges. *Biophysics Reports*, 4(1), 1–16. DOI: 10.1007/s41048-017-0045-8 PMID: 29577065

Yahaya, A.-F., Khalid, A. A., Lim, K., Poh, H. X., Aziz, A., Khalid, K., Lim, H. X., & Poh, C. L. (2023). Development of Next Generation Vaccines against SARS-CoV-2 and Variants of Concern. *Viruses, 15*(3), 624. DOI: 10.3390/v15030624

Yang, A., Zhang, W., Wang, J., Yang, K., Han, Y., & Zhang, L. (2020, September). Review on the Application of Machine Learning Algorithms in the Sequence Data Mining of DNA. *Frontiers in Bioengineering and Biotechnology*, 8, 4. DOI: 10.3389/fbioe.2020.01032 PMID: 33015010

Yang, C., Chowdhury, D., Zhang, Z., Cheung, W. K., Lu, A., Bian, Z., & Zhang, L. (2021). A review of computational tools for generating metagenome-assembled genomes from metagenomic sequencing data. *Computational and Structural Biotechnology Journal*, 19, 6301–6314. DOI: 10.1016/j.csbj.2021.11.028 PMID: 34900140

Yang, J., Cai, Y., Zhao, K., Xie, H., & Chen, X. (2022, November 1). Concepts and applications of chemical fingerprint for hit and lead screening. *Drug Discovery Today*, 27(11), 103356. DOI: 10.1016/j.drudis.2022.103356 PMID: 36113834

Yang, J., Yan, R., Roy, A., Xu, D., Poisson, J., & Zhang, Y. (2015). The I-TASSER Suite: Protein structure and function prediction. *Nature Methods*, 12(1), 7–8. DOI: 10.1038/nmeth.3213 PMID: 25549265

Yang, L., Zhang, Z., Hong, S., Xu, R., Zhao, Y., Shao, Y., Zhang, W., Yang, M., & Cui, B. (2022). Diffusion models: A comprehensive survey of methods and applications. *ACM Computing Surveys*, 56, 1–39. DOI: 10.1145/3505243

Yang, R., Xia, Y., Jin, X., Yu, H., Yan, B., & Cheng, B. (2022). Identification of potential dual farnesol x receptor/retinoid x receptor α agonists based on machine learning models, admet prediction and molecular docking. *ChemistrySelect*, 7(28), e202200715. Advance online publication. DOI: 10.1002/slct.202200715

Yang, T.-J., Yu, P.-Y., Chang, Y.-C., Liang, K.-H., Tso, H.-C., Ho, M.-R., Chen, W.-Y., Lin, H.-T., Wu, H.-C., & Hsu, S.-T. D. (2021). Effect of SARS-CoV-2 B.1.1.7 mutations on spike protein structure and function. *Nature Structural & Molecular Biology*, 28(9), 731–739. DOI: 10.1038/s41594-021-00652-z PMID: 34385690

Yang, X., Wang, Y., Byrne, R., Schneider, G., & Yang, S. (2019, September 25). Concepts of Artificial Intelligence for Computer-Assisted Drug Discovery. *Chemical Reviews*, 119(18), 10520–10594. DOI: 10.1021/acs.chemrev.8b00728 PMID: 31294972

Yan, Y., Tao, H., He, J., & Huang, S.-Y. (2020). The HDOCK server for integrated protein–protein docking. *Nature Protocols*, 15(5), 1829–1852. DOI: 10.1038/s41596-020-0312-x PMID: 32269383

Ye. (2003). Flexible structure alignment by chaining aligned fragment pairs allowing twists. Bioinformatics, 19(2), ii246-ii255.

Zaha, A., Ferreira, H. B., & Passaglia, L. M. P. (2014). *Biologia Molecular Básica—5. ed.* Artmed.

Zaki, M. J., & Meira, W.Jr. (2020). *Data mining and machine learning: fundamental concepts and algorithms* (2nd ed.). Cambridge University Press. DOI: 10.1017/9781108564175

Zarei, A. E., Almehdar, H. A., & Redwan, E. M. (2016). Hib Vaccines: Past, Present, and Future Perspectives. *Journal of Immunology Research*, 2016(1), 7203587. DOI: 10.1155/2016/7203587 PMID: 26904695

Zeevi, D., Korem, T., Zmora, N., Israeli, D., Rothschild, D., Weinberger, A., Ben-Yacov, O., Lador, D., Avnit-Sagi, T., Lotan-Pompan, M., Suez, J., Mahdi, J. A., Matot, E., Malka, G., Kosower, N., Rein, M., Zilberman-Schapira, G., Dohnalová, L., Pevsner-Fischer, M., & Segal, E. (2015). Personalized nutrition by prediction of glycemic responses. *Cell*, 163(5), 1079–1094. DOI: 10.1016/j.cell.2015.11.001 PMID: 26590418

Zeng, Z., Shi, H., Wu, Y., & Hong, Z. (2015). Survey of natural language processing techniques in bioinformatics. *Computational and Mathematical Methods in Medicine*, 674296, 1–10. Advance online publication. DOI: 10.1155/2015/674296 PMID: 26525745

Zerbino, D. R., Frankish, A., & Flicek, P. (2020). Progress, challenges, and surprises in annotating the human genome. *Annual Review of Genomics and Human Genetics*, 21(1), 55–79. DOI: 10.1146/annurev-genom-121119-083418 PMID: 32421357

Zhang, L., Chen, F., Zeng, Z., Xu, M., Sun, F., Yang, L., & Hao, H. (2021). Advances in metagenomics and its application in environmental microorganisms. *Frontiers in Microbiology*, 12, 766364. DOI: 10.3389/fmicb.2021.766364 PMID: 34975791

Zhang, L., Jackson, C. B., Mou, H., Ojha, A., Peng, H., Quinlan, B. D., & Suthar, M. S. (2020). SARS-CoV-2 spike-protein D614G mutation increases virion spike density and infectivity. *Nature Communications*, 11(1), 6013. DOI: 10.1038/s41467-020-19808-4 PMID: 33243994

Zhang, S., Fan, R., Liu, Y., Chen, S., Liu, Q., & Zeng, W. (2023). Applications of transformer-based language models in bioinformatics: A survey. *Bioinformatics Advances*, 3(1), vbad001. Advance online publication. DOI: 10.1093/bioadv/vbad001 PMID: 36845200

Zhang, T., Aipire, A., Li, Y., Guo, C., & Li, J. (2023). Antigen cross-presentation in dendric cells: From bench to bedside. *Biomedicine and Pharmacotherapy*, 168, 115758. DOI: 10.1016/j.biopha.2023.115758 PMID: 37866002

Zhang, X., Shen, C., Zhang, H., Kang, Y., Hsieh, C.-Y., & Hou, T. (2024). Advancing Ligand Docking through Deep Learning: Challenges and Prospects in Virtual Screening. *Accounts of Chemical Research*, 57(10), 1500–1509. DOI: 10.1021/acs.accounts.4c00093 PMID: 38577892

Zhang, Y. (2008). I-TASSER server for protein 3D structure prediction. *BMC Bioinformatics*, 9(1), 40. DOI: 10.1186/1471-2105-9-40 PMID: 18215316

Zhao, J., Qin, B., Nikolay, R., Spahn, C. M. T., & Zhang, G. (2019). Translatomics: The Global View of Translation. *International Journal of Molecular Sciences*, 20(1), 212. Advance online publication. DOI: 10.3390/ijms20010212 PMID: 30626072

Zhao, Y., Zheng, K., Guan, B., Guo, M., Song, L., Gao, J., Qu, H., Wang, Y., Shi, D., & Zhang, Y. (2020). DLDTI: A learning-based framework for drug-target interaction identification using neural networks and network representation. *Journal of Translational Medicine*, 18(1), 434. DOI: 10.1186/s12967-020-02602-7 PMID: 33187537

Zheng, L., Wang, H., Lin, J., Zhou, Y., Xiao, J., & Li, K. (2023). Population genomics provides insights into the genetic diversity and adaptation of the Pieris rapae in China. *PLoS One*, 18(11), e0294521. DOI: 10.1371/journal.pone.0294521 PMID: 37972203

Zhou, X., Li, Y., Liu, S., Yang, Q., Su, X., Zhou, L., Tang, M., Fu, R., Li, J., & Huang, Q. (2013). Ultra-deep sequencing enables high-fidelity recovery of biodiversity for bulk arthropod samples without PCR amplification. *GigaScience*, 2(1), 4. DOI: 10.1186/2047-217X-2-4 PMID: 23587339

Zhu, X. J. (2005). Semi-supervised learning literature survey. Academic Press.

Zubair, M., Wang, J., Yu, Y., Faisal, M., Qi, M., Shah, A. U., Feng, Z., Shao, G., Wang, Y., & Xiong, Q. (2022). Proteomics approaches: A review regarding an importance of proteome analyses in understanding the pathogens and diseases. *Frontiers in Veterinary Science*, 9, 1079359. DOI: 10.3389/fvets.2022.1079359 PMID: 36601329

Zwanzig, R., Szabo, A., & Bagchi, B. (1992, January). Levinthal's paradox. *Proceedings of the National Academy of Sciences of the United States of America*, 89(1), 20–22. DOI: 10.1073/pnas.89.1.20 PMID: 1729690

# About the Contributors

**Paulo Fazendeiro** received a PhD in Informatics Engineering (2010), held Pedagogical Aptitude and Scientific Capacity Tests (2001), and holds a Degree in Mathematics and Informatics (1995, best student). Both degrees were obtained at the University of Beira Interior, Portugal. He is currently an assistant professor at the Faculty of Engineering at the University of Beira Interior. He belongs to the Informatics Department where he teaches predominantly subjects in the areas of Artificial Intelligence, Bioinformatics and Programming. He is course director of the 2nd cycle of studies in Computer Science and Engineering (MSc), having previously served as director of the 1st cycle of studies (BSc) in Computer Science and Engineering and also of the 1st cycle of studies in Bioengineering. His research interests include Computational Intelligence, Granular Computing and Fuzzy Systems, Knowledge Discovery and Data Mining, Evolutionary Algorithms, as well as clustering techniques with applications for different areas. He is a member of the Pattern Recognition and Image Analysis group (PIA-Cv) of the Instituto de Telecomunicações (IT).

**Carmelina Leite** has a PhD degree in Bioinformatics, Federal University of Minas Gerais, Brasil. She developed an algorithm for the ligand-based virtual screening based on linear algebra - The Milk-Way algorithm. This algorithm can be used not only to discover new drugs but also to reposition commercialized drugs. This structure-based screening resulted in a suggestion of treatment for COVID-19, the tetrachlorodecaoxide. Her dissertation comprised four years of work, giving rise to two deposit patents. She also holds an MSc in Bioinformatics and in Pharmaceutical Sciences. Currently, Carmelina is an entrepreneur in an innovation project of nutraceuticals and drug development using data mining techniques.

* * *

**Janaíne Aparecida de Paula** has a degree in Clinical Pathology (Clinical Analysis) from the Federal University of Triângulo Mineiro (UFTM). She has been working in the Bioinformatics and Omic Sciences Laboratory of the Microbiology, Immunology and Parasitology department at UFTM since 2021, where she carries out research related to immunoinformatics and structural bioinformatics, with viruses as the microorganisms of interest, especially HIV and HTLV. She is currently a master's student in the Interunit Graduate Program in Bioinformatics at the Federal University of Minas Gerais (UFMG), working at the Laboratory of Cellular and Molecular Genetics (LGCM) at UFMG.

**Sheila Araujo** graduated with a degree in Natural Sciences from the School of Arts, Sciences and Humanities at the University of São Paulo (EACH-USP) in 2012. Subsequently, she completed a master's degree in Science and Technology, focusing on Chemistry at the Federal University of ABC (UFABC) in 2015. Following this, she pursued and achieved a PhD in Science and Technology from UFABC, with a sandwich internship at the Vancouver Prostate Center - University of British Columbia, Canada, from 2016 to 2021, focusing on neglected diseases, particularly Chagas disease. Her diverse educational background is reflected in her work as a Science and Chemistry teacher in primary education (Elementary II and High School) in both public and private schools in São Paulo from 2012 to 2018. During this time, she also taught at the high school level as a teacher employed by the Department of Education of São Paulo. She also served as a tutor in the mandatory internship subject at the Open University of Brazil (UAB) at the Jardim Miragaia - Vila Curuçá hub linked to the Federal Institute of São Paulo (IFSP) in the Pedagogy and EPT course. Furthermore, she has been a voluntary collaborator in the UFABC extension project and the scientific dissemination blog "Science Enthusiast's Guide." Her research projects have revolved around Theoretical Chemistry, focusing primarily on medicinal, computational chemistry, molecular modeling, and bioinformatics. She is currently a postdoctoral student in the UFMG and UFSJ bioinformatics program, where her studies encompass biophysics, molecular dynamics of enzymes and macromolecules, virtual screening, and theoretical and medicinal chemistry.

**Vasco Ariston de Carvalho Azevedo** is visiting full professor at the School of Veterinary Medicine and Zootechnics at UFBA, retired full professor and volunteer at UFMG, CNPq 1A researcher, full member of the Brazilian Academy of Sciences and corresponding member of the Mathematics and Natural Sciences class of the Göttingen Academy of Sciences and Humanities. He is a Commander of the Order of Scientific Merit of the MCTI, a member of the Genetics Advisory Committee and the COBRG/CNPq Working Group on Public Policies in Biotechnology and

Genetic Resources, coordinator of the Bact-infl Associated International Laboratory of INRA and UFMG and was vice-president of the Brazilian Association of Bioinformatics and Computational Biology (AB3C) until December 2023. He holds a degree in Veterinary Medicine from the Faculty of Veterinary Medicine of the Federal University of Bahia (1986), a master's degree (1989) and a doctorate (1993) in Microorganism Genetics from the Institut National Agronomique Paris Grignon. Post-doctorate from the Department of Microbiology, School of Medicine, University of Pennsylvania (USA, 1994). Full Professor at the Institute of Biomedical Sciences of the University of São Paulo (2004) and Ph.D. in Bioinformatics from UFMG (2017). He has experience in Genetics and Bioinformatics, with an emphasis on Molecular Genetics and Microorganisms, working mainly on the following topics: Genomics; transcriptomics; proteomics, metabolomics; Functional analysis of genes; Study of virulence and pathogenicity, Molecular Biotechnology and Biosafety. He works on vaccine development, diagnostics and the development of state-of-the-art probiotics.

**Luana Luiza Bastos** has a degree in Biomedicine from Faculdade de Minas-BH (2019) and a master's degree in Bioinformatics from the Federal University of Minas Gerais (2021). She is currently a doctoral student in the postgraduate program in bioinformatics at UFMG, working mainly on the following topics: contacts, rational design of enzymes, structural alignment, structural bioinformatics, protein-protein docking, protein-protein interactions.

**Gabriel Camargos Gomes** graduated in Biological Sciences (Licenciatura) in 2023 from the Federal University of Minas Gerais (UFMG) and is studying for a master's degree in Bioinformatics at UFMG.

**Marcus Canário Viana** has a degree in Biological Sciences from the State University of Southwest Bahia (2010), a master's degree in Genetics, Biodiversity and Conservation from the State University of Southwest Bahia (2013), a PhD in Genetics (2018) and in Bioinformatics (2022) from the Federal University of Minas Gerais. He is currently a postdoctoral researcher in Bioinformatics at the Federal University of Minas Gerais. Experience in Bioinformatics and Molecular and Microorganism Genetics, working mainly on the following topics: genome assembly and annotation, comparative genomics and molecular phylogenetics.

**Sangramjit Chakraborty** was born in 1999 in West Bengal, India. He has completed his undergraduate B.Tech degree in Computer Science and Engineering from Dr. B.C. Roy Engineering College, Durgapur. His research interests include swarm intelligence, computer vision, and artificial intelligence.

**Alessandra Cioletti** has a PhD in Chemistry, post-doctorate in Carbohydrate and Peptide Chemistry, specialization in Data Science and Big Data, currently a master's student in Computer Science. Researching in the field of Artificial Intelligence applied to Bioinformatics, Medicinal Chemistry and Organic Electrochemistry

**Mariana Costa Pedrosa** is an undergraduate student in Biomedicine at the Pontifical Catholic University (PUC). She is a scientific researcher at the LGCM (Laboratory of Cellular and Molecular Genetics) in the area of Bioinformatics at the Federal University of Minas Gerais.

**Siomar de Castro Soares** graduated in Biomedicine from the University of Uberaba (2002-2006). Further training in Bioinformatics (2007). Master's degree in Genetics from UFMG (2008-2009). PhD in Genetics from UFMG (2009-2013) with a 1-year sandwich PhD at the Center for Biotechnology (CeBiTec) of the Universität Bielefeld (2011-2012). PhD in Bioinformatics from UFMG (2015-2019). Post-doctorate in Bioinformatics from UFMG (2013-2014). Senior Bioinformatics Researcher at the Official Central Laboratory of the Ministry of Fisheries and Aquaculture (2014-2015). Secretary of the Southeast region of the Brazilian Genetics Society (2017-2018), Substitute Coordinator of the Biomedicine undergraduate course at UFTM (2017-2019), Director of the Department of Research Development and Technological Innovation at UFTM (2019-2022), Affiliate Member of the Brazilian Academy of Sciences (2018-2022). He is currently a Professor at the Federal University of the Triângulo Mineiro - UFTM (2015-Present). Areas of expertise: molecular genetics, genomic sequencing and comparative genomics of microorganisms, with a focus on pan-genomics, genomic plasticity in the identification of pathogenicity islands and virulence factors, phylogenomics, molecular epidemiology, reverse vaccinology and software development (perl and java languages).

**Ana de Oliveira Vinhal** is currently attending Biological Sciences degree in Universidade Federal de Minas Gerais. Since 2022 she has been part of Laboratório de Genética Molecular e Celular, first as a volunteer and now as scholarship holder. Experienced at bioinformatics.

**Fernanda Diniz Prates** has a degree in Biomedicine from Universidade FUMEC (2010), a Master's degree in Microbiology (CAPES 7) from Universidade Federal de Minas Gerais, where she developed a project entitled: "Effect of the number of EPIYA-C sites of the CagA protein of Helicobacter pylori on the gastric concentration of cytokines associated with the Th17 response in patients with gastritis and gastric carcinoma". She also has an MBA in Strategic Project Management from Universidade FUMEC (2018) and a specialization in Bioinformatics from Hospital

Israelita Albert Einstein (2022). She has experience in microbiology, molecular biology, immunology and bioinformatics. She is currently a technical area supervisor in the Microbiology sector of the Pardini Group and a PhD student in Bioinformatics at the Federal University of Minas Gerais.

**Talita Gomes** is a PhD student in Microbiology at the Postgraduate Program in Microbiology at the Federal University of Minas Gerais. She holds a Master's degree in Microbiology and is a Veterinary Physician at the Federal University of Bahia. Member of the National Institute of Science and Technology (INCT) in Interdisciplinary and Transdisciplinary Studies in Ecology and Evolution (IN-TREE). He has taught the subjects "Ethics, citizenship and human rights" and "Career planning and professional development" to technical students. Taught the subject "Biology" to pre-university students. She has experience in microbiology, clinical microbiology, infectious diseases, spirochetes (Leptospira and Treponema), molecular biology and biofilms.

**Ajaygiri K. Goswami** has completed Masters and Doctorate in Chemical engineering and technology and presently working as Head Pharmaceutical Technology, KBCNorth Maharashtra University, Jalgaon. He has presented more than fifty research articles in International Conferences and also published research papers more than hundred in National and international reputed Journals. He is working in various academic bodies of the University like BOS,BOG and Coordinators for IIC, Startup, KCIL cell etc. He has more than 25 years Teaching and research experience in Engineering College and University Departments and is research area are Pharmaceutical technology, Chemical engineering, Oil technology, Bio fuel, Nano catalyst, etc.

**Vinay Kabara** has completed his graduation in Food Technology at the Department of Food Technology, University Institute of Chemical Technology, Kavayitri Bahinabai Chaudhari North Maharashtra University, Jalgaon. His academic focus has been predominantly on bioinformatics for food and nutritional science, which is crucial for improving nutritional profiles in consumer diets. Mr. Kabara has actively participated in numerous national and international conferences and workshops, including those organized by CFTRI (Mysuru), IIT Kharagpur, NIFTEM etc. These engagements have not only broadened his knowledge base but also sparked a deeper interest in pursuing further research in the field of food technology. He is keen to contribute to advancing the industry through innovative research, particularly in areas that intersect with food safety and nutritional enhancement.

**Sunil J. Kulkarni** has done his PhD in Chemical Engineering with focus on adsorption for wastewater treatment. He has done his Masters in chemical engineering from Tatyasaheb Kore Institute of Engineering and Technology, Warananagar. He has worked as an Assistant Professor in Chemical Engineering Department of Datta Meghe College of Engineering, Airoli, Navi Mumbai, India for 13 years. He has total 25 years of teaching and research experience. Dr. Sunil is involved with different academic activities across the universities in Maharashtra. His area of investigation in doctoral studies is phenol and heavy metal removal from waste water. He has also worked on kinetics and isotherms of various sorption operations. His area of research includes adsorption, clean technology and environmental biotechnology. He is on the reviewer board of many international journals and reviewed many international papers. He is also governing body member, promotional editor and executive member of many reputed international journals. Currently he is working as Associate Professor in chemical engineering and and Dean Research and Development at GIT, Lavel, India.

**Diego Mariano** is Ph.D. and M.S. in Bioinformatics by the Universidade Federal de Minas Gerais. He is currently a post-doctoral fellow at the Department of Computer Science at UFMG (Brazil). He has experience in the development of Web systems, Data Science and data visualization, Bioinformatics, and machine learning. Has knowledge in the following languages: PHP, JavaScript, Python, R, Perl, HTML, CSS, and SQL.

**Luiza Pereira Reis** is an undergraduate student in Biomedicine at the Pontifical Catholic University of Minas Gerais. She is a scientific researcher at the LGCM (Laboratory of Cellular and Molecular Genetics) in the area of bioinformatics at the Federal University of Minas Gerais.

**Moumita Pradhan** received her BTech degree in Computer Science and Engineering from JIS College of Engineering College, Kalyani (under West Bengal University of Technology), India, in 2005 and MTech degree from NIT, Durgapur, West Bengal, India in Operation Research and Business management in 2009. She have completed her PhD in Computer science and engineering from N.I.T Durgapur. Her area of research includes hydrothermal scheduling, economic load dispatch, soft computing techniques and NLP. Presently she is working as an Assistant Professor in the Department of Information Technology, Dr. B.C. Roy Engineering College, Durgapur, India.

**Lucas Rodrigues Gomes** is a PhD student at the Graduate Program in Bioinformatics - UFMG, interested in immunoinformatics and reverse vaccinology. Bachelor's degree in Biological Sciences from the Federal University of Minas Gerais (UFMG) and Master's degree in Bioinformatics from the Graduate Program in Bioinformatics - UFMG.

**Mohsen Sheykhhasan** is a dedicated researcher in the fields of stem cell biology and medical biotechnology. He has made significant contributions to the understanding and development of novel therapeutic approaches, particularly in the areas of exosomes and drug delivery systems. Throughout his career, Mohsen has conducted extensive research on exosomes derived from adipose-derived stem cells, focusing on their potential applications in treating neurodegenerative disorders such as Alzheimer's disease. His PhD thesis explored innovative drug delivery methods for Alzheimer's models, which has contributed to the advancement of targeted therapies for the disease. Dr. Sheykhhasan has refereed over 550 articles for prestigious journals, according to Web of Science, demonstrating his expertise and commitment to maintaining high standards in scientific research. His research interest's span various areas, including exosomes, adipose-derived stem cells, and drug delivery systems. In addition to his research endeavors, Dr. Sheykhhasan has authored and co-authored several publications in esteemed journals such as Biomedicine & Pharmacotherapy, Reviews in Medical Virology, Expert Reviews in Molecular Medicine, Cancer Gene Therapy, and Cellular Oncology. He has also contributed to the understanding of the therapeutic roles of chimeric antigen receptor (CAR) T cells in cancer immunotherapy and explored the potential of exosomes as vehicles for delivering miRNA to cancer cells.

**Eduarda Sousa** is a biomedical student at the Federal University of Triângulo Mineiro. Developed a scientific initiation project in the Immunology laboratory, with an emphasis on autoimmune diseases. Coordinator of Fertiliga, an in vitro reproduction league, the UFTM Genetics League and Research Coordinator of the LACEC Extracorporeal Circulation League. Took part in the digital writing of the UFTM Biomedicine Journal and in the Nossa Jornada extension project as an Event and Training Analyst for Meu Futuro. Has worked on bioinformatics projects in the department of microbiology, immunology and parasitology, working in the area of reverse vectionology at UFTM. Now PhD student in Postgraduate Program in Bioinformatics at UFMG.

**Samruddhi Vaidya** is B.Tech graduate in Food Technology from Department of Food Technology, University Institute of Chemical Technology, Kavayitri Bahinabai Chaudhari North Maharashtra University, Jalgaon. During her academic journey, Samruddhi embarked on a six-month internship journey at ITC Ltd. Haridwar. Where she gained invaluable insights into the corporate world. Her tenure at ITC provided her with hands-on experience in the projects of Quality assurance, honing her skills in smooth production process, Quality assurance, Problem identification and solving etc. She actively participated in various academic events, lectures, seminars and conferences, including presenting a poster at a AEEAR – 2024 national conference.

Currently, Samruddhi continues to pursue her professional endeavours, driven by a desire to make meaningful contributions in food processing and preservation sector.

# Index

## R



## S



## V